Fracture Mechanics

Nestor Perez

Fracture Mechanics

Second Edition

 Springer

Nestor Perez
Department of Mechanical Engineering
University of Puerto Rico
Mayaguez, Puerto Rico

Additional material to this book can be downloaded from http://extras.springer.com

ISBN 978-3-319-79718-2   ISBN 978-3-319-24999-5   (eBook)
DOI 10.1007/978-3-319-24999-5

Printed on acid-free paper

This Springer imprint is published by Springer Nature
The registered company is Springer International Publishing AG Switzerland

To my wife Neida
To my daughters Jennifer and Roxie
To my son Christopher

Foreword

First Edition of Fracture Mechanics by Dr. Nestor Perez

I had the pleasure and the privilege of reading the original book, particularly some of the important chapters of the book on fracture mechanics by Dr. Nestor Perez. I find the book, in general, very well written for the academicians as well as for the practitioners in the field of engineering fracture mechanics. The language is simple, clear, and straight to the point. Each chapter is developed in such a way that the mathematical treatment supports the main physical mechanism of fracture. The equations are arranged in an orderly fashion in harmony with the descriptive concepts of a phenomenon that is highly complex, very nonlinear, and often unpredictable. The main objective of a mathematical analysis is to explain and clarify a physical phenomenon and definitely not to jeopardize it by undue and unwanted complexity at the cost of brevity. If this is true for a textbook or any treatise for that matter, Dr. Perez's book has done the job quite elegantly.

This book is suitable as a textbook for senior undergraduate and graduate students of a one-semester course in mechanical, civil, chemical, and industrial engineering, materials science, as well as applied physics and applied chemistry programs.

Each chapter is self-contained and self-sufficient in descriptive details, but it keeps a smooth continuity with its preceding and following chapters. The analytic and the algebraic illustrations are just in place with theoretical analyses and examples. Unlike many voluminous works on fracture mechanics, in this book, mathematics does not overburden the physics of fracture mechanics and thus shows a more realistic route to solve a particular problem. Hence, practicing engineers in consulting firms and design offices can use this book in a very handy and straightforward fashion. Also it is a good reference book in the personal library of many retired professionals and professors who still like to keep in touch with the reality as a hobby, pastime, or pleasure.

I highly recommend this book to any technical publishing house for the timely birth of this solid but simple work on engineering fracture mechanics.

Journal of Mechanical Behavior of Materials Editor (2003)

Jay K. Banerjee, Ph.D., P.E., M.Ed.

Second Edition of Fracture Mechanics by Dr. Nestor Perez

It is just over a decade since Dr. Nestor Perez published his book on fracture mechanics. During these 10 years, many emerging fields have gained momentum and expanded their boundaries related to the failure of materials through instability and fracture. Some of them are the rupture of cell membranes and tissues in tissue mechanics, the plastic flow and fracture of bones and tendons in the knees and elbows, as well as the allied areas of biomechanics and sports medicine. I am sure that the fundamental concepts of flow and fracture in all "engineerable" materials as developed and presented by Dr. Perez in the first edition of his book, Fracture Mechanics, has served in many of the emerging areas of materials science and technology as mentioned above.

In continuation with this spirit, Dr. Perez has added new concepts in the second edition of his book. In each chapter, additional approaches are introduced in simple but convincing manners through a variety of examples. Besides, new numerical problems are added as examples on how to apply the advanced theories in practice. Such examples do help not only the students and their instructors in a classroom environment but also the researchers and their mentors in a laboratory of fracture mechanics.

In sum, in the second edition, Dr. Perez not only reinforces the basic concepts of fracture mechanics by presenting new examples on their applications but also adds and hence expands them into the novel and emerging branches of materials science. This new edition will help enhance both teaching at the undergraduate- and graduate-level courses as well as will facilitate the researchers to advance their theoretical concepts and apply them in practice in the realm of the failure of materials. These are the improvements in the second edition worth navigating and exploring into the unknown of the matter: the materials research.

Professor of Manufacturing Processes
and Systems
Mechanical Engineering Department
University of Puerto Rico at Mayaguez
Mayaguez, PR, USA
2016

Jay K. Banerjee, Ph.D., P.E., M.Ed.

Preface

This second edition of the book retains all the features of the previous edition while new ones are added. The main work in this edition includes refining text in each chapter, expansion of some sections in several chapters, and addition of examples, problems, and new sections, such as conformal mapping and mechanical behavior of wood.

The purpose of this book is to present, in a closed form, analytical methods in deriving stress and strain functions related to fracture mechanics. This book contains a compilation of work available in the literature in a scatter form and, to a certain extent, selected experimental data of many researchers to justify the theoretical fracture mechanics models in solving crack problems. It is a self-contained and detailed book for the reader (senior and graduate students and engineers) involved in the analysis of failure using a mathematical approach for designing against fracture. However, it is important that the reader understands the concept of modeling, problem solving, and interpreting the meaning of mathematical solution for a particular engineering problem or situation. Once this is accomplished, the reader should be able to think mathematically, foresee metallurgically the significance of microstructural parameters on properties, analyze the mechanical behavior of materials, and recognize realistically how dangerous a crack is in a stressed structure, which may fail catastrophically.

In spite of the advances in fracture mechanics, many principles remain the same. Dynamic fracture mechanics is included through the field of fatigue and Charpy impact testing. The material included in this book is based upon the development of analytical and numerical procedures pertinent to particular fields of linear elastic fracture mechanics (LEFM) and plastic fracture mechanics (PFM), including mixed-mode-loading interaction. The mathematical approach undertaken herein is coupled with a brief review of several fracture theories available in cited references.

Fracture mechanics of engineering materials deals with fracture of solids undergoing large deformation (ductile materials) and/or fracture (brittle materials) when subjected to extreme loading. The analysis of a solid responding to loads is concerned partly with microscopic mechanisms of fracture, establishing fracture criteria, and predicting the fracture stress from a macroscopic approach.

Chapter 1 includes definitions of variables such as force, load, stress, strain, and displacement. These are vital for the understanding of state properties of solid materials and for characterizing the mechanical behavior of crack-free or cracked solids.

Chapter 2 deals with the introduction to fracture mechanics. It also includes the close form of Griffith crack theory and the strain-energy release rate associated with fracture.

Chapter 3 is devoted to solid bodies under quasi-static stress modes containing cracks. The theory of linear elastic fracture mechanics (LEFM) is integrated in this chapter using an analytical approach for deriving field equations ahead of a crack tip.

Chapter 4 includes the derivation of elastic field equations for mode I (tension), mode II (sliding), and mode III (tearing) loadings.

Chapter 5 is devoted to crack tip plasticity and relevant configuration. The yielding phenomenon is analyzed for a better understanding of the plastic deformation ahead of a crack tip.

Chapter 6 deals with the energy principles for assessing the elastic behavior of solids containing cracks. The energy terms included in this chapter are the energy release rate and the J-integral which are used to define fracture criteria.

Chapter 7 includes the theoretical concepts of plastic fracture mechanics for deriving the HRR field equations using the J-integral approach. An engineering approach is also included for determining the plastic J-integral.

Chapter 8 deals with a realistic engineering problem related to mixed-mode fracture mechanics. This is the case for a crack in a component being subjected to a mixed-mode loading, such as tension and torsion (mode I and II stress loading). A closed-form analytical approach is used in this chapter for deriving the field elastic equations.

Chapter 9 is devoted to fatigue crack growth since fatigue in solid materials being subjected to repeated cyclic loading is a cumulative damage phenomenon. Fatigue crack initiation is modeled using a crystallographic approach, and the fatigue crack growth rate is determined as a function of the change of the stress intensity factor. Thus, a fatigue life formula is derived for predicting fracture.

Chapter 10 is devoted to fracture toughness correlations, including indentation-induced cracking, Charpy impact energy, and dynamic effects.

A solution manual is available for educators or teachers upon the consent of the book publisher. Also, all images, pictures, or data taken from reliable sources are included in this book for educational purposes and academic support only. Additional material to this book can be downloaded from http://extras.springer.com.

Mayaguez, Puerto Rico   Nestor Perez
2016

Contents

Theory of Elasticity

1

1.1   Introduction

The definition of variables such as force, load, stress, strain, and displacement is vital for the understanding of state properties of solid materials and for characterizing the mechanical behavior of crack-free or cracked solids. Clearly, the latter have a different mechanical behavior than the former, and it is characterized according to the principles of fracture mechanics, which are divided into two areas. Linear-elastic fracture mechanics (LEFM) considers the fundamentals of linear elasticity theory, and elastic-plastic fracture mechanics (EPFM) characterizes plastic behavior of cracked ductile solids. In order to characterize cracked solids, knowledge of the aforementioned variables is necessary. The term force in the fracture mechanics field is an applied mechanical load being fixed, quasi-static or dynamics. In physics, a dynamic force, $F = ma$, depends on the acceleration (a) of a moving mass (m). However, if this mass is stationary and susceptible to be deformed, a quasi-static force or mechanical force must be defined. Both dynamic and mechanical forces have the same units, but different physical meaning. Moreover, this mechanical force is analogous to load (P). Obviously, this is the point of departure in this chapter for defining an important engineering parameter called elastic stress, $\sigma = P/A_o$, where A_o is the original cross-sectional area of a specimen.

Now, the strain is defined as $\epsilon = d\mu/dx$ where $d\mu$ is the change of displacement, say, in the x-direction. The intent here is to indicate how certain parameters or variables are related to one another. Nevertheless, if two variables are known, the third one can be estimated or predicted. This is one of the benefits of mathematics for solving engineering problems, which have their own constraints for dictating the magnitude of a particular variable. In fact, one or more variables may define a material property, while a property depends on the microstructure of a solid material.

© Springer International Publishing Switzerland 2017
N. Perez, *Fracture Mechanics*, DOI 10.1007/978-3-319-24999-5_1

1.2   Definitions

This section is concerned with some definitions the reader needs to assimilate before the fracture mechanics theories and mathematical definitions are introduced in a progressive manner. It is important to have a clear and precise definition of vital concepts in the field of applied mechanics so that the learning process for understanding fracture mechanics becomes obviously easy. However, basic concepts such as stress, strain, safety factor, deformation, and the like are important in characterizing the mechanical behavior of solid materials subjected to forces or loads in service [1, 2]. Hence,

Deformation: The movement of points in a solid body relative to each other. Deformation is also the change in shape of objects due to applied forces and it may be elastic or plastic.

Displacement: The movement of a point in a vector quantity in a body subjected to a loading mode.

Strain: This is a geometric quantity, which depends on the relative movement of two or three points in a body. Strain is also a measure of deformation of the material based on a reference length.

Stress: A stress at a point on a body represents the internal resistance of the body due to an external force. Thus, the load (P) and the cross-sectional area (A) are related as indicated by the equation of equilibrium of forces. This implies that stress is defined as force per unit area.

$$\sum F_y = P - \sigma A = 0 \qquad (1.1)$$

$$\sigma = \frac{P}{A} \qquad (1.2)$$

If A is the original cross-sectional area (A_o), then σ is an engineering stress; otherwise, it is a local stress. The theory of elasticity deals with isotropic materials subjected to elastic stresses, strains, and displacements. In engineering, the elasticity behavior of a material is characterized by the tensile modulus of elasticity and the elastic limit. The latter is just the transition stress between elastic and plastic deformation.

Safety Factor: This is a parameter used in designing structural components to assure structural integrity. Simply stated, the safety factor is a design factor defined by

$$S_F = \frac{Strength}{Stress} > 1 \qquad (1.3)$$

Here, the strength may represent a material's property, such as the yield strength σ_{ys}, and the stress σ is the variable to be applied to a structure. The role of

S_F in this simple relationship is to control the design stress so that $\sigma < \sigma_{ys}$ in designing applications, where it is desirable to have a prolong design life for assuring structural integrity.

Usually, the safety factor is in the order of two, but its magnitude depends on the designer's experience or on a design code.

1.3 Stress State

According to the theory of elasticity, the field equations are based on the normal strains (ϵ_i) and the shear strains (γ_{ij}). These are related to displacements (u_i), which are illustrated in Fig. 1.1 for an element being distorted due to an external applied load.

Consequently, the reference point P becomes P^* and the element gets distorted at angle θ. In general, the strains associated with distortion in three dimensions are defined by

$$
\epsilon_x = \frac{\partial \mu_x}{\partial x} \qquad \gamma_{xy} = \frac{\partial \mu_y}{\partial x} + \frac{\partial \mu_x}{\partial y} = \frac{\tau_{xy}}{G}
$$

$$
\epsilon_y = \frac{\partial \mu_y}{\partial y} \qquad \gamma_{yz} = \frac{\partial \mu_z}{\partial y} + \frac{\partial \mu_y}{\partial z} = \frac{\tau_{yz}}{G} \qquad (1.4)
$$

$$
\epsilon_z = \frac{\partial \mu_z}{\partial z} \qquad \gamma_{zx} = \frac{\partial \mu_x}{\partial z} + \frac{\partial \mu_z}{\partial x} = \frac{\tau_{zx}}{G}
$$

where G = Shear modulus of rigidity (MPa)

According to the theory of elasticity, stresses and strains are generalized as $\epsilon_{ij} = f(\sigma_{ij})$, $\gamma_{ij} = f(\tau_{ij})$, $\sigma_{ij} = f(\epsilon_{ij})$, and $\tau_{ij} = f(\gamma_{ij})$. These quantities are treated as second-rank tensors, and the matching mathematical framework of tensor analysis can be found elsewhere [3,4]. It is not intended herein to review the theory of elasticity, but include simplified forms of stresses and strains so that the reader is reminded about the use of these second-rank tensors as powerful tools for solving engineering problems or situations.

Fig. 1.1 Displacement of a point

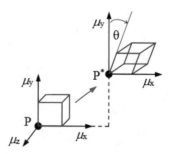

In addition, the relationships between stresses and strains are known as constitutive equations, which are classified as equilibrium equations, compatibility equations, and boundary equations [1].

Problems, whose solutions require the analysis of stresses, strains, and displacements, are normally encountered in engineering structures, which are susceptible to develop dangerous cracks during service. Thus, it is important to visualize stresses and strains as three-dimensional entities that develop around discontinuities in microstructures, such as dislocations.

Triaxial Sate of Stress Consider a homogeneous and isotropic solid body in equilibrium subjected to an external loading. As a result, elastic deformation takes place to an extent, provided that the applied external stress (σ) does not exceed the yield strength (σ_{ys}) of the body.

In order to understand the level and distribution of the resultant elastic internal stress, the use of a three-dimensional element helps visualizing the type and location of acting stresses on the element. Figure 1.2 shows an ideal model for defining the element of a solid being elastically deformed.

In general, the stresses acting on the entire solid body are continuously distributed over the surface and are the same acting on the element in equilibrium. At equilibrium, the shear stresses are related as $\tau_{xy} = \tau_{yx}$, $\tau_{yz} = \tau_{zy}$, and $\tau_{zx} = \tau_{xz}$, and the tensile stresses are σ_x, σ_y, and σ_z.

According to Hooke's law for isotropic solid materials, the strain components and related elastic stresses (Fig. 1.2) are defined below, without proof, for elastic bodies being elastically deformed in tension mode. Hence, the three-dimensional entities in Cartesian coordinates are

$$\epsilon_x = \frac{\sigma_x}{E} \qquad \epsilon_y = \epsilon_z = -\frac{\nu\sigma_x}{E}$$

$$\epsilon_y = \frac{\sigma_y}{E} \qquad \epsilon_x = \epsilon_z = -\frac{\nu\sigma_y}{E} \tag{1.5}$$

Fig. 1.2 Three-dimensional stress element

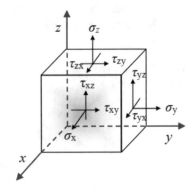

$$\epsilon_z = \frac{\sigma_z}{E} \qquad \epsilon_x = \epsilon_y = -\frac{\nu\sigma_z}{E}$$

$$\nu = -\frac{\epsilon_y}{\epsilon_x} = -\frac{\epsilon_z}{\epsilon_x}$$

where ν = Poisson's ratio

Using the **principle of superposition** (process of adding stress or strain solutions for an identical geometry), the strains and stresses in matrix form are

$$\begin{bmatrix} \epsilon_x \\ \epsilon_y \\ \epsilon_z \end{bmatrix} = \frac{1}{E} \begin{bmatrix} \sigma_x - \nu\left(\sigma_y + \sigma_z\right) \\ \sigma_y - \nu\left(\sigma_x + \sigma_z\right) \\ \sigma_z - \nu\left(\sigma_x + \sigma_y\right) \end{bmatrix} \tag{1.6}$$

For convenience, the stresses as functions of strains may be defined in a matrix form as indicated below. The stresses along the principal axes are

$$\begin{bmatrix} \sigma_x \\ \sigma_y \\ \sigma_z \end{bmatrix} = \frac{E}{(1+\nu)(1-2\nu)} \begin{bmatrix} (1-\nu)\,\epsilon_x + \nu\left(\epsilon_y + \epsilon_z\right) \\ (1-\nu)\,\epsilon_y + \nu\left(\epsilon_x + \epsilon_z\right) \\ (1-\nu)\,\epsilon_z + \nu\left(\epsilon_x + \epsilon_y\right) \end{bmatrix} \tag{1.7}$$

and the shear stresses on planes are

$$\begin{bmatrix} \tau_{xy} \\ \tau_{yz} \\ \tau_{zx} \end{bmatrix} = G \begin{bmatrix} \gamma_{xy} \\ \gamma_{yz} \\ \gamma_{zx} \end{bmatrix} = \frac{E}{2\,(1+\nu)} \begin{bmatrix} \gamma_{xy} \\ \gamma_{yz} \\ \gamma_{zx} \end{bmatrix} \tag{1.8}$$

since the shear modulus of elasticity is

$$G = \frac{E}{2\,(1+\nu)} \tag{1.9}$$

where E = Elastic modulus of elasticity (MPa)

Biaxial Sate of Stress If $\sigma_z = \tau_{zx} = \tau_{zy} = \gamma_{xz} = \gamma_{yz} = 0$, then Eq. (1.6) gives the strain entities for the biaxial state. These are defined in matrix form below

$$\begin{bmatrix} \epsilon_x \\ \epsilon_y \\ \epsilon_z \end{bmatrix} = \frac{1}{E} \begin{bmatrix} \sigma_x - \nu\sigma_y \\ \sigma_y - \nu\sigma_x \\ -\nu\left(\sigma_x + \sigma_y\right) \end{bmatrix} \tag{1.10}$$

$$\begin{bmatrix} \gamma_{xy} \\ \gamma_{xz} \\ \gamma_{yz} \end{bmatrix} = \frac{1}{G} \begin{bmatrix} \tau_{xy} \\ 0 \\ 0 \end{bmatrix} \tag{1.11}$$

and the stresses are

$$
\begin{bmatrix} \sigma_x \\ \sigma_y \\ \sigma_z \end{bmatrix} = \frac{E}{(1 - v^2)} \begin{bmatrix} \epsilon_x + v\epsilon_y \\ \epsilon_y + v\epsilon_x \\ 0 \end{bmatrix} \tag{1.12}
$$

$$
\begin{bmatrix} \tau_{xy} \\ \tau_{yz} \\ \tau_{zx} \end{bmatrix} = \frac{E}{2(1 + v)} \begin{bmatrix} \gamma_{xy} \\ 0 \\ 0 \end{bmatrix} \tag{1.13}
$$

In addition, the principle stresses and strains occur on the main axes, and their maximum and minimum values can be predicted using the Mohr's circle on a point. Mohr allows the determination of the normal and shear stress in a two-dimensional plane.

Hence, if $\sigma_z = 0$, then the principle stresses and strains can be predicted from the following expressions:

$$
\sigma_{1,2} = \frac{\sigma_x + \sigma_y}{2} \pm \sqrt{\left(\frac{\sigma_x - \sigma_y}{2}\right)^2 + \tau_{xy}^2} \tag{1.14}
$$

$$
\sigma_{1,2} = \frac{\sigma_x + \sigma_y}{2} \pm \tau_{max} \tag{1.15}
$$

Recall that the third principal stress is perpendicular to the outward plane of the paper implying that $\sigma_3 = \sigma_z$. In addition, if the shear stress $\tau_{xy} = 0$, then σ_1 and σ_2 are principle stresses, which are related to their principal directions. The angle between the principal directions is 90°.

Conversely, the principle strains are strains in the direction of the principle stresses. For a two-dimensional analysis, the principle strains are determined using the following quadratic expression:

$$
\epsilon_{1,2} = \frac{\epsilon_x + \epsilon_y}{2} \pm \sqrt{\left(\frac{\epsilon_x - \epsilon_y}{2}\right)^2 + \left(\frac{\gamma_{xy}}{2}\right)^2} \tag{1.16}
$$

The principle stresses and principle strains are the components of the stress tensor and strain tensor when the shear stress components become zero. A tensor stands for an entity that has nine components (σ_{ij} or ϵ_{ij}) defining the stress state, and it is defined as vector $\boldsymbol{T}^{(n)} = \boldsymbol{n} \cdot \boldsymbol{\sigma}$ or $T^{(n)} = n\sigma_{ij}$, where $\boldsymbol{n}$ is the unit length vector.

Example 1.1. *Compute the principle stresses acting on a steel machine part if the rectangular stress components are*

$$
\sigma_x = 280 \text{ MPa} \quad \sigma_y = -120 \text{ MPa} \quad \sigma_z = 140 \text{ MPa}
$$

$$
\tau_{xy} = 280 \text{ MPa} \quad \tau_{yz} = 0 \quad \tau_{zx} = 0
$$

Solution. *From Eq. (1.14),*

$$\sigma_{1,2} = \frac{280 - 120}{2} \pm \sqrt{\left(\frac{280 + 120}{2}\right)^2 + 280^2}$$

$$\sigma_{1,2} = 80 \text{ MPa} \pm \sqrt{1.184 \times 10^5}$$

$$\sigma_{1,2} = 80 \text{ MPa} \pm 344 \text{ MPa}$$

$$\sigma_1 = 80 \text{ MPa} + 344 \text{ MPa} = 424 \text{ MPa}$$

$$\sigma_2 = 80 \text{ MPa} - 344 \text{ MPa} = -264 \text{ MPa}$$

$$\sigma_3 = \sigma_z = 140 \text{ MPa}$$

1.4   Engineering Stress State

This section includes the engineering stress that is based on specimen bulk dimensions and the applied external loading mode, while the engineering strain depends on specimen bulk dimensions during or after deformation of a solid body. On the other hand, the theory of elasticity provides internal or surface elastic stresses, strains, and displacements in two or three dimensions that develop during elastic deformation.

For uniaxial tension testing, the state of stress and the state of strain are described by the uniaxial relationships. From an engineering point of view, the tensile or longitudinal strain is defined as elongation or stretching, which is related to Hooke's law of elastic deformation. For a uniaxial tensile test on a crack-free specimen shown in Fig. 1.3, the strain and Hooke's law are

$$\epsilon_t = \int_{l_o}^{l} \frac{dl}{l} = \ln\left(\frac{l}{l_o}\right) \qquad \text{(natural or true strain)} \qquad (1.17)$$

$$\epsilon = \frac{\Delta l}{l_o} \qquad \text{(engineering strain)} \qquad (1.18)$$

$$\epsilon = \frac{\sigma}{E} \qquad \text{(Hooke's Law)} \qquad (1.19)$$

Here, Δl is the change in gage length of a line segment between two points on a solid, and l_o is the original gage length. It is clear that Hooke's law gives a linear stress-strain relationship. Most structural materials have some degree of plasticity, which is not defined by Hooke's law.

In general, the mechanical behavior of a material under a stress-loading mode depends on the microstructure, strain rate, and environment. The behavior of an initially crack-free material is characterized by one of the typical stress-strain curves shown in Fig. 1.4.

Fig. 1.3 Schematic tensile
crack-free round specimen

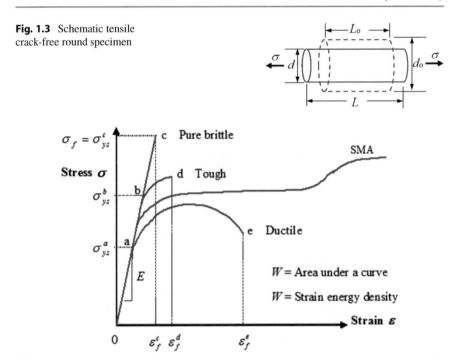

Fig. 1.4 Schematic stress-strain curves

Typical tension properties, such as yield strength, tensile strength, ductility, and the modulus of elasticity, are obtained from these curves. The strength refers to a property and stress is a parameter related to an applied loading mode.

Nevertheless, the area under the curve is a measure of fracture toughness in terms of strain energy density, which is not a common variable used by engineers in structural analysis, but it may be used as a controlling parameter in classifying structural materials. Particularly, the *SMA* curve in Fig. 1.4 is for a shape memory alloy, such as 55Ni-45Ti (nitinol), which exhibits significant high strain to failure (superelastic) and high total strain energy density [5, 6].

In fact, the strain energy density W (Joules/m^3) is the energy required to deform the material. According to Fig. 1.4, this energy is the area under the curve. For elastic behavior (up to the yield point), fracture toughness is the elastic strain energy density known as resilience, and it is defined as

$$W = \int_0^\epsilon \sigma \, d\epsilon \tag{1.20}$$

This expression represents an elastic behavior up to the yield strain for points a, b, c, and d in Fig. 1.4. Hooke's law, Eq. (1.19), is used to solve the integral given by Eq. (1.20). Thus, the elastic strain energy density becomes

$$W = \int_o^{\epsilon_{ys}} E\epsilon \, d\epsilon = \frac{1}{2} E\epsilon_{ys}^2 = \frac{\sigma_{ys}^2}{2E} \tag{1.21}$$

On the other hand, **tough materials** have fracture toughness based on *oe* and SMA curves. Thus, the strain energy density for curve *oe* takes the form

$$W = \int_o^{\epsilon_f^e} \sigma \, d\epsilon = \int_o^{\epsilon_{ys}} \sigma \, d\epsilon + \int_{\epsilon_{ys}}^{\epsilon_f^e} \sigma \, d\epsilon \tag{1.22}$$

$$W = \frac{\sigma_{ys}^2}{2E} + \int_{\epsilon_{ys}}^{\epsilon_f^e} \sigma \, d\epsilon \tag{1.23}$$

This integral can be solved once a stress function in terms of strain, $\sigma = f(\epsilon)$, is available. The most common **plastic stress functions** applicable from the yield strength σ_{ys} (*YS*) to the ultimate tensile strength or simply the tensile strength σ_{ts} (*TS*) as the maximum strength on a stress-strain curve, Fig. 1.4, are known as Ramberg–Osgood [7] and Hollomon [8] equations. These functions are defined as

$$\sigma = \sigma_{ys} \left(\frac{\epsilon}{\alpha' \epsilon_{ys}} \right)^{1/n'} \quad \text{for } \sigma \geq \sigma_{ys} \quad \text{(Ramberg–Osgood)} \tag{1.24}$$

$$\sigma = k\epsilon^n \quad \text{for } \sigma \geq \sigma_{ys} \quad \text{(Hollomon)} \tag{1.25}$$

where n, n' = Strain-hardening exponents
   k = Strength coefficient or proportionality constant (MPa)
   σ = Plastic stress (MPa)
   σ_{ys} = Yield strength (MPa)
   ϵ = Plastic strain
   α' = Constant

Both Eqs. (1.24) and (1.25) give effective or true plastic stresses and plastic strains within the convex shape (parabolic) of a stress-strain curve. Conversely, the strain-hardening exponent n or n' measures the rate at which a metallic solid body becomes strengthened or hardened as a result of plastic straining. The main mechanism for plastic straining arises due to dislocation interactions in most polycrystalline structures. This physical phenomenon can also be referred to as dislocation strengthening which is very important in cold-work hardening or simply work hardening. However, plastic deformation due to martensitic phase transformation (stainless steels), mechanical twinning, strain rate, and testing temperature can affect the convex shape of a stress-strain curve, and therefore, Eqs. (1.24) and (1.25) may not describe the plastic behavior of a metal or an alloy. The magnitude of the strain-hardening exponent depends on the nature of the material. In fact, $n < 1$ and $n' > 1$ due to the nature of the empirical mathematical models.

In addition, Eq. (1.24) predicts that $\sigma \rightarrow \infty$ as $\epsilon \rightarrow \infty$ for $0 < n < 1$, but experimental observations reveal that σ must have a finite magnitude at high strains. On the other hand, if $\sigma \rightarrow k$ when $n \rightarrow 0$, then the material approaches a true plastic behavior with vanishing strain or work-hardening capability.

For a strain hardenable material, the Hollomon or power-law equation may be used as an effective stress expression in Eq. (1.23) so that the integral can easily be solved. Inserting Eq. (1.25) into (1.23) and integrating yields the total strain energy density up to the tensile strength (TS) point

$$W = \frac{\sigma_{ys}^2}{2E} + \frac{k\epsilon^{n+1}}{n+1}\bigg|_{\epsilon_{ys}}^{\epsilon_{ts}} \tag{1.26}$$

$$W = \frac{\sigma_{ys}^2}{2E} + \frac{k\left(\epsilon_{ts}^{n+1} - \epsilon_{ys}^{n+1}\right)}{n+1} \tag{1.27}$$

$$W = W_e + W_p \tag{1.28}$$

Here, the first and second terms are the elastic (W_e) and plastic (W_p) strain energy densities. Since Hooke's law applies up to the yield point, the total energy as per Eq. (1.28) takes the form

$$W = \frac{\sigma_{ys}^2}{2E} - \frac{k}{n+1}\left(\frac{\sigma_{ys}}{E}\right)^{n+1} + \frac{k\epsilon_{ts}^{n+1}}{n+1} \tag{1.29}$$

An ideal tough material must exhibit high strength and ductility. Despite that ductile materials are considered tough; they have low strength and high ductility. However, if a notched tensile specimen made of a ductile material is loaded in tension, the plastic flow is shifted upward since a triaxial state of stress is developed at the root of the notch. This is a constraint against plastic flow, but it enhances the magnitude of the elastic stresses at the notch root [9].

In summary, the yield strength (material property) and the fracture toughness in terms of total strain energy density of crack-free materials can be compared using the inequalities shown below

$$\sigma_{ys}^{ductile} < \sigma_{ys}^{tough} < \sigma_{ys}^{brittle} \tag{1.30}$$

$$W_{brittle} < W_{tough} < W_{ductile} \tag{1.31}$$

This analogy implies that the yield strength decreases and the total strain energy density increases with increasing strength and decreasing strain to failure. These expressions can be used for classifying solid materials. However, an ideal material for practical engineering applications should be characterized according to the above inequalities, but slightly modified as indicated below for certain applications. Therefore,

$$\sigma_{ys}^{ductile} << \sigma_{ys}^{tough} \approx \sigma_{ys}^{brittle} \tag{1.32}$$

$$W_{brittle} << W_{tough} \approx W_{ductile} \tag{1.33}$$

since high ductility is not desired.

1.4.1   Plane Conditions

One important material's condition for characterizing the mechanical behavior of either a cracked or a crack-free specimen is its thickness. Thus, plane conditions are classified below.

Plane Stress: This is a stress condition used for thin bodies (plates), in which the specimen thickness must be $B \ll w$, where w is the width. Therefore, the negligible stresses through thickness are

$$\sigma_z = \tau_{yz} = \tau_{zx} = 0 \tag{1.34}$$

Most solid bodies under a quasi-static or dynamic loading are subjected to monotonic or fracture mechanics testing, where $\sigma_z = 0$ at the surface and $\sigma_z \simeq 0$ at the mid-thickness plane.

Plane Strain: This particular condition is for thick bodies, which develop a triaxial state of local stress at the crack tip. The through-thickness stress in Cartesian coordinates is

$$\sigma_z \simeq v\left(\sigma_x + \sigma_y\right) \tag{1 35}$$

which is a controlling stress entity.

1.5   Equilibrium Equations

The objective of this section is to show the equilibrium field equations used for analytically deriving solutions to the unknown elastic stresses σ_x, σ_y, and τ_{xy}. Subsequently, this requires an elementary treatment of the theory of elasticity.

Rectangular Coordinates Consider a small two-dimensional element of unit thickness in equilibrium as shown in Fig. 1.5.

Fig. 1.5 Stresses in Cartesian coordinates

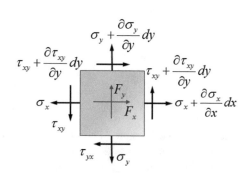

For a two-dimensional stress analysis as per Fig. 1.5, the sums of the forces in the x-direction and y-direction in Cartesian coordinates are respectively

$$\sum F_x = 0 \tag{a}$$

$$0 = \left(\sigma_x + \frac{\partial \sigma_x}{\partial x}dx\right)dy - \sigma_x dy \tag{b}$$

$$+ \left(\tau_{yx} + \frac{\partial \tau_{xy}}{\partial y}dy\right)dx - \tau_{yx}dx + F_x dxdy$$

$$\sum F_y = 0 \tag{c}$$

$$0 = \left(\sigma_y + \frac{\partial \sigma_y}{\partial x}dx\right)dy - \sigma_y dy \tag{d}$$

$$+ \left(\tau_{xy} + \frac{\partial \tau_{xy}}{\partial y}dy\right)dx - \tau_{xy}dx + F_y dxdy$$

Divide Eqs. (b) and (d) by $dxdy$; let $dx \to 0$ and $dy \to 0$ and $\tau_{yx} = \tau_{xy}$ due to symmetry. Hence, the equilibrium equations become

$$\frac{\partial \sigma_x}{\partial x} + \frac{\partial \tau_{xy}}{\partial y} + F_x = 0 \tag{1.36}$$

$$\frac{\partial \tau_{xy}}{\partial y} + \frac{\partial \sigma_y}{\partial x} + F_y = 0 \tag{1.37}$$

Here, F_x and F_y are body-force intensities. The resultant stress equations in plane Cartesian coordinates along with body forces are defined components of the stress tensor [1, 10, 11]. Hence,

$$\sigma_x = \frac{\partial^2 \phi}{\partial y^2} + F_x$$

$$\sigma_y = \frac{\partial^2 \phi}{\partial x^2} + F_y \tag{1.38}$$

$$\tau_{xy} = -\frac{\partial^2 \phi}{\partial x \partial y}$$

The body forces F_x and F_y commonly arise from the gravitational field, which has insignificant effects on the elastic stresses given by (1.38). In such a case, $F_x = 0$ and $F_y = 0$ for computational purposes.

Fig. 1.6 Stresses in polar coordinates

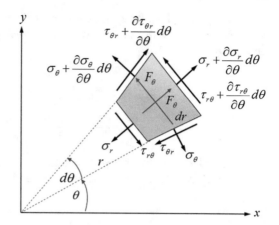

Polar Coordinates Consider a small two-dimensional element in equilibrium of unit thickness shown in Fig. 1.6.

For a two-dimensional stress analysis as per Fig. 1.6, the sum of the forces in the radial direction (r-direction) and tangential direction (θ-direction) in polar coordinates are, respectively

$$\sum F_r = 0 \tag{a}$$

$$0 = \left(\sigma_r + \frac{\partial \sigma_r}{\partial r} dr \right) (r + dr) \, d\theta - \sigma_r r d\theta - \tau_{\theta r} dr \cos (d\theta/2) \tag{b}$$

$$+ \left(\tau_{r\theta} + \frac{\partial \tau_{\theta r}}{\partial \theta} d\theta \right) dr \cos (d\theta/2) - \left(\sigma_\theta + \frac{\partial \sigma_\theta}{\partial \theta} d\theta \right) dr \sin (d\theta/2)$$

$$- \sigma_\theta dr \sin (d\theta/2) + F_r r dr d\theta$$

and

$$\sum F_\theta = 0 \tag{c}$$

$$0 = \left(\tau_{r\theta} + \frac{\partial \tau_{r\theta}}{\partial r} dr \right) (r + dr) \, d\theta - \tau_{r\theta} r d\theta - \tau_{r\theta} dr \sin (d\theta/2) \tag{d}$$

$$+ \left(\tau_{r\theta} + \frac{\partial \tau_{r\theta}}{\partial \theta} d\theta \right) dr \sin (d\theta/2) - \left(\sigma_\theta + \frac{\partial \sigma_\theta}{\partial \theta} d\theta \right) dr \cos (d\theta/2)$$

$$- \sigma_\theta dr \cos (d\theta/2) + F_\theta r dr d\theta$$

Rearranging and dividing Eqs. (b) and (d) by $drd\theta$, letting $\tau_{r\theta} = \tau_{\theta r}$ and, $dr \to 0$ and $d\theta \to 0$ in an infinitely small element gives the equilibrium equations defined by

$$\frac{\partial \sigma_r}{\partial r} + \frac{1}{r}\frac{\partial \tau_{r\theta}}{\partial \theta} + \frac{1}{r}(\sigma_r - \sigma_\theta) + F_r = 0 \tag{1.39}$$

$$\frac{1}{r}\frac{\partial \sigma_\theta}{\partial \theta} + \frac{\partial \tau_{r\theta}}{\partial r} + \frac{2\tau_{r\theta}}{r} + F_\theta = 0 \tag{1.40}$$

Here, F_r and F_θ are also body-force intensities. The resultant stress equations in polar coordinates are defined by Timoshenko and Goodier [1], Sadd [11], Dally and Riley [10]

$$
\begin{aligned}
\sigma_r &= \frac{1}{r}\frac{\partial \phi}{\partial r} + \frac{1}{r^2}\frac{\partial^2 \phi}{\partial \theta^2} \\[2mm]
\sigma_\theta &= \frac{\partial^2 \phi}{\partial r^2} \\[2mm]
\tau_{r\theta} &= \frac{1}{r^2}\frac{\partial \phi}{\partial \theta} - \frac{1}{r}\frac{\partial^2 \phi}{\partial r \partial \theta} = -\frac{\partial}{\partial r}\left(\frac{1}{r}\frac{\partial \phi}{\partial \theta}\right)
\end{aligned}
\tag{1.41}
$$

In general, body forces are categorized as gravitational forces (gravitational forces), electromagnetic forces (electromagnetic field), and inertial forces (motion). For example, gravitational forces arise when a single fixed-end cantilever beam is subjected to its own mass-weight force, while inertial forces arise from a rotating shaft with an accelerating force direction acting on it with opposite direction [11]. The mass of the shaft is in motion and acquires an acceleration of low magnitude. Consequently, inertial forces are induced by motion and are considered fictitious forces. Recall that inertia is the resistance of a body to changes in its state of motion. On the other hand, internal forces arise within a body being subjected to external forces and are induced uniformly and continuously within the body.

Normally, the rate of work done on an elastic solid by the external loads determines the state of elastic deformation as a quasi-static or dynamics. Subsequently, the related Airy elastic stresses are determined without the body-force components, which are considered vanish within the solid.

1.6 Biharmonic Equation

The goal in this section is to show how to transform the rectangular biharmonic equation into polar coordinates using the Airy stress function ϕ. The limitation of any arbitrary ϕ function is that it must satisfy the biharmonic equation. Once this is accomplished, the stresses given by Eq. (1.41) are also satisfied.

The starting point is to find a relation between rectangular and polar coordinates. This is given by

$$r^2 = x^2 + y^2, \quad r = \sqrt{x^2 + y^2}$$
$$x = r\cos\theta, \quad y = r\sin\theta \tag{1.42}$$
$$\theta = \arctan\left(\frac{y}{x}\right)$$

from which

$$\frac{\partial r}{\partial x} = \frac{x}{r} = \cos\theta, \qquad \frac{\partial\theta}{\partial x} = -\frac{y}{r^2} = -\frac{1}{r}\sin\theta \tag{1.43}$$

$$\frac{\partial r}{\partial y} = \frac{y}{r} = \sin\theta, \qquad \frac{\partial\theta}{\partial y} = -\frac{x}{r^2} = +\frac{1}{r}\cos\theta$$

If $\phi = f(r, \theta)$, then

$$\frac{\partial\phi}{\partial x} = \frac{\partial\phi}{\partial r}\frac{\partial r}{\partial x} + \frac{\partial\phi}{\partial\theta}\frac{\partial\theta}{\partial x} = \frac{\partial\phi}{\partial r}\cos\theta - \frac{1}{r}\frac{\partial\phi}{\partial\theta}\sin\theta \tag{1.44}$$

$$\frac{\partial\phi}{\partial y} = \frac{\partial\phi}{\partial r}\frac{\partial r}{\partial y} + \frac{\partial\phi}{\partial\theta}\frac{\partial\theta}{\partial y} = \frac{\partial\phi}{\partial r}\sin\theta + \frac{1}{r}\frac{\partial\phi}{\partial\theta}\sin\theta$$

Thus,

$$\frac{\partial^2\phi}{\partial x^2} = \left(\frac{\partial}{\partial r}\cos\theta - \frac{1}{r}\sin\theta\frac{\partial}{\partial\theta}\right)\left(\frac{\partial\phi}{\partial r}\cos\theta - \frac{1}{r}\frac{\partial\phi}{\partial\theta}\sin\theta\right)$$

$$= \frac{\partial^2\phi}{\partial r^2}\cos^2\theta - \frac{2}{r}\frac{\partial^2\phi}{\partial\theta\partial r}\cos\theta\sin\theta + \frac{1}{r}\frac{\partial\phi}{\partial r}\sin^2\theta \tag{1.45}$$

$$+ \frac{2}{r^2}\frac{\partial\phi}{\partial\theta}\cos\theta\sin\theta + \frac{1}{r^2}\frac{\partial^2\phi}{\partial\theta^2}\sin^2\theta$$

and

$$\frac{\partial^2\phi}{\partial y^2} = \left(\frac{\partial}{\partial r}\sin\theta - \frac{1}{r}\sin\theta\frac{\partial}{\partial\theta}\right)\left(\frac{\partial\phi}{\partial r}\sin\theta + \frac{1}{r}\frac{\partial\phi}{\partial\theta}\sin\theta\right)$$

$$= \frac{\partial^2\phi}{\partial r^2}\sin^2\theta + \frac{2}{r}\frac{\partial^2\phi}{\partial\theta\partial r}\cos\theta\sin\theta + \frac{1}{r}\frac{\partial\phi}{\partial r}\cos^2\theta \tag{1.46}$$

$$- \frac{2}{r^2}\frac{\partial\phi}{\partial\theta}\cos\theta\sin\theta + \frac{1}{r^2}\frac{\partial^2\phi}{\partial\theta^2}\cos^2\theta$$

Adding Eqs. (1.45) and (1.46) gives

$$\frac{\partial^2\phi}{\partial x^2} + \frac{\partial^2\phi}{\partial y^2} = \frac{\partial^2\phi}{\partial r^2} + \frac{1}{r}\frac{\partial\phi}{\partial r} + \frac{1}{r^2}\frac{\partial^2\phi}{\partial\theta^2} \tag{1.47}$$

Take the fourth derivatives of Eq. (1.47) to get the biharmonic equations in rectangular and polar coordinates. Hence,

$$\nabla^4 \phi\,(x, y) = \left(\frac{\partial^2}{\partial x^2} + \frac{\partial^2}{\partial y^2} \right) \left(\frac{\partial^2 \phi}{\partial x^2} + \frac{\partial^2 \phi}{\partial y^2} \right)$$

$$\nabla^4 \phi\,(x, y) = \frac{\partial^4 \phi}{\partial x^4} + 2\frac{\partial^4 \phi}{\partial x^2 \partial y^2} + \frac{\partial^4 \phi}{\partial y^4} = 0 \tag{1.48}$$

and that for polar coordinates becomes

$$\nabla^4 \phi\,(r, \theta) = \left(\frac{\partial^2}{\partial r^2} + \frac{1}{r}\frac{\partial}{\partial r} + \frac{1}{r^2}\frac{\partial^2}{\partial \theta^2} \right) \left(\frac{\partial^2 \phi}{\partial r^2} + \frac{1}{r}\frac{\partial \phi}{\partial r} + \frac{1}{r^2}\frac{\partial^2 \phi}{\partial \theta^2} \right) = 0 \tag{1.49}$$

Therefore, $\nabla^4 \phi\,(x, y) = \nabla^4 \phi\,(r, \theta) = 0$.

Expanding and simplifying Eq. (1.49) requires a step-by-step algebraic manipulation of the partial derivatives of any Airy stress function ϕ. The algebraic manipulation is carried out by the nine multiplication steps. They are:

$$\text{(1)}\quad \frac{\partial}{\partial r^2} \left(\frac{\partial^2 \phi}{\partial r^2} \right) = \frac{\partial^4 \phi}{\partial r^4}$$

$$\text{(2)}\quad \frac{\partial^2}{\partial r^2} \left(\frac{1}{r}\frac{\partial \phi}{\partial r} \right) = \frac{2}{r^3}\frac{\partial \phi}{\partial r} - \frac{2}{r^2}\frac{\partial^2 \phi}{\partial r^2} + \frac{1}{r}\frac{\partial^3 \phi}{\partial r^3}$$

$$\text{(3)}\quad \frac{\partial^2}{\partial r^2} \left(\frac{1}{r^2}\frac{\partial^2 \phi}{\partial \theta^2} \right) = \frac{6}{r^4}\frac{\partial^2 \phi}{\partial \theta^2} - \frac{4}{r^3}\frac{\partial^3 \phi}{\partial r \partial \theta^2} + \frac{1}{r^2}\frac{\partial^4 \phi}{\partial r^2 \partial \theta^2}$$

$$\text{(4)}\quad \frac{1}{r}\frac{\partial}{\partial r} \left(\frac{\partial^2 \phi}{\partial r^2} \right) = \frac{1}{r}\frac{\partial^3 \phi}{\partial r^3} \tag{1.50}$$

$$\text{(5)}\quad \frac{1}{r}\frac{\partial}{\partial r} \left(\frac{1}{r}\frac{\partial \phi}{\partial r} \right) = \frac{1}{r^2}\frac{\partial^2 \phi}{\partial r^2} - \frac{1}{r^3}\frac{\partial \phi}{\partial r}$$

$$\text{(6)}\quad \frac{1}{r}\frac{\partial}{\partial r} \left(\frac{1}{r^2}\frac{\partial^2 \phi}{\partial \theta^2} \right) = \frac{1}{r^3}\frac{\partial^3 \phi}{\partial r \partial \theta^2} - \frac{2}{r^4}\frac{\partial^2 \phi}{\partial \theta^2}$$

$$\text{(7)}\quad \frac{1}{r^2}\frac{\partial^2}{\partial \theta^2} \left(\frac{\partial^2 \phi}{\partial r^2} \right) = \frac{1}{r}\frac{\partial^4 \phi}{\partial \theta^2 \partial r^2}$$

$$\text{(8)}\quad \frac{1}{r^2}\frac{\partial^2}{\partial \theta^2} \left(\frac{1}{r}\frac{\partial \phi}{\partial r} \right) = \frac{1}{r^3}\frac{\partial^3 \phi}{\partial \theta^2 \partial r}$$

$$\text{(9)}\quad \frac{1}{r^2}\frac{\partial^2}{\partial \theta^2} \left(\frac{1}{r^2}\frac{\partial^2 \phi}{\partial \theta^2} \right) = \frac{1}{r^4}\frac{\partial^4 \phi}{\partial \theta^4}$$

The subsequent step for completing this analytical procedure is the determination of the elastic stresses using a particular Airy stress function ϕ. This is accomplished in the next section.

1.7   The Airy Stress Function

The Airy stress function approach [12] is used in order to analytically determine the unknown stresses σ_x, σ_y, and τ_{xy} in two-dimensional elasticity problem. The use of the type of coordinates depends on the nature of the problem and the complexity of the needed analytical approach, such as the equilibrium, compatibility, and governing biharmonic equation. Hence, the objective of this section is to describe the method for finding solutions of engineering problems using the Airy stress function ϕ, which must also satisfy boundary conditions.

In a two-dimensional analysis, the general mathematical definition of the elastic stress field in terms of the Airy stress function in Cartesian coordinates is

$$\sigma_x = \frac{\partial^2 \phi}{\partial y^2} + \Omega$$

$$\sigma_y = \frac{\partial^2 \phi}{\partial x^2} + \Omega \qquad (1.51)$$

$$\tau_{xy} = -\frac{\partial^2 \phi}{\partial x \partial y}$$

Here, $\Omega = \Omega(x, y)$ is the body-force field that includes forces due to gravity, water pressure in porous materials, and centrifugal forces in rotating machine parts. Once an Airy stress function ϕ is selected, the stress solutions may not necessarily satisfy the equilibrium equation. Instead, the biharmonic equation is used in order to verify if the stress definitions meet the equilibrium requirements.

The body-force intensities or the body force of magnitudes are

$$F_x = -\frac{\partial \Omega}{\partial x} \qquad F_y = -\frac{\partial \Omega}{\partial y} \qquad (1.52)$$

The stress compatibility equation is [1, 10, 11]

$$\nabla^2 \left(\sigma_x + \sigma_y \right) = -\frac{\beta}{\nu} \left(\frac{\partial F_x}{\partial x} + \frac{\partial F_y}{\partial y} \right) \qquad (1.53)$$

where $\beta = 1$ for plane stress condition
$\beta = 1/(1 - \nu)$ for plane strain condition
For a crack-free hollow cylinder is of the type [10]

$$\phi = a_1 + a_2 \ln(r) + a_3 r^2 + a_4 r^2 \ln(r) \qquad (1.54)$$

and for a cracked plate, ϕ may be of the form [13]

$$\phi = r^{\lambda+1} f(\theta) \tag{1.55}$$

$$\phi = g(r) f(\theta) \tag{1.56}$$

where λ is an eigenvalue and $f(\theta)$ is an unknown trigonometric function.

1.7.1　Airy Power Series

Theoretically, any Airy stress function ϕ used for finding the solution to plane elasticity engineering problems with no body forces should satisfy a respective biharmonic equation. However, the choice of coordinates depends on the boundary conditions induced by the loading mode imposed on a particular specimen geometry. Excluding any environmental effects on a stressed body, the governing equations of interest must include primarily the stresses, which are summarized below.

- Rectangular coordinates: $\phi = \phi(x, y)$ for σ_x, σ_y, and τ_{xy} if $\nabla^4 \phi(x, y) = 0$
- Polar coordinates $\phi = \phi(r, \theta)$ for σ_r, σ_θ, and $\tau_{r\theta}$ if $\nabla^4 \phi(r, \theta) = 0$.

It should be mentioned that the Airy stress function ϕ is also used for determining the stress and strain fields around edge dislocations in isotropic and continuous media. A mathematical and theoretical treatment on this particular subject is given in a book written by Meyers and Chawla [3] in which isostress contours indicate the maximum tension, compression, and shear stresses.

Let the Airy stress function ϕ be defined as an Airy power series having a_i the polynomial coefficients

$$\phi = \sum_{i=1}^{m} a_i x^{m-i} y^{i-1} \tag{1.57}$$

For convenience, the first-order derivatives of this function, Eq. (1.57), with respect to two-dimensional Cartesian coordinates are

$$\frac{\partial \phi}{\partial x} = \sum_{I=1}^{m} (m-i) a_i x^{m-i-1} y^{i-1} \tag{1.58}$$

$$\frac{\partial \phi}{\partial y} = \sum_{I=1}^{m} (i-1) a_i x^{m-i} y^{i-2} \tag{1.59}$$

A polynomial described by Eq. (1.57) must satisfy the biharmonic expression, Eq. (1.48). For instance, let the order of the polynomial be

$$\phi = \sum_{i=1}^{5} a_i x^{5-i} y^{i-1} \tag{1.60}$$

$$\phi = a_1 x^4 + a_2 x^3 y + a_3 x^2 y^2 + a_4 xy^3 + a_5 y^4 \tag{1.61}$$

Thus, the pertinent derivatives are

$$\frac{\partial^4 \phi}{\partial x^4} = 24a_1$$

$$\frac{\partial^4 \phi}{\partial y^4} = 24a_5 \tag{1.62}$$

$$\frac{\partial^4 \phi}{\partial x^2 \partial y^2} = 4a_3$$

Substituting Eq. (1.62) into (1.48) yields a non-satisfactory result

$$\nabla^4 \phi = \frac{\partial^4 \phi}{\partial x^4} + 2\frac{\partial^4 \phi}{\partial x^2 \partial y^2} + \frac{\partial^4 \phi}{\partial y^4} \tag{1.63}$$

$$\nabla^4 \phi = 24a_1 + 8a_3 + 24a_5 \neq 0 \tag{1.64}$$

This problem can be solved by letting $\nabla^4 \phi = 0$ in Eq. (1.64) so that

$$24a_1 + 8a_3 + 24a_5 = 0 \tag{a}$$

Solving Eq. (a) for a_5 yields

$$a_5 = -(a_1 + a_3/3) \tag{b}$$

Now, substituting a_5 in Eq. (1.62) gives the redefined Airy stress function as

$$\phi = a_1\left(x^4 - y^4\right) + a_2 x^3 y + a_3\left(x^2 y^2 - \frac{1}{3}y^4\right) + a_4 xy^3 \tag{1.65}$$

from which the fourth-order derivatives are

$$\frac{\partial^4 \phi}{\partial x^4} = 24a_1$$

$$\frac{\partial^4 \phi}{\partial y^4} = -24a_1 - 8a_3 \tag{1.66a}$$

$$\frac{\partial^4 \phi}{\partial x^2 \partial y^2} = 4a_3$$

Substituting these fourth-order derivatives, Eq. (1.66a), into the biharmonic expression, Eq. (1.48), yields $\nabla^4\phi = 0$. The reader should verify these partial results as an exercise.

Therefore, Eq. (1.65) satisfies the condition $\nabla^4\phi = 0$. Subsequently, using Eqs. (1.51) and (1.65) gives the expected stress equations along with zero body-force field as

$$\sigma_x = \frac{\partial^2\phi}{\partial y^2} = -12a_1y^2 + 2a_3\left(x^2 - 2y^2\right) + 6a^4xy$$

$$\sigma_y = \frac{\partial^2\phi}{\partial x^2} = 12a_1x^2 + 6a_2xy + 2a_3y^2 \tag{1.66b}$$

$$\tau_{xy} = -\frac{\partial^2\phi}{\partial x\partial y} = -3a_2x^2 - 4a_3xy - 3a_4y^2$$

This particular exercise shows that a fourth-order polynomial does not satisfy the biharmonic equation unless the above adjustment is made. Apparently, a third-order polynomial can give a straightforward solution. This is left as an exercise.

An example can make this procedure sufficiently clear how to determine the elastic stresses using the Airy stress function for a cantilever beam.

Example 1.2. *Consider a cantilever beam of width b and height h being supported at its ends and uniformly loaded under plane stress condition by a load P per unit length. This is shown in the figure below.*

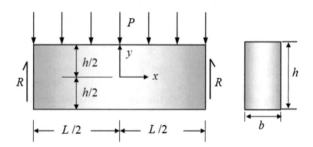

Find the stresses in Cartesian coordinates using an Airy stress function defined by (From ref. [10])

$$\phi = a_1x^2 + a_2x^2y + a_3\left(x^2y^3 - \frac{1}{5}y^5\right) + a_4y^3 \tag{a7}$$

Loading conditions:

$$\sigma_y = \tau_{xy} = 0 \quad \text{at } y = -h/2 \tag{a1}$$

$$\sigma_y = -P/b \quad \text{at } y = h/2 \tag{a2}$$

$$\tau_{xy} = 0 \qquad \text{at } y = h/2 \tag{a3}$$

$$R = \int_{-h/2}^{h/2} \tau_{xy} dy = \frac{PL}{2} \quad \text{at } x = \pm \frac{L}{2} \tag{a4}$$

$$\int_{-h/2}^{h/2} \sigma_x dy = 0 \quad \text{at } x = \pm \frac{L}{2} \tag{a5}$$

$$\int_{-h/2}^{h/2} \sigma_x y\, dy = 0 \quad \text{at } x = \pm \frac{L}{2} \tag{a6}$$

Solution. *Equations (a4) and (a5) indicate that there is no longitudinal force and bending coupling at the ends of the beam.*

Now, use the given polynomial to derive the derivatives of the fourth order

$$\phi = a_1 x^2 + a_2 x^2 y + a_3 \left(x^2 y^3 - \frac{1}{5} y^5 \right) + a_4 y^3 \tag{a7}$$

from which

$$\frac{\partial^4 \phi}{\partial x^4} = 0$$

$$\frac{\partial^4 \phi}{\partial y^4} = -24 a_3 y \tag{a8}$$

$$\frac{\partial^4 \phi}{\partial x^2 \partial y^2} = 12 a_3 y$$

These expressions satisfy the condition $\nabla^4 \phi = 0$. Thus, the elastic stresses take the form along with $\Omega = 0$

$$\sigma_x = \frac{\partial^2 \phi}{\partial y^2} = a_3 \left(6x^2 y - 4y^3 \right) + 6a_4 y \tag{a9}$$

$$\sigma_y = \frac{\partial^2 \phi}{\partial x^2} = 2a_1 + 2a_2 y + 2a_3 y^3 \tag{a10}$$

$$\tau_{xy} = -\frac{\partial^2 \phi}{\partial x \partial y} = -2a_2 x - 6a_3 x y^2 \tag{a11}$$

Let us use the given loading conditions in order to determine the polynomial constants a_i. Using Eq. (a10) along with the loading condition given by Eqs. (a1) and (a2) gives

$$\sigma_y = 2a_1 + 2a_2 y + 2a_3 y^3 = 0 \qquad \text{at } y = -h/2 \tag{b1}$$

$$\sigma_y = 2a_1 + 2a_2 y + 2a_3 y^3 = -p/b \quad \text{at } y = +h/2 \tag{b2}$$

or

$$a_1 - a_2 \left(\frac{h}{2}\right) - a_3 \left(\frac{h^3}{8}\right) = 0 \tag{b3}$$

$$a_1 - a_2 \left(\frac{h}{2}\right) - a_3 \left(\frac{h^3}{8}\right) = -P/2b \tag{b4}$$

Solving Eqs. (b3) and (b4) simultaneously for a_1 yields

$$a_1 = -\frac{P}{4b} \tag{b5}$$

Furthermore, using Eq. (a11) along with $y = -h/2$ yields

$$\tau_{xy} = -2a_2 x - 6a_3 x y^2 = 0 \tag{c1}$$

$$0 = -a_2 - 3a_3 y^2 \tag{c2}$$

$$0 = a_2 + 3a_3 \left(-\frac{h}{2}\right)^2 \tag{c3}$$

$$a_2 = -\frac{3}{4}a_3 h^2 \tag{c4}$$

Substituting Eqs. (b5) and (c4) into (b3) gives

$$a_3 = \frac{P}{bh^3} \tag{c5}$$

Combining Eqs. (a6) and (a9) yields

$$\int_{-h/2}^{h/2} \sigma_x y\, dy = 0 \quad @\ x = \frac{L}{2} \tag{d1}$$

$$\int_{-h/2}^{h/2} \left[a_3 \left(6x^2 y - 4y^3\right) + 6a_4 y\right] y\, dy = 0 \quad @\ x = \frac{L}{2} \tag{d2}$$

$$\int_{-h/2}^{h/2} \left[a_3 \left(6x^2 y^2 - 4y^4\right) + 6a_4 y^2\right] dy = 0 \quad @\ x = \frac{L}{2} \tag{d3}$$

$$\int_{-h/2}^{h/2} \left[a_3 \left(\frac{3}{2}L^2 y^2 - 4y^4\right) + 6a_4 y^2\right] dy = 0 \tag{d4}$$

Solving this integral and using Eq. (c5) gives

$$a_4 = a_3 \left(\frac{h^2}{10} - \frac{L^2}{4}\right) \tag{d5}$$

$$a_4 = \frac{P}{20bh^3}\left(2h^2 - 5L^2\right) \tag{d6}$$

If the moment of inertia of the beam having a rectangular cross section is $I = bh^3/12$, then Eq. (d6) becomes

$$a_4 = \frac{P}{240I}\left(2h^2 - 5L^2\right) \tag{d7}$$

Inserting a_1 through a_4 into Eqs. (a9) through (a11) yields the required stresses

$$\sigma_x = \frac{P}{8I}\left(4x^2 - L^2\right)y + \frac{P}{60I}\left(3h^2y - 20y^3\right) \tag{e1}$$

$$\sigma_y = \frac{P}{24I}\left(4y^2 - 3h^2y - h^3\right) \tag{e2}$$

$$\tau_{xy} = \frac{Px}{8I}\left(h^2 - 4y^2\right) \tag{e3}$$

Conventional strength of materials gives the stress in the x-direction as

$$\sigma_x = \frac{My}{I} = \frac{P}{8I}\left(4x^2 - L^2\right)y \tag{e4}$$

Here, M is the moment. Furthermore, Eq. (e4) implies that the second term can be interpreted as a correction term. Therefore, Eq. (e1) is more accurate than Eq. (e4) because of the extra term in the expression.

In conclusion, it has been shown that the use of an appropriate Airy stress function gives suitable results.

Example 1.3. **(a)** *A large beam is subjected to a line of uniform distribution of load as indicated below. Determine the elastic stresses in polar coordinates when (a) $\alpha = 90°$;* **(b)** *plot the resultant radial stress equation when $\theta = 30°$ and $\theta = 60°$, and $P = 100\,MPa\,mm$. The upward load reaction is defined as $R = \int_o^\pi \sigma_r r \sin\theta\, d\theta$. Assume that the analysis of the line of point forces included in this example provides accurate results.*

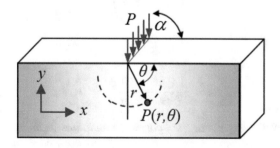

Solution.

$$\sum F_y = R - P = 0 \tag{a}$$

$$\int_o^\pi \sigma_r r \sin\theta d\theta - P = 0 \tag{b}$$

(a) *For $\alpha = 90°$ and $P = P_y$, let the Airy stress function be*

$$\phi = ar\theta \cos\theta \tag{c}$$

Using Eqs. (1.63) on (1.62) follows that

$$\nabla^4\phi = \left[\frac{\partial^4\phi}{\partial r^4}\right] + \left[\frac{2}{r^3}\frac{\partial\phi}{\partial r} - \frac{2}{r^2}\frac{\partial^2\phi}{\partial r^2}\right] + \left[\frac{6}{r^4}\frac{\partial^2\phi}{\partial\theta^2} - \frac{4}{r^3}\frac{\partial^3\phi}{\partial r\partial\theta^2} + \frac{1}{r^2}\frac{\partial^4\phi}{\partial r^2\partial\theta^2}\right]$$

$$+ \left[\frac{1}{r}\frac{\partial^3\phi}{\partial r^3}\right] + \left[\frac{1}{r}\left(-\frac{1}{r^2}\frac{\partial\phi}{\partial r} + \frac{1}{r}\frac{\partial^2\phi}{\partial r^2}\right)\right] + \frac{1}{r}\left(-\frac{2}{r^3}\frac{\partial^2\phi}{\partial\theta^2} + \frac{1}{r^2}\frac{\partial^3\phi}{\partial r\partial\theta^2}\right)$$

$$+ \left[\frac{1}{r}\frac{\partial^4\phi}{\partial r^2\partial\theta^2}\right] + \left[\frac{1}{r^3}\frac{\partial^3\phi}{\partial\theta^2\partial r}\right] + \left[\frac{1}{r^4}\frac{\partial^4\phi}{\partial\theta^4}\right] \tag{d}$$

and

$$\nabla^4\phi = 0 + \frac{2a\theta}{r^3}\cos\theta - \frac{1}{r^3}(4a\sin\theta + 2a\theta\cos\theta)$$

$$+ 0 - \frac{a}{r^3}\theta\cos\theta + \frac{1}{r^3}(2a\sin\theta + a\theta\cos\theta)$$

$$+ 0 - \frac{1}{r^3}(2a\sin\theta + a\theta\cos\theta) + \frac{1}{r^4}(4ar\sin\theta + ar\theta\cos\theta) \tag{e}$$

Therefore, Eq. (c) satisfies $\nabla^4\phi = 0$. Furthermore, the elastic stresses can be defined in terms of trigonometric functions. According to Eq. (1.58), the elastic stresses in polar coordinates are

$$\sigma_r = \frac{1}{r}\frac{\partial\phi}{\partial r} + \frac{1}{r^2}\frac{\partial^2\phi}{\partial\theta^2} = -\frac{2a}{r}\sin\theta \tag{f}$$

$$\sigma_\theta = \frac{\partial^2\phi}{\partial r^2} = 0 \tag{g}$$

$$\tau_{r\theta} = \frac{1}{r^2}\frac{\partial\phi}{\partial\theta} - \frac{1}{r}\frac{\partial^2\phi}{\partial r\partial\theta} = 0 \tag{h}$$

These results imply that the boundary conditions were correct. Combining Eqs. (b) and (f) provides the final expression for the load P and the constant a. Thus,

$$P = -\int_{o}^{\pi} \sigma_r r \sin\theta \, d\theta = -2a \int_{o}^{\pi} \sin^2\theta \, d\theta \tag{i1}$$

$$P = -\pi a \tag{i2}$$

$$a = -P/\pi \tag{i3}$$

Substituting Eq. (i3) into (f) yields the radial stress

$$\sigma_r = \frac{2P}{\pi r} \sin\theta \tag{j}$$

(b) *For $P = 100\,MPa\,mm$, $\sin(30°) = 1/2$, and $\sin(60°) = \sqrt{3}/2$, Eq. (j) gives the radial stress distribution shown in the figure below.*

$$\sigma_r = (31.831 MPa\,mm)/r \quad \text{for } \theta = 30° \tag{k1}$$

$$\sigma_r = (55.133\ MPa\,mm)/r \quad \text{for } \theta = 60° \tag{k2}$$

Thus, the distribution of σ_r is depicted in the figure below.

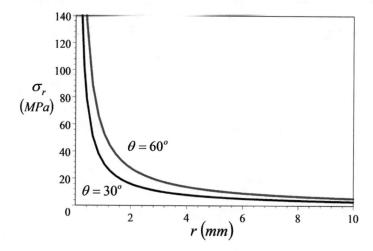

Denote that the radial stress, σ_r, for both selected angles decreases very rapidly from the upper edge of the plate for small radius r. These results imply that σ_r reaches a greater magnitude as the angle θ approaches $90° = \pi/2$ despite that the loading mode is vertically downward.

1.8   Complex Variable Theory

The application of complex variable theory for solving many plane elasticity problems [11, 14] is briefly introduced in this section to gain sufficient theoretical background on the subject matter. Hence, one will be able to find and accurately interpret solutions to problems using the powerful method of complex variable theory, which is usually an inherently two-dimensional approach where z is the complex variable. Figure 1.7 shows the complex plane of an infinite solid in Cartesian coordinates, where a point z and its conjugate $\bar{z}$ are located in the domain D. The letter C in this figure represents an arbitrary and continuous contour enclosing the complex z-plane.

For any complex function, the independent variable x and the dependent variable y may be separated into real and imaginary parts. Thus, the complex variable z defines a point as (x, y) in the domain D in the z-plane and has an image point called complex conjugate variable $\bar{z}$. These points are defined as

$$z = x + iy \quad \text{and} \quad \bar{z} = x - iy \tag{1.67a}$$

where $i = \sqrt{-1}$ is the imaginary unit ($i^2 = -1$, $i^3 = -i$, $i^4 = 1$), Re $(z) = x$ is the real part of z, and Im $(z) = y$ is the imaginary part of z.

Converting the complex variable z to polar coordinates yields the Euler's formula. Thus,

$$z = re^{i\theta} = r\left(\cos\theta + i\sin\theta\right) \tag{1.67b}$$

$$\bar{z} = re^{-i\theta} = r\left(\cos\theta - i\sin\theta\right) \tag{1.67c}$$

$$z^m = r^m e^{\pm im\theta} = r^m\left(\cos m\theta \pm i\sin m\theta\right) \tag{1.67d}$$

where

$$|z| = r = \sqrt{x^2 + y^2} = \sqrt{(x + iy)(x - iy)} \tag{1.67e}$$

$$|z| = r = \sqrt{z\bar{z}} \quad \text{and} \quad r^2 = z\bar{z} \tag{1.67f}$$

Fig. 1.7 Complex z-plane within a contour C

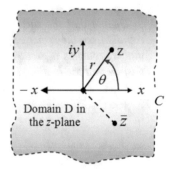

1.8.1 Cauchy–Riemann Equations

This section describes the Cauchy–Riemann equations (criterion) for characterizing a complex function $f(z)$ with respect to its analyticity on some domain D and its differentiability in D. Let

$$f(z) = u(x, y) + iv(x, y) \tag{1.68}$$

which is analytic if it satisfies the Cauchy–Riemann equations

$$\frac{\partial u}{\partial x} = \frac{\partial v}{\partial y} \tag{1.69}$$

$$\frac{\partial u}{\partial y} = -\frac{\partial v}{\partial x}$$

where $u(x, y)$ and $v(x, y)$ are real-valued continuous functions.

Analytic Function: Let

$$f(z) = z^2 = x^2 - y^2 + 2ixy \tag{1.70}$$

$$u = u(x, y) = x^2 - y^2 \tag{1.71a}$$

$$v = v(x, y) = 2xy \tag{1.71b}$$

The continuous first-order partial derivatives are

$$\frac{\partial u}{\partial x} = 2x \quad \& \quad \frac{\partial v}{\partial x} = 2y \tag{1.72}$$

$$\frac{\partial u}{\partial y} = -2y \quad \& \quad \frac{\partial v}{\partial y} = 2x \tag{1.73}$$

Thus,

$$\frac{\partial u}{\partial x} = \frac{\partial v}{\partial y} = 2x \quad \text{(satisfied)} \tag{1.74}$$

$$\frac{\partial u}{\partial y} = -\frac{\partial v}{\partial x} = -2y \quad \text{(satisfied)} \tag{1.75}$$

Therefore, $f(z) = z^2$ is analytic everywhere because it satisfies the Cauchy–Riemann criterion. The sum of the second partial derivatives yields the Laplace equations

$$\frac{\partial^2 u}{\partial x^2} + \frac{\partial^2 u}{\partial y^2} = 2 - 2 = 0 \tag{1.76}$$

$$\frac{\partial^2 v}{\partial x^2} + \frac{\partial^2 v}{\partial y^2} = 0 + 0 = 0 \tag{1.77}$$

These expressions, Eqs. (1.76) and (1.77), imply that u and v are harmonic.
Not Analytic Function: Let

$$f(z) = \bar{z} = x - iy \tag{1.78}$$

$$u = x \quad \& \quad v = y \tag{1.79}$$

from which

$$\frac{\partial u}{\partial x} = 1 \quad \& \quad \frac{\partial u}{\partial y} = 0 \tag{1.80}$$

$$\frac{\partial v}{\partial x} = 0 \quad \& \quad \frac{\partial v}{\partial y} = -1 \tag{1.81}$$

and

$$\frac{\partial u}{\partial x} \neq \frac{\partial v}{\partial y} \quad \text{(not satisfied)} \tag{1.82}$$

$$\frac{\partial u}{\partial y} = -\frac{\partial v}{\partial x} = 0 \quad \text{(satisfied)}$$

Therefore, $f(z) = \bar{z} = x - iy$, Eq. (1.78), is not analytic because $\partial u / \partial x \neq \partial v / \partial y$.

1.8.2   Complex Potential Functions

Consider a two-dimensional approach in plane elasticity for assessing the elastic behavior of isotropic materials using complex variable theory. The mathematical treatment is concerned with the determination of two complex potential, γz and ψ, for defining the state of stress in a domain on the z-plane. The driving force for deriving the state of stress at particular point on the complex domain is the connection between the complex analytic potential functions γz and ψ. Eventually, these complex potentials should give the stress distribution on the z-plane at suitable distance r from the chosen origin of the rectangular or polar coordinates

This section treats the Airy stress function ϕ in terms of functions of a complex variable z; $\phi = \phi(z, \bar{z})$. The aim is to find the general solution to a particular plane elasticity problem, involving two complex potential, $\gamma(z)$ and $\psi(z)$, within a infinite domain D (Fig. 1.7). A comprehensive and detail analysis of this topic can be found elsewhere [11, 15]. According to Muskhelishvili [15, 16], the complexity of a problem in plane theory of elasticity can be simplified very significantly by finding γ (z) and ψ (z), which must satisfy the problem boundary conditions.

Thus, one can generalize the Airy stress function as $\phi(z, \bar{z}) = f[\gamma(z), \psi(z)]$. For instance, consider the existence of several small regions (holes) within a large region in the z-plane. This is known as an infinite multiply connected system. Now, assume that the z-plane represents a plate subjected to external stresses which must remain bounded at infinity. The goal is to derive the elastic displacements and stresses at a point in the z-plane as a representation of the mechanical behavior of a solid. The complex potentials or complex stress functions that may provide a solution to the elasticity problem can be expressed as [11]

$$\gamma(z) = -\frac{\sum_{j=1}^{m} F_j}{2\pi(1+\kappa)} \log(z) + \left(\frac{\sigma_x^\infty + \sigma_y^\infty}{4}\right) z + \gamma^*(z) \tag{1.83}$$

$$\psi(z) = \frac{\kappa \sum_{j=1}^{m} \overline{F}_j}{2\pi(1+\kappa)} \log(z) + \left(\frac{\sigma_y^\infty - \sigma_x^\infty + 2i\tau_{xy}^\infty}{2}\right) z + \psi^*(z) \tag{1.84}$$

Here, m represents internal boundaries; $F_j - \Gamma_x + i\Gamma_y$ or $\overline{F}_j - \Gamma_x - i\Gamma_y$ is the resultant force on a contour, where the overbar in $\overline{F}_j$ indicates the conjugate function of F_j obtained by replacing i with $-i$. Both $\gamma^*(z)$ and $\psi^*(z)$ are arbitrary analytic functions outside the region enclosing all contours, and the κ is the plane elasticity factor defined as

$$\kappa = 3 - 4v \quad \text{for plane strain} \tag{1.85}$$

$$\kappa = \frac{3-v}{1+v} \quad \text{for plane stress}$$

Here, v is the Poisson's ratio. Using power series theory, the analytic functions, $\gamma^*(z)$ and $\psi^*(z)$ given in Eqs. (1.83) and (1.84), can be expressed as complex polynomials [11]

$$\gamma^*(z) = \sum_{n=1}^{\infty} a_n z^{-n} \tag{1.86}$$

$$\psi^*(z) = \sum_{n=1}^{\infty} b_n z^{-n}$$

The number of terms in the series is determined by the boundary conditions. Now, let us derive a complex expression for the Airy stress function $\phi = \phi(z, \bar{z})$. Firstly, develop the following operators:

$$\frac{\partial \phi}{\partial x} = \frac{\partial \phi}{\partial z} + \frac{\partial \phi}{\partial \bar{z}} \tag{a}$$

$$\frac{\partial \phi}{\partial z} = \frac{1}{2} \left(\frac{\partial \phi}{\partial x} - i \frac{\partial \phi}{\partial y} \right) \tag{b}$$

$$\frac{\partial \phi}{\partial y} = i \left(\frac{\partial \phi}{\partial z} - \frac{\partial \phi}{\partial \bar{z}} \right) \tag{c}$$

$$\frac{\partial \phi}{\partial \bar{z}} = \frac{1}{2} \left(\frac{\partial \phi}{\partial x} + i \frac{\partial \phi}{\partial y} \right) \tag{d}$$

Secondly, use the differential operators, Eqs. (a) through (d), repeatedly to get the harmonic ($\nabla^2 \phi$) and biharmonic ($\nabla^4 \phi$) operators [11]

$$\nabla^2 \phi = 4 \frac{\partial^2 \phi}{\partial z \partial \bar{z}} \tag{e}$$

$$\nabla^4 \phi = 16 \frac{\partial^4 \phi}{\partial z^2 \partial \bar{z}^2} \tag{f}$$

Thus, the governing biharmonic equation defined by Eq. (1.41) becomes

$$\frac{\partial^4 \phi}{\partial z^2 \partial \bar{z}^2} = 0 \tag{1.87}$$

Integrating Eq. (1.87) yields the general Airy stress function in terms of arbitrary complex potentials of the complex variable z and $\bar{z}$ [11, 14, 16, 17]

$$\phi(z, \bar{z}) = \frac{1}{2} \left[\bar{z} \overline{\gamma(z)} + \bar{z} \gamma(z) + \chi(z) + \overline{\chi(z)} \right] \tag{1.88}$$

$$\phi(z, \bar{z}) = \text{Re} \left[\bar{z} \gamma(z) + \chi(z) \right] \tag{1.89}$$

Using Eq. (1.40) without body forces and (1.89) yields the stress components in terms of the complex potentials [11, 15]

$$\sigma_x = \text{Re} \left[2\gamma'(z) - \bar{z} \gamma''(z) - \chi''(z) \right]$$
$$\sigma_y = \text{Re} \left[2\gamma'(z) + \bar{z} \gamma''(z) + \chi''(z) \right] \tag{1.90}$$
$$\tau_{xy} = \text{Re} \left[\bar{z} \gamma''(z) + \chi''(z) \right]$$

where the primes indicate differentiation with respect to z. Furthermore, the arbitrary function $\chi(z)$ can be expressed as

$$\chi(z) = \int \psi(z)\, dz \quad \text{or} \quad \psi(z) = \chi'(z) = \frac{d\chi(z)}{dz} \tag{1.91}$$

The complex displacement function $U = u + iv$ and its boundary condition give [11, 15]

$$2GU = \kappa\gamma(z) - z\overline{\gamma'(z)} - \overline{\psi(z)} \tag{1.92}$$

where κ is defined by Eq. (1.85) and G is the shear modulus of elasticity, which is related to the tensile modulus of elasticity E and Poisson's ratio v in the following form:

$$2G = \frac{E}{1 + v} \tag{1.93}$$

In addition, the Airy stress function ϕ, as defined by Eq. (1.89), can be used to determine stress relationships by adding to it its conjugate, which in turn will give twice the real part.

The procedure to carry out the analytical work can be found in Timoshenko and Goodier book, Chap. 7 [1]. The resultant function is

$$2\phi(z, \bar{z}) = \bar{z}\gamma(z) + \chi(z) + \overline{z\gamma(z)} + \overline{\chi(z)} \tag{1.94}$$

Applying Eq. (1.58) to (1.94) yields the general stress relationships, including Eq. (1.92) for convenience

$$\sigma_x + \sigma_y = 2\left[\gamma'(z) + \overline{\gamma'(z)}\right] = 4\,\mathrm{Re}\left[\gamma'(z)\right]$$

$$\sigma_y - \sigma_x + 2i\tau_{xy} = 2\left[\bar{z}\gamma''(z) + \psi'(z)\right] \tag{1.95}$$

$$2GU = \kappa\gamma(z) - z\overline{\gamma'(z)} - \overline{\psi(z)}$$

Extracting and combining the real parts of the first two expression in Eq. (1.95) gives the combination of stresses as

$$\sigma_x + \sigma_y = 4\,\mathrm{Re}\left[\gamma'(z)\right] \tag{1.95a}$$

$$\sigma_y - \sigma_x = 2\,\mathrm{Re}\left[\bar{z}\gamma''(z) + \psi'(z)\right] \tag{1.95b}$$

Add these two equations to get

$$\sigma_y = 2\,\mathrm{Re}\left[\gamma'(z)\right] + \mathrm{Re}\left[\bar{z}\gamma''(z) + \psi'(z)\right] \tag{1.95c}$$

$$\sigma_x = 2\,\mathrm{Re}\left[\gamma'(z)\right] - \mathrm{Re}\left[\bar{z}\gamma''(z) + \psi'(z)\right] \tag{1.95d}$$

For a case when the solid is half-space in the region D, where $y \geq 0$ with the system boundary at $y = 0$, Eq. 1.95 becomes [18]

$$\sigma_x + \sigma_y = 4 \operatorname{Re} \left[\gamma' (z) \right]$$
$$\sigma_y - i\tau_{xy} = \gamma' (z) + \overline{\gamma' (z)} + (z - \overline{z}) \overline{\gamma'' (z)} \qquad (1.96a)$$
$$2GU = \kappa \gamma (z) - z\overline{\gamma' (z)} - \overline{\psi (z)}$$

where $\overline{z} = 1/z$.

The stresses and displacements in polar coordinates are [11]

$$\sigma_r + \sigma_\theta = \sigma_x + \sigma_y$$
$$\sigma_\theta - \sigma_r + 2i\tau_{r\theta} = \left(\sigma_y - \sigma_x + 2i\tau_{xy} \right) e^{i2\theta} \qquad (1.96b)$$
$$u_r + iu_\theta = (u + iv) e^{-i\theta}$$

Recall that σ_r is the normal stress component in the radial direction and σ_θ is the normal stress component in the circumferential direction. Furthermore, the general stress relationship in terms of complex potentials for a circular domain with arbitrary edge loading may be expressed as [11]

$$\sigma_r - i\tau_{r\theta} = \gamma' (z) + \overline{\gamma' (z)} - e^{2i\theta} \left[\overline{z} \gamma'' (z) + \psi' (z) \right] \qquad (1.97)$$

This equation is very useful for determining power series coefficients. A classical example can illustrate the application of the complex variable theory for determining the stresses and displacements [11, 17].

Example 1.4. *This example illustrates the usefulness of Eqs. (1.95) and (1.96b) for determining stress and displacement equations with unknown coefficients a_n and b_n. Consider the following complex potential functions for deriving equations for stresses and displacements.*

$$\gamma (z) = (a_1 + ia_2) z$$
$$\psi (z) = (b_1 + ib_2) z$$

Solution. *Rectangular coordinates:*   The derivatives are

$$\gamma' (z) = (a_1 + ia_2) ; \ \psi' (z) = (b_1 + ib_2) ; \ \gamma'' (z) = 0 \ \& \ \psi'' (z) = 0$$

From Eq. (1.95),

$$\sigma_x + \sigma_y = 2 \left[\gamma' (z) + \overline{\gamma' (z)} \right] = 2 \left[(a_1 + ia_2) + (a_1 - ia_2) \right] \qquad (a)$$
$$= 4 \operatorname{Re} \gamma' (z) = 4a_1$$
$$\sigma_y - \sigma_x + 2i\tau_{xy} = 2 \left[\overline{z} \gamma'' (z) + \psi' (z) \right] = 2 \left[0 + (b_1 + ib_2) \right] \qquad (b)$$
$$= 2 \operatorname{Re} \psi' (z) + 2 \operatorname{Im} \psi' (z) = 2b_1 + 2ib_2$$

Equating real and imaginary parts gives

$$\sigma_y - \sigma_x = 2b_1 \tag{c1}$$

$$2i\tau_{xy} = 2ib_2 \ \ or \ \ \tau_{xy} = b_2 \tag{c2}$$

Adding Eqs. (a) and (c1) and substituting the resultant expression into (c1) give

$$\sigma_y = 2a_1 + b_1 \tag{c3}$$

$$\sigma_x = 2a_1 - b_1 \tag{c4}$$

Polar coordinates: *From Eq. (1.96b) and the above results, the stress relation-ships are*

$$\sigma_r + \sigma_\theta = \sigma_x + \sigma_y = 4a_1 \tag{d1}$$

$$\sigma_\theta - \sigma_r + 2i\tau_{r\theta} = \left(\sigma_y - \sigma_x + 2i\tau_{xy}\right) e^{i2\theta} = (2b_1 + 2ib_2)(\cos 2\theta + i \sin 2\theta)$$
$$= 2b_1 \cos 2\theta + 2ib_1 \sin 2\theta + 2ib_2 \cos 2\theta - 2b_2 \sin 2\theta$$
$$= 2(b_1 \cos 2\theta - b_2 \sin 2\theta) + 2i(b_1 \sin 2\theta + b_2 \cos 2\theta) \tag{d2}$$

Separating and equating real and imaginary terms yields

$$\sigma_\theta - \sigma_r = 2(b_1 \cos 2\theta - b_2 \sin 2\theta) \tag{d3}$$

$$\tau_{r\theta} = b_1 \sin 2\theta + b_2 \cos 2\theta \tag{d4}$$

Combine Eqs. (d1) and (d3) to get the circumferential and radial stress compo-nents, respectively:

$$\sigma_\theta = 2a_1 + b_1 \cos 2\theta - b_2 \sin 2\theta \tag{d5}$$

$$\sigma_r = 2a_1 - b_1 \cos 2\theta + b_2 \sin 2\theta \tag{d6}$$

Displacements: *From Eq. (1.92),*

$$2G(u + iv) = \kappa\gamma(z) - z\overline{\gamma'(z)} - \overline{\psi(z)} \tag{e1}$$
$$= \kappa(a_1 + ia_2)z - z(a_1 - ia_2) - (b_1 - ib_2)z$$
$$= [\kappa(a_1 + ia_2) - (a_1 - ia_2) - (b_1 - ib_2)]z$$
$$= [\kappa(a_1 + ia_2) - (a_1 - ia_2) - (b_1 - ib_2)](x + iy)$$
$$= [a_1(\kappa - 1)x - a_2(\kappa + 1)y - b_1x - b_2y]$$
$$+ i[a_1(\kappa - 1)y + a_2(\kappa + 1)x - b_1y + b_2x]$$

Then,

$$(u + iv) = \frac{1}{2G} [a_1 (\kappa - 1) x - a_2 (\kappa + 1) y - b_1 x - b_2 y] \tag{e2}$$

$$+ i \left(\frac{1}{2G} \right) [a_1 (\kappa - 1) y + a_2 (\kappa + 1) x - b_1 y + b_2 x]$$

Equating the real and imaginary parts yields

$$u = \frac{1}{2G} [a_1 (\kappa - 1) x - a_2 (\kappa + 1) y - b_1 x - b_2 y] \tag{e3}$$

$$v = \frac{1}{2G} [a_1 (\kappa - 1) y + a_2 (\kappa + 1) x - b_1 y + b_2 x] \tag{e4}$$

$$u_r + i u_\theta = (u + iv) e^{-i\theta} \tag{e5}$$

$$= (u + iv) (\cos \theta - i \sin \theta) \tag{e6}$$

$$= (u \cos \theta + v \sin \theta) + i (v \cos \theta - u \sin \theta) \tag{e7}$$

Equating the real and imaginary parts yields

$$u_r = u \cos \theta + v \sin \theta \tag{f1}$$

$$u_\theta = v \cos \theta - u \sin \theta \tag{f2}$$

The directions of these displacements in the interval $0 \leq \theta \leq \pi/2$ are:

- *If $\theta = 0$, then $u_r = u$ and $u_\theta = v$ in the positive direction.*
- *If $\theta = \pi/2$, then $u_r = v$ in the positive direction and $u_\theta = -u$ in negative direction.*
- *If $0 < \theta < \pi/2$, then u_r and u_θ fluctuate depending on the values of u and v.*

Example 1.5. *Consider an infinite elastic plate containing a circular hole subjected to a uniform tensile far field (remote). The loading condition and the boundary conditions are $\sigma_x^\infty = S$, $\sigma_y^\infty = 0$, $\tau_{xy}^\infty = 0$ at $r = \infty$, and $\sigma_r - i\tau_{r\theta} = 0$ at $r = a$, respectively:* **(a)** *expand the complex power series up to three terms,* **(b)** *derive the Airy stress function $\phi = \phi(r, \theta)$,* **(c)** *derive the stress expressions for σ_r, σ_θ, and $\tau_{r\theta}$ in polar coordinates and evaluate them at the edge of the hole when $\theta = \pi/2 = 3\pi/2$, and* **(d)** *calculate the values of these stresses if $S = 100\,MPa$, $a = 0.025\,m$, $r = 2a$, and $\theta = \pi/4$.*

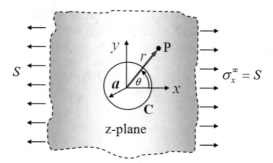

Solution. *Loading condition:*

$$\sigma_x^\infty = S, \ \sigma_y^\infty = 0 \ and \ \tau_{xy}^\infty = 0 \ at \ r = \infty$$

Boundary conditions:

$$\sigma_r - i\tau_{r\theta} = 0 \ at \ r = a$$

(a) **Power series expansion:** *An examination of Eqs. (1.83) and (1.84) indicates that* $\log(z)$ *vanishes due to the discontinuity represented by the stress-free hole. From Eqs. (1.83) and (1.84) and (1.87) and (1.88) and the loading condition, the complex potentials become*

$$\gamma(z) = \frac{Sz}{4} + \gamma^*(z) = \frac{Sz}{4} + \sum_{n=1}^{2} a_n z^{-n} \tag{a1}$$

$$\psi(z) = -\frac{Sz}{2} + \gamma^*(z) = -\frac{Sz}{2} + \sum_{n=1}^{2} b_n z^{-n} \tag{a2}$$

and

$$\gamma(z) = \frac{Sz}{4} + \frac{a_1}{z} + \frac{a_2}{z^2} + \frac{a_3}{z^3} \tag{a3}$$

$$\psi(z) = -\frac{Sz}{2} + \frac{b_1}{z} + \frac{b_2}{z^2} + \frac{b_3}{z^3} \tag{a4}$$

From Eq. (1.91), the integration of the complex potential $\psi(z)$*, Eq. (a4), yields a new complex potential. Thus,*

$$\chi(z) = \int \psi(z)\,dz = \int \left(-\frac{Sz}{2} + \frac{b_1}{z} + \frac{b_2}{z^2} + \frac{b_3}{z^3} \right) dz \tag{a5}$$

$$\chi(z) = -\frac{Sz^2}{4} + b_1 \log z - \frac{b_2}{z} - \frac{b_3}{2z^2} \tag{a6}$$

(b) *Denote that* log *(z) in Eq. (a6) is just the result of integration and therefore, it does not vanish. Substituting Eqs. (a3) and (a6) into (1.89) yields the Airy stress function in terms of the complex variable z*

$$\phi\,(z,\bar{z}) = \mathrm{Re}\left[\bar{z}\left(\frac{Sz}{4} + \frac{a_1}{z} + \frac{a_2}{z^2} + \frac{a_3}{z^3}\right) - \frac{Sz^2}{4} + b_1 \log\,(z) - \frac{b_2}{z} - \frac{b_3}{2z^2}\right]$$

(a7)

where

$$z = re^{i\theta} = r\,(\cos\theta + i\sin\theta)$$

$$\bar{z} = re^{-i\theta} = r\,(\cos\theta - i\sin\theta) \qquad (a8)$$

$$z^2 = r^2 e^{i2\theta} = r^2\,(\cos 2\theta + i\sin 2\theta)$$

$$z^3 = r^3 e^{i3\theta} = r^3\,(\cos 3\theta + i\sin 3\theta)$$

Inserting Eq. (a8) into (a7) gives $\phi\,(z,\bar{z}) \rightarrow \phi\,(r,\theta)$ *and collecting* $\cos m\theta \pm i\sin m\theta = e^{\pm im\theta}$ *terms yields*

$$\phi\,(r,\theta) = \mathrm{Re}\,\frac{1}{r^2}\left(\frac{1}{4}Sr^4 - \frac{1}{4}Sr^4 e^{2i\theta} + r^2 a_1 e^{-2i\theta} + ra_2 e^{-3i\theta} + a_3 e^{-4i\theta}\right)$$

$$+ \mathrm{Re}\,\frac{1}{r^2}\left[r^2\,(\log r)\,b_1 e^{i\theta} - rb_2 e^{-i\theta} - \frac{1}{2}b_3 e^{-2i\theta}\right]$$

Extracting the real parts along with $\sin^2\theta = (1 - \cos 2\theta)/2$ *gives the Airy stress function,* $\phi = \phi\,(r,\theta)$, *in a simplified form*

$$\phi = \frac{1}{r^2}\left(\frac{1}{2}Sr^4 \sin^2\theta + r^2 a_1 \cos 2\theta + ra_2 \cos 3\theta + a_3 \cos 4\theta\right)$$

(a9)

$$+ \frac{1}{r^2}\left(r^2 b_1\,(\log r) - rb_2 \cos\theta - \frac{1}{2}b_3 \cos 2\theta\right)$$

(c) *This expression, Eq. (a9) satisfies* $\nabla^4\phi = 0$. *Applying Eq. (a9) to (1.41) gives the elastic stresses in polar coordinates as*

$$\sigma_r = \frac{1}{r}\frac{\partial\phi}{\partial r} + \frac{1}{r^2}\frac{\partial^2\phi}{\partial\theta^2}$$

$$\sigma_r = \frac{1}{r^4}\left(\frac{1}{2}Sr^4 \cos 2\theta - 4r^2 a_1 \cos 2\theta - 10ra_2 \cos 3\theta - 18a_3 \cos 4\theta\right)$$

$$+ \frac{1}{r^4}\left(r^2 b_1 + 2rb_2 \cos\theta + 3b_3 \cos 2\theta\right)$$

$$\sigma_\theta = \frac{\partial^2 \phi}{\partial r^2} = \frac{1}{r^4} \left(\frac{1}{2} Sr^4 - \frac{1}{2} Sr^4 \cos 2\theta + 2ra_2 \cos 3\theta + 6a_3 \cos 4\theta \right)$$

$$- \frac{1}{r^4} \left(r^2 b_1 + 2rb_2 \cos \theta + 3b_3 \cos 2\theta \right) \tag{a10}$$

$$\tau_{r\theta} = \frac{1}{r^2} \frac{\partial \phi}{\partial \theta} - \frac{1}{r} \frac{\partial^2 \phi}{\partial r \partial \theta} = -\frac{\partial}{\partial r} \left(\frac{1}{r} \frac{\partial \phi}{\partial \theta} \right)$$

$$\tau_{r\theta} = -\frac{1}{r^4} \left(\frac{1}{2} Sr^4 \sin 2\theta + 2r^2 a_1 \sin 2\theta + 6ra_2 \sin 3\theta + 12a_3 \sin 4\theta \right)$$

$$+ \frac{1}{r^4} \left(2rb_2 \sin \theta + 3b_3 \sin 2\theta \right)$$

Using $\sigma_r - i\tau_{r\theta} = 0$ at $r = a$, one can collect $(\cos m\theta \pm i \sin m\theta) = e^{\pm im\theta}$ terms and equate like powers of $e^{\pm im\theta}$. The resultant equation is

$$0 = \frac{1}{2} Sr^4 + r^2 b_1 + \frac{1}{2} Sr^4 (\cos 2\theta + i \sin 2\theta) + 2rb_2 (\cos \theta - i \sin \theta)$$

$$+ 3b_3 (\cos 2\theta - i \sin 2\theta) - 4r^2 a_1 \cos 2\theta + 2ir^2 a_1 \sin 2\theta \tag{a11}$$

$$- 18a_3 \cos 4\theta + 12ia_3 \sin 4\theta - 10ra_2 \cos 3\theta + 6ira_2 \sin 3\theta$$

Let $-4r^2 a_1 \cos 2\theta + 2ir^2 a_1 \sin 2\theta = -3r^2 a_1 \cos 2\theta - r^2 a_1 \cos 2\theta + 3ir^2 a_1 \sin 2\theta - ir^2 a_1 \sin 2\theta$ so that

$$0 = 2rb_2 e^{-i\theta} + \frac{1}{2} Sr^4 + r^2 b_1 + 3b_3 e^{-i2\theta} + \frac{1}{2} Sr^4 e^{i2\theta}$$

$$- 3r^2 a_1 (\cos 2\theta - i \sin 2\theta) - r^2 a_1 (\cos 2\theta + i \sin 2\theta) \tag{a12}$$

$$- 18a_3 \cos 4\theta + 12ia_3 \sin 4\theta - 10ra_2 \cos 3\theta + 6ira_2 \sin 3\theta$$

$$0 = 2rb_2 e^{-i\theta} + \frac{1}{2} Sr^4 + r^2 b_1 + 3b_3 e^{-i2\theta} + \frac{1}{2} Sr^4 e^{i2\theta} - 3r^2 a_1 e^{-i2\theta} - r^2 a_1 e^{i2\theta}$$

$$- 18a_3 \cos 4\theta + 12ia_3 \sin 4\theta - 10ra_2 \cos 3\theta + 6ira_2 \sin 3\theta$$

Equating like powers of $e^{\pm im\theta}$ and letting $r = a$ yields the coefficients a_n and b_n. Hence,

$\frac{1}{2} Sr^4 = -r^2 b_1$	$-r^2 a_1 e^{i2\theta} = \frac{1}{2} Sr^4 e^{i2\theta}$	$-3r^2 a_1 e^{-i2\theta} = 3b_3 e^{-i2\theta}$
$b_1 = -\frac{1}{2} Sa^2$	$a_1 = -\frac{1}{2} Sa^2$	$b_3 = -a^2 a_1$
$b_2 = 0$	$a_2 = a_3 = 0$	$b_3 = -\frac{1}{2} Sa^4$

Inserting the above coefficients into Eq. (a9), the Airy stress function becomes

$$\phi = \frac{1}{r^2}\left(\frac{1}{2}Sr^4\sin^2\theta + \frac{1}{2}Sa^2r^2\cos 2\theta - \frac{1}{2}Sa^2r^2\left(\log r\right) - \frac{1}{4}Sa^4\cos 2\theta\right)$$

(a13)

Applying Eq. (1.58) or (a10) to (a13) yields the elastic stresses as reported in Timoshenko and Goodier book, page 80 [1]

$$\sigma_r = \frac{S}{2}\left(1 - \frac{a^2}{r^2}\right) + \frac{S}{2}\left(1 - \frac{4a^2}{r^2} + \frac{3a^4}{r^4}\right)\cos 2\theta$$

$$\sigma_\theta = \frac{S}{2}\left(1 + \frac{a^2}{r^2}\right) - \frac{S}{2}\left(1 + \frac{3a^4}{r^4}\right)\cos 2\theta$$

(a14)

$$\tau_{r\theta} = -\frac{S}{2}\left(1 + \frac{2a^2}{r^2} - \frac{3a^4}{r^4}\right)\sin 2\theta$$

The stresses at the edge of the hole, r = a, become

$$\sigma_r = \tau_{r\theta} = 0$$

$$\sigma_\theta = S - 2S\cos 2\theta = S\left(1 - 2\cos 2\theta\right)$$

(a15)

$$\sigma_{max} = \sigma_\theta = 3S \ @ \ \theta = \pi/2 = 3\pi/2$$

If $\theta = \pi/2 = 3\pi/2$ and r = a, then

$$\sigma_{max} = \sigma_\theta = 3S$$

which gives the maximum value of the concentration factor as

$$K_t = \frac{\sigma_{max}}{S} = 3$$

(a16)

(d) *If $S = 100\,MPa$, $a = 0.025\,m$, $r = 2a$, and $\theta = \pi/4$, then the elastic stresses are*

$$\sigma_r = 37.50\,MPa$$

$$\sigma_\theta = 62.50\,MPa$$

(a17)

$$\tau_{r\theta} = -65.63\,MPa$$

For comparison purposes, assume that the loaded plate has no hole and the origin of the polar coordinates remains in the center of the plate. Let the Airy stress function be

$$\phi = \frac{S}{2}y^2 = \frac{1}{2}S\left(r\sin\theta\right)^2$$

(b1)

$$\phi = \frac{S}{4}r^2\left(1 - \cos 2\theta\right)$$

(b2)

As a result, the elastic stress becomes

$$\sigma_r = \frac{S}{2}(1 + \cos 2\theta)$$

$$\sigma_\theta = \frac{S}{2}(1 - \cos 2\theta) \qquad \text{(b3)}$$

$$\tau_{r\theta} = -\frac{S}{2}\sin 2\theta$$

1.9 Conformal Mapping

The connection between harmonic and complex functions provides the driving force for conformal mapping, or transformation of a point (z_o) in the z-plane onto a point (ζ_o) in the ζ-plane can be accomplished since harmonic functions are infinitely differentiable on a z-plane. In fact, a complex-valued function $f(z)$ depends on a single complex variable $z = x + iy$, defined by Eq. (1.64). Thus, conformal (angle preserving) mapping is a mathematical technique used to convert or map one mathematical problem and its solution onto another complex plane, usually, in two dimensions. Thus, a complex mapping function takes every point in one complex plane (z-plane) and maps it onto another complex plane (ζ-plane) as shown in Fig. 1.8.

In fact, when the boundary conditions in the original domain D (z-plane) are not suitable for solving a plane elasticity problem, it is convenience and advantageous to use a conformal mapping function of the form $z = f(\zeta)$ and its inverse $\zeta = f^{-1}(z)$ to solve the problem.

The function $z = f(\zeta)$ maps the point (x, y) of the z-plane onto the ζ-plane as point (ξ, η). Conversely, $\zeta = f(z)$ maps the point (ξ, η) of the ζ-plane back onto the z-plane as point (x, y). Denote that the z-plane in Fig. 1.8 has an arbitrary region D containing four points to be mapped onto a unit circle in the ζ-plane. Finding the function $z = f(\zeta)$ for an arbitrary region D is complicated task; however, one can find in the literature mapping functions for regular shapes, such as round hole, elliptical hole, elliptical crack, straight crack, and the like. Then the mapping is mostly done onto a unit circle as depicted in Fig. 1.8.

Fig. 1.8 Conformal mapping of points (x,y) of an arbitrary z-plane into a ζ-plane on an infinite plate.

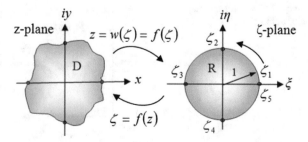

In general, carrying out a conformal mapping requires that the angle of intersection of any two lines or curves remains unchanged between a given set of elements in a domain D of the z-plane and another set of elements in a codomain R of the ζ-plane. Thus, mapping of complex functions preserves angles if and only if $f(z)$ is analytic or $\overline{f(z)}$ is anti-analytic (conjugate). Thus, $df/dz = f'(z) \neq 0$ and $\overline{f'(z)} \neq 0$ [19–21].

Let z and ζ be complex variables so that

$$z = w(\zeta) = f(\zeta) \quad \& \quad \zeta = f(z) \tag{1.98}$$

The mapping function $z = w(\zeta)$ is very common in the literature. Instead, $z = f(\zeta)$ is adopted in this section, Nonetheless, $f(\zeta)$ and $f(z)$ are analytic and $z = f(\zeta)$ is the conformal mapping function, which is to be found using a suitable analytical procedure that includes some boundary conditions. Nonetheless, the analytical procedure requires that there must be a one-to-one mapping of points between the z-plane and the ζ-plane. This simply means that points on the z-plane are mapped onto the ζ-plane by means of the function $\zeta = f(z)$ as indicated in Fig. 1.18 [22]. Thus, $\zeta = f(z)$ is the transformation function of interest.

Basically, the significance of conformal mapping is to determine a mapping function that will transform a complex region in the z-plane into a simple region in the ζ-plane. Let us illustrate the application of conformal mapping theory by using a simple transformation found elsewhere [19].

In general, transformation can be envisioned rather readily by letting $\zeta = \xi + i\eta$ be a **single-valued function** of $z = x + iy$ so that $\xi + i\eta = f(x + iy)$. This means that $\eta = f(x, y)$ and $\xi = f(x, y)$. For instance, given a point P located at $z_o = (x_o, y_o)$ in the z-plane, there corresponds a point P' at $\zeta_o = (\xi_o, \eta_o)$ in the ζ-plane. This is illustrated in Fig. 1.9 for an arbitrary PQ trajectory being mapped as an arbitrary $P'Q'$ trajectory. Thus, point P' at $\zeta_o = (\xi_o, \eta_o)$ is the image of P at $z_o = (x_o, y_o)$. Similarly, Q' is the image of Q.

On the other hand, a multiple-valued function is commonly used for mapping a region onto a circle or half-plane because the analytical procedure generally provides simple expressions of the boundary conditions. Apparently mapping onto a unit circle is most common since complex potentials can be expressed by power series. For instance, Fig. 1.10 schematically illustrates the conformal mapping of an

Fig. 1.9 Conformal mapping of points P and Q

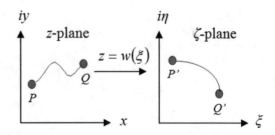

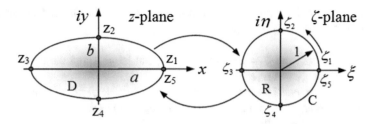

Fig. 1.10 Conformal mapping of an elliptic region D into a unit circle C.

elliptical region D onto a unit circle R having a contour C. For the sake of clarity, marked points on the elliptical contour are mapped as ζ points on the contour C.

The elliptical region in Fig. 1.10 with semimajor axis a and semiminor axis b may represent an elliptical hole in a large plate or an embedded elliptical crack in a solid material subjected to an external loading mode. This elliptical region can be mapped using a mapping function defined by Sadd [22]

$$z = c\left(\zeta + \frac{d}{\zeta}\right) \tag{1.99}$$

where the constants c and d take the form

$$c = \frac{(a+b)}{2} \tag{1.100}$$

$$d = \frac{(a-b)}{(a+b)} \tag{1.101}$$

The inverse mapping function of Eq. (1.99) is

$$\zeta = f^{-1}(z) = \frac{1}{2c}\left(z + \sqrt{z^2 - 4dc^2}\right) \quad \text{(exterior with } |\zeta| > 1 \text{)} \tag{1.102}$$

$$\zeta = f^{-1}(z) = \frac{1}{2c}\left(z - \sqrt{z^2 - 4dc^2}\right) \quad \text{(interior with } |\zeta| < 1 \text{)} \tag{1.103}$$

Either inverse function maps the ellipse onto the unit circle. However, one has to select the exterior or the interior mapping based on the absolute value of $|\zeta|$. Usually, exterior mapping the ellipse onto the exterior of the unit circle is common when the elliptical hole or crack is treated as a traction-free discontinuity.

In general, the literature offers many examples of analytical and numerical solutions to hole and crack problems being independently influenced by a simple or biaxial loading mode. Most of these solutions do not include environmental effects and some include dislocation models to describe the influence of dislocations near a crack tip [23].

For the sake of clarity, the graphical procedure to map the ellipse onto the circle is as follows:

- First: If $z = re^{i\theta}$ and $\zeta = z^{1/2}$, then $\bar{z} = 1/z$ at $|z| = 1$ and the first points are $z_1 = r_1 e^{i\theta_1}$ and $\zeta_1 = \sqrt{r_1} e^{i\theta_1/2}$.
- Second: If $z_2 = r_2 e^{i\theta_2} = r_1 e^{i\theta_2}$ and $\zeta_2 = \sqrt{r_2} e^{i\theta_2/2} = \sqrt{r_1} e^{i\theta_2/2}$ since $r = r_1 = r_2$. The five points shown in Fig. 1.10 are determined as follows:
- If $\theta_1 = 0$ and $r_1 = 1$, then

$$|\zeta_1| = e^{i0} = 1 \text{ and } |\zeta_1| = 1 = \xi + i\eta$$

- If $\theta_2 = \pi/2$ and $r_2 = 1$, then

$$\zeta_2 = \zeta_1 e^{i\pi/2} \text{ and } e^{i\pi/2} = \xi + i\eta$$

- If $\theta_3 = \pi$ and $r_3 = 1$, then

$$\zeta_3 = \zeta_1 e^{i\pi} \text{ and } e^{i\pi} = \xi + i\eta$$

- For $\theta_4 = 3\pi/2$ and $r_4 = 1$,

$$\zeta_4 = \zeta_1 e^{i3\pi/2} \text{ and } e^{i3\pi/2} = \xi + i\eta$$

- For $\theta_5 = 2\pi$ and $r_5 = 1$,

$$\zeta_5 = \zeta_1 e^{i2\pi} \text{ and } e^{i2\pi} = \xi + i\eta$$

Denote that $z_1 = z_5$ in Fig. 1.10. At any rate, the above procedure can also be presented as

$$\theta_1 = 0 \text{ and } r_1 = 1 \text{ so that } |\zeta_1| = e^{i0} = 1 \text{ and } |\zeta_1| = 1 = \xi + i\eta$$
$$\theta_2 = \pi/2 \text{ and } r_2 = 1 \text{ so that } \zeta_2 = \zeta_1 e^{i\pi/2} \text{ and } e^{i\pi/2} = \xi + i\eta$$
$$\theta_3 = \pi \text{ and } r_3 = 1 \text{ so that } \zeta_3 = \zeta_1 e^{i\pi} \text{ and } e^{i\pi} = \xi + i\eta$$
$$\theta_4 = 3\pi/2 \text{ and } r_4 = 1 \text{ so that } \zeta_4 = \zeta_1 e^{i3\pi/2} \text{ and } e^{i3\pi/2} = \xi + i\eta$$
$$\theta_5 = 2\pi \text{ and } r_5 = 1 \text{ so that } \zeta_5 = \zeta_1 e^{i2\pi} \text{ and } e^{i2\pi} = \xi + i\eta$$

The main goal on conformal mapping hereby is to expand the theoretical work in order to determine complex potentials for deriving displacement and stress field equations.

1.9.1 Cauchy Integral Formula

So far the complex functions $\gamma(z)$ and $\psi(z)$ do not have useful mathematical forms. However, using the powerful Cauchy integral theorem (CIT), and the Cauchy integral formula (CIF), one can derive useful complex expressions to solve a particular plane elasticity problem. Thus, [15, 19, 22, 24]

- (1) **The Cauchy integral theorem (CIF)** states that if a function $f(z)$ is analytic in a region D at all points interior to and on a closed contour C, then

$$\int_C f(z)dz = 0 \tag{1.104}$$

- (2) **The Morera's integral theorem** states if $f(z)$ is a continuous function of z in the region D and

$$\int_C f(z)dz = 0 \tag{1.105}$$

about any closed contour C in D, then $f(z)$ represents an analytic function in D.
- (3) **The Cauchy integral formula** states that if $f(z)$ is analytic everywhere within a region D and on the boundary of a simply connected contour C, and for any point z_o inside D, there holds

$$f(z_o) = \frac{1}{2\pi i} \int_C \frac{f(z)}{z - z_o} dz \tag{1.106}$$

$$f^{(n)}(z_o) = \frac{n!}{2\pi i} \int_C \frac{f(z)}{(z - z_o)^{n+1}} dz \tag{1.107}$$

where $f^{(n)}(z_o)$ represents derivatives of all orders in D that are analytic functions, $f^{(n)'}(z_o) \neq 0$, in D. Here, $n = 0, 1, 2, \ldots$ and $n! = 0!, 1!, 2!, \ldots (n$ factorial). According to **calculus of residues** along with $\bar{\zeta} = 1/\zeta$ and $z = \zeta = 1$, the following conditions hold:

$$\frac{1}{2\pi i} \int_C \frac{1}{\zeta^n (\zeta - z)} d\zeta = \begin{cases} 0 \text{ for } n > 0 \\ 1 \text{ for } n = 0 \end{cases} \tag{1.108}$$

From Eq. (1.106) [11, 15],

$$\left[\frac{1}{2\pi i} \int_C \frac{\gamma(\zeta)}{\zeta - z} d\zeta \right]_1 + \left[\frac{1}{2\pi i} \int_C \frac{z\overline{\gamma'(\zeta)}}{\zeta - z} d\zeta \right]_2 + \left[\frac{1}{2\pi i} \int_C \frac{\overline{\psi(\zeta)}}{\zeta - z} d\zeta \right]_3$$

$$= \left[\frac{1}{2\pi i} \int_C \frac{h(\zeta)}{\zeta - z} d\zeta \right]_4 \tag{1.109}$$

Each square bracket has a subscript for proper identification.

The evaluation of Eq. (1.109) is most suited using the Taylor series for determining the coefficients a_n and b_n that contribute to the stresses and displacements and to the constant boundary function $h(\zeta) = h$, which, in turn, depends on the remote or infinity loading condition.

Procedure: The procedure to find the proper expressions for $\gamma(z)$ and $\psi(z)$ is described by Muskhelishvili [15] in Chap. 15, and it is included in this section for convenience. Thus, the reader may have a better insight on the application of Cauchy integral formula for solving some plane elasticity problems. The procedure requires proper evaluation of the integrals defined by Eq. (1.109).

- **The First Integral**: It is simply the Cauchy integral formula term in Eq. (1.109)

$$\gamma(z) = \left[\frac{1}{2\pi i} \int_C \frac{\gamma(\zeta)}{\zeta - z} d\zeta \right]_1 = \sum_{n=o}^{\infty} a_n z^n \qquad (1.110)$$

Let $\gamma(z)$ be a polynomial from which $\gamma'(z)$ and $\overline{\gamma'(z)}$ are easily obtained. Hence,

$$\gamma(z) = \sum_{n=o}^{\infty} a_n z^n = a_o + a_1 z + a_2 z^2 + \dots . \qquad (1.111)$$

$$\gamma'(z) = \sum_{n=o}^{\infty} a_n n z^{n-1} = a_1 + 2a_2 z \qquad (1.112)$$

$$\overline{\gamma'(z)} = \sum_{n=o}^{\infty} \bar{a}_n n \bar{z}^{n-1} = \overline{a_1} + 2\overline{a_2}\bar{z} \qquad (1.113)$$

- **The Second Integral**: Using Eq. (1.113) in the second integral gives

$$\left[\frac{1}{2\pi i} \int_C \frac{z\overline{\gamma'(\zeta)}}{\zeta - z} d\zeta \right]_2 = \overline{a_1} z + 2\overline{a_2} \qquad (1.114)$$

where $z\bar{z} = z/z = 1$ and $z\overline{\gamma'(z)} = z(\overline{a_1} + 2\overline{a_2}\bar{z}) = \overline{a_1} z + 2\overline{a_2}.$

- **The Third Integral**: In order to evaluate the third integral some requirements must be met. Let

$$\gamma(z) + z\overline{\gamma'(z)} + \overline{\psi(z)} = h(z) \qquad (1.115a)$$

$$\overline{\gamma(z)} + \bar{z}\gamma'(z) + \psi(z) = h(z) \qquad (1.115b)$$

From Eq. (1.115b), the Cauchy integral formula for the conjugate complex potential $\overline{\gamma(z)}$ takes the form

$$\frac{1}{2\pi i}\int_C \frac{\overline{\gamma(\zeta)}}{\zeta - z}d\zeta = \overline{\gamma(z)} = \overline{\gamma(z=0)} = 0 \tag{1.115c}$$

Handling the above integral requires that z be replaced by ζ inside the integral and set $z = 0$ in the solution so that $\overline{\gamma(z)} = \overline{\gamma(0)} = 0$. Now, the Cauchy integral formula for the term $\bar{z}\gamma'(z)$ in Eq. (1.115b) is

$$\frac{1}{2\pi i}\int_C \frac{\bar{\zeta}\gamma'(\zeta)}{\zeta - z}d\zeta = \frac{1}{2\pi i}\int_C \frac{\gamma'(\zeta)}{\zeta(\zeta - z)}d\zeta = \frac{1}{2\pi i}\int_C \frac{\gamma'(\zeta)}{\zeta}\frac{d\zeta}{\zeta - z}$$

$$= \frac{\gamma'(z)}{z} - \frac{a_1}{z} \tag{1.116}$$

Solving Eq. (1.115b) for $\psi(z)$, the solution for the third integral follows

$$\psi(z) = \frac{1}{2\pi i}\int_C \frac{\overline{h(\zeta)}}{\zeta - z}d\zeta - \frac{1}{2\pi i}\int_C \frac{\bar{\zeta}\gamma'(\zeta)}{\zeta - z}d\zeta - \frac{1}{2\pi i}\int_C \frac{\overline{\gamma(\zeta)}}{\zeta - z}d\zeta$$

$$\psi(z) = \frac{1}{2\pi i}\int_C \frac{\overline{h(\zeta)}}{\zeta - z}d\zeta - \frac{\gamma'(z)}{z} + \frac{a_1}{z} - \overline{\gamma(0)} \tag{1.117}$$

- **The Fourth Integral**: The remaining procedure is for defining the constants $a_n \neq 0$. In fact, $a_n = 0$ for $n > 2$ as implicitly suggested in Eq. (1.111) [15, 22]. Then, Eq. (1.109) becomes

$$[\gamma(z)]_1 + \left[\frac{1}{2\pi i}\int_C \frac{z\gamma'(\zeta)}{\zeta - z}d\zeta\right]_2 + \left[\overline{\gamma(0)}\right]_3 = \left[\frac{1}{2\pi i}\int_C \frac{\overline{h(\zeta)}}{\zeta - z}d\zeta\right]_4$$

$$\gamma(z) + \overline{a_1}z + 2\overline{a_2} = \left[\frac{1}{2\pi i}\int_C \frac{\overline{h(\zeta)}}{\zeta - z}d\zeta\right]_4 \tag{1.118}$$

Let [15]

$$\frac{1}{2\pi i}\int_C \frac{h(\zeta)\,d\zeta}{\zeta - z} = \frac{1}{2\pi i}\int_C \frac{h(\zeta)\,d\zeta}{\zeta(1 - z/\zeta)}$$

$$= \frac{1}{2\pi i}\int_C h(\zeta)\left(1 + \frac{z}{\zeta} + \frac{z^2}{\zeta^2}\right)\frac{d\zeta}{\zeta}$$

$$\frac{1}{2\pi i}\int_C \frac{h(\zeta)\,d\zeta}{\zeta - z} = \frac{1}{2\pi i}\int_C \frac{h(\zeta)\,d\zeta}{\zeta} + \frac{z}{2\pi i}\int_C \frac{h(\zeta)\,d\zeta}{\zeta^2} \tag{1.119}$$

$$+ \frac{z^2}{2\pi i}\int_C \frac{h(\zeta)\,d\zeta}{\zeta^3}$$

so that Eqs. (1.110), (1.114), and (1.119) yield

$$\gamma(z) + \overline{a_1}z + 2\overline{a_2} = \frac{1}{2\pi i} \int_C \frac{h(\zeta)\,d\zeta}{\zeta} + \frac{z}{2\pi i} \int_C \frac{h(\zeta)\,d\zeta}{\zeta^2} \qquad (1.120)$$

$$+ \frac{z^2}{2\pi i} \int_C \frac{h(\zeta)\,d\zeta}{\zeta^3}$$

The first derivative of Eq. (1.120) with respect to z gives

$$\gamma'(z) + \overline{a_1} = \frac{1}{2\pi i} \int_C \frac{h(\zeta)\,d\zeta}{\zeta^2} + \frac{z}{\pi i} \int_C \frac{h(\zeta)\,d\zeta}{\zeta^3} \qquad (1.121)$$

Letting $z = 0$ Eq. (1.112) yields $\gamma'(0) = \sum_{n=o}^{\infty} a_n n z^{n-1} = a_1$ and consequently, Eq. (1.121) becomes

$$\gamma'(0) + \overline{a_1} = \frac{1}{2\pi i} \int_C \frac{h(\zeta)\,d\zeta}{\zeta^2}$$

$$a_1 + \overline{a_1} = \frac{1}{2\pi i} \int_C \frac{h(\zeta)\,d\zeta}{\zeta^2} \qquad (1.122)$$

Take the second derivative of Eq. (1.121) with respect to z to get

$$\gamma''(z) = \frac{1}{\pi i} \int_C \frac{h(\zeta)\,d\zeta}{\zeta^3} \qquad (1.121)$$

From Eq. (1.112) at $z = 0$, $\gamma''(z) = \sum_{n=1}^{\infty} a_n(n-1) z^{n-2} = 2a_2$. Then, Eq. (1.121) becomes

$$\gamma''(0) = \frac{1}{\pi i} \int_C \frac{h(\zeta)\,d\zeta}{\zeta^3} = 2a_2$$

$$a_2 = \frac{1}{2\pi i} \int_C \frac{h(\zeta)\,d\zeta}{\zeta^3} \qquad (1.123)$$

The final results based on the unit disk or unit circle mapping are summarized along with $h = h(\zeta)$ as

$$\gamma(z) = \frac{1}{2\pi i} \int_C \frac{h(\zeta)}{\zeta - z}\,d\zeta - \overline{a_1}z \qquad (1.124)$$

$$\psi(z) = \frac{1}{2\pi i} \int_C \frac{\overline{h(\zeta)}}{\zeta - z}\,d\zeta - \frac{\gamma'(z)}{z} + \frac{a_1}{z} \qquad (1.125)$$

$$a_1 + \overline{a_1} = \frac{1}{2\pi i} \int_C \frac{h(\zeta)}{\zeta^2}\,d\zeta \qquad (1.126)$$

$$a_2 = \frac{1}{2\pi i} \int_C \frac{h(\zeta)}{\zeta^3}\,d\zeta \qquad (1.127)$$

from which the complex potentials $\gamma(z)$ and $\psi(z)$ can be determined according to a particular plane elasticity problem and its loading conditions. This implies that the constants a_1 and $\overline{a_1}$ and the boundary function $h(\zeta) = h$ also depend on the loading condition.

The application of this technique can be made clear using the classic example of a circular disk under concentrated forces applied to its boundary. This boundary value problem is described in Muskhelishvili's classic book [15], and it is also found solved in Sadd's modern book [11].

For the sake of clarity, the example below illustrates the usefulness of the Cauchy integral formula for determining the elastic stress equations of crack-free planes. In addition, the above procedure is used in Chap. 4 for determining the stress and displacement field equations near a crack tip in the z-plane.

Among the applications of the Cauchy integral formula based on the complex function $f(z)$, the Laurent Series of the form

$$f(z) = \sum_{n=0}^{\infty} a_n (z - z_0)^n \tag{1.128}$$

can be solved using the Cauchy integral formula for defining the power series coefficients. Hence,

$$a_n = \frac{1}{2\pi i} \oint \frac{f(z)}{(z - z_o)^{n+1}} dz \quad \text{for } n = 0, 1, 2, 3, \ldots \tag{1.129}$$

where $f(z)$ is analytic at all points on a closed curve C and z_0 is a point interior to C on the z-plane. If $f(z)$ is not analytic, then $f^{(n+1)}(z_o) = 0$ and the result is called a singular point [11].

Example 1.6. *Consider a unit circular disk subjected to a uniform compressive pressure as shown in the figure below to determine the hydrostatic state of stress.*

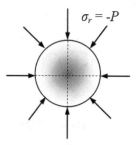

Solution. *The boundary traction is*

$$T_x + iT_y = (\sigma_r + i\tau_{r\theta}) z \quad \text{where } \tau_{r\theta} = 0$$

$$T_x + iT_y = \sigma_r z = -Pz = -Pe^{i\theta}$$

Let

$$h(z) = i \int_0^\theta \left(T_x + i T_y \right) d\theta = -i \int_0^\theta P e^{i\theta} d\theta$$

$$h(z) = -Pz$$

Substituting this result into Eq. (1.124) yields

$$\gamma(z) = \frac{1}{2\pi i} \int_C \frac{h(\zeta)}{\zeta - z} d\zeta - \overline{a_1} z = -\frac{1}{2\pi i} \int_C \frac{P\zeta}{\zeta - z} d\zeta - \overline{a_1} z$$

Evaluate the Cauchy integral formula such that

$$\frac{1}{2\pi i} \int_C \frac{P\zeta}{\zeta - z} d\zeta = [f(\zeta)]_{\zeta=z} = [P\zeta]_{\zeta=z} = Pz$$

Then,

$$\gamma(z) = -Pz - \overline{a_1} z$$

$$\gamma'(z) = -P - \overline{a_1}$$

From Eq. (1.126),

$$a_1 + \overline{a_1} = -\frac{1}{2\pi i} \int_C \frac{P\zeta}{\zeta^2} d\zeta = -\frac{1}{2\pi i} \int_C \frac{P}{\zeta} d\zeta = -P$$

From Eq. (1.125),

$$\psi(z) = -\frac{1}{2\pi i} \int_C \frac{P\overline{\zeta}}{\zeta - z} d\zeta + \frac{P + \overline{a_1}}{z} + \frac{a_1}{z}$$

$$\psi(z) = -\frac{1}{2\pi i} \int_C \frac{P}{\zeta(\zeta - z)} d\zeta + \frac{P}{z} + \frac{1}{z}(\overline{a_1} + a_1)$$

$$\psi(z) = -\frac{1}{2\pi i} \int_C \frac{P}{\zeta(\zeta - z)} d\zeta + \frac{P}{z} - \frac{P}{z} = 0$$

From Eqs. (1.95) and (1.96b), respectively,

$$\sigma_x + \sigma_y = 2 \left[\gamma'(z) + \overline{\gamma'(z)} \right] = -2P$$

$$\sigma_y - \sigma_x + 2i\tau_{xy} = 2 \left[\overline{z}\gamma''(z) + \psi'(z) \right] = 0$$

$$\sigma_r + \sigma_\theta = \sigma_x + \sigma_y = 2 \left[\gamma'(z) + \overline{\gamma'(z)} \right] = -2P$$

$$\sigma_\theta - \sigma_r + 2i\tau_{r\theta} = \left(\sigma_y - \sigma_x + 2i\tau_{xy} \right) e^{i2\theta} = 0$$

Thus,

$$\sigma_\theta = \sigma_r = -P; \quad \sigma_x = \sigma_y = -P \quad \& \quad \tau_{xy} = \tau_{r\theta} = 0$$

These represent the hydrostatic state of stress of the disk under uniform compression.

1.10   Problems

1.1. A thin sheet made of an aluminum alloy having $E = 67\,\text{GPa}$, $G = 25.125\,\text{GPa}$, and $v = 1/3$ was used for two-dimensional surface strain measurements. The measurements provided the strains as $\epsilon_x = 10.5 \times 10^{-5}$, $\epsilon_y = -20 \times 10^{-5}$, and $\gamma_{xy} = 240 \times 10^{-5}$. Determine the corresponding stresses in Cartesian coordinates. An element is shown below. [Solution: $\sigma_x = 2.89\,\text{MPa}$, $\sigma_y = -12.44\,\text{MPa}$, and $\tau_{xy} = 60.45\,\text{MPa}$.]

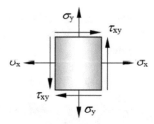

1.2. Determine **(a)** the principle stresses and strains and **(b)** the maximum shear stress for the case described in Problem 1.1.

1.3. Calculate **(a)** the diameter of a 1-m long wire that supports a weight of 200 Newton. If the wire stretches 2 mm, determine **(b)** the strain and the stress induced by the weight. Let the modulus of elasticity be $E = 207\,\text{GPa}$. [Solution: $\epsilon = 0.20\,\%$, $\sigma = 414\,\text{MPa}$, and $d = 0.78\,\text{mm}$.]

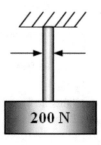

200 N

1.4. Derive an expression for the local uniform strain across the neck of a round bar being loaded in tension. Then, determine its magnitude if the original diameter is reduced to 80 %.

1.5. The torsion of a bar containing a longitudinal sharp groove may be characterized by a warping function of the type [after F.A McClintock, Proc. Inter. Conf. On
Fracture of Metals, Inst. of Mechanical Eng., London, (1956) 538]

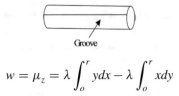

Groove

$$w = \mu_z = \lambda \int_0^r y\, dx - \lambda \int_0^r x\, dy$$

The displacements are $\mu_x = 0$ and $\mu_y = rz\lambda$, where λ and r are the angle of twist
per unit length and the crack tip radius, respectively. The polar coordinates have
the origin at the tip of the groove, which has a radius (R). Determine w, the shear
strains γ_{rz} and $\gamma_{z\theta}$. In addition, predict the maximum of the shear strain [Solution:
If $r \to 0$, then $\gamma_{rz} = \gamma_{max} \to -\infty$].

1.6. A steel cantilever beam having a cross-sectional area of $1.5\,cm^2$ is fixed at the
left-hand side and loaded with a 100 N downward vertical force at the extreme end
as shown in the figure shown below.

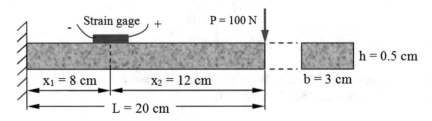

(a) Derive the stress equations in Cartesian coordinates if the Airy stress function
is $\phi = a_1xy + a_2xy^3$. The loading conditions are

$$\int_{-h/2}^{h/2} \tau_{xy}\, dy = \frac{P}{b}$$

$$\tau_{xy} = 0 \text{ at } y = \pm\frac{h}{2}$$

(b) Calculate the stresses and the strains at 8 cm from the fixed end of the shown
steel cantilever beam. The steel modulus of elasticity of is $E = 207\,GPa$.
[Solution: (a) $\sigma_x = 96\,MPa$, $\sigma_y = \tau_{xy} = 0$, and (b) $\epsilon = 4.64 \times 10^{-4}$.]

1.7. The stress-strain behavior of an annealed low-carbon steel ($\sigma_{ys} = 900\,MPa$
and $E = 207\,GPa$) obeys the Hollomon equation with $k = 1200\,MPa$ and $n = 0.25$.
(a) Plot the true and the engineering stress-strain curves. Calculate **(b)** the tensile
strength (σ_{ts}) and **(c)** the strain energy density up to the instability point.

1.8. The figure below shows a schematic cross-sectional view of a pressure vessel (hollow cylinder) subjected to internal and external pressures.

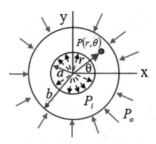

Determine the stresses at a point $P(r, \theta)$ in polar coordinates when **(a)** $P_i \neq 0$ and $P_o \neq 0$, **(b)** $P_o = 0$, **(c)** $P_i = 0$, **(d)** $a = 0$ so that the hollow cylinder becomes a solid cylinder. **(e)** Plot $\sigma_r = f(r)$ and $\sigma_\theta = f(r)$. Let $a = 450\,\text{mm}$, $b = 800\,\text{mm}$, $P_o = 0$ and $P = 40\,\text{MPa}$. The valid radius range must be $0.45\,\text{m} \le r \le 0.80\,\text{m}$. Use the following Airy stress function

$$\phi = c_1 + c_2 \ln(r) + c_3 r^2$$

along with the boundary conditions $\sigma_r = -P_i$ & $\tau_{r\theta} = 0$ @ $r = a$ and $\sigma_r = -P_o$ & $\tau_{r\theta} = 0$ @ $r = b$. This problem can be found elsewhere [13].

1.9. Consider an infinite plate with a central hole subjected to a remote uniform stress as shown in Example 1.5. The boundary conditions for this loaded plate are (1) $\sigma_x^\infty = \sigma_x = S$ and $\sigma_y = \tau_{xy} = 0$ at $r = \infty$ and (2) $\sigma_r = \tau_{r\theta} = 0$ at $r = a$. Use the following complex potentials [14]:

$$\psi'(z) = \frac{S}{4}\left(1 - \frac{2a^2}{z^2}\right) \quad \text{and} \quad \chi''(z) = -\frac{S}{2}\left(1 - \frac{a^2}{z^2} + \frac{3a^4}{z^4}\right)$$

to determine σ_r, σ_θ and $\tau_{r\theta}$ (in polar coordinates).

1.10. Use the Cauchy–Riemann condition to show that **(a)** $f(z) = 1/z$ is analytic and **(b)** its derivative is $f(z) = -1/z^2$.

1.11. Solve the Cauchy integral formula given below for the complex function $f(z) = e^z$ at $z = z_o$ and $z = 2$.

$$f(z_o) = \frac{1}{2\pi i} \int \frac{f(z)}{z - z_o} dz$$

1.12. Evaluate the Cauchy integral formula given below for the complex function when $z_o = \pi$.

$$f(z_o) = \frac{1}{2\pi i} \int \frac{\cos z}{z^2 - 1} dz$$

References

1. S. Timoshenko, J.N. Goodier, *Theory of Elasticity*, Chaps. 1, 2, 7. 2nd edn. (McGraw-Hill, New York, 1951)
2. J.E. Shigley, C.R. Mischke, *Mechanical Engineering Design*, 5th edn, vol. 14 (McGraw-Hill, New York, 1989)
3. M.A. Meyers, K.K. Chawla, *Mechanical Metallurgy: Principles and Applications* (Prentice-Hall, Englewood Cliffs, 1984)
4. A. Kelly, G.W. Groves, *Crystallography and Crystal Defects* (Addison-Wesley Publishing Co, California, 1970)
5. R.L. Ellis, Ph.D. Dissertation, Virginia Polytechnic Institute and State University, 1996
6. H. Jia, Ph.D. Dissertation, Virginia Polytechnic Institute and State University, 1998
7. W. Ramberg, W.R. Osgood, Description of stress-strain curves by three parameters, Technical Note No. 902, National Advisory Committee For Aeronautics, Washington, DC, 1943
8. J.H. Hollomon, Tensile deformation. Trans. Am. Inst. Min. Metall. Pet. Eng. **162**, 268–290 (1945)
9. W.S. Peeling, Design options for selection of fracture control procedures in the modernization of codes, rules, and standards, in *Proceedings: Joint United States–Japan Symposium on Applications of Pressure Component Codes*, Tokyo, 1973
10. J.W. Dally, W.F. Riley, *Experimental Stress Analysis*, 3rd edn. (McGraw-Hill, New York, 1991)
11. M.H. Sadd, *Elasticity: Theory, Applications and Numerics*, Chaps. 3 and 10, 2nd edn. (Elsevier, New York, 2009)
12. G.B. Airy, On the strains in the interior of beams. Philos. Trans. R. Soc. Lond. **153**, 49–80 (1863)
13. K. Hellan, *Introduction to Fracture Mechanics* (McGraw-Hill, New York, 1984)
14. A.S. Saada, *Elasticity Theory and Applications*, Chap. 19, 2nd edn. (J. Ross Publishing, Florida, 2009)
15. N.I. Muskhelishvili, *Some Basic Problems of the Mathematical Theory of Elasticity*, Chaps. 12–15. 2nd English edn. (Noordhoff International Publishing, Layden, 1977)
16. N.I. Muskhelishvili, Study on the boundary value problems related to the biharmonic equation and equations of elasticity in two dimensions (translated). Math. Ann. **107**(2), 282–312 (1932)
17. E. Goursat, The existence of the integral functions of a system of partial differential equations (translated). Bull. Soc. Math. Fr. **26**, 129–134 (1898)
18. A.F. Bower, *Applied Mechanics of Solids* (CRC Press, New York, 2009)
19. M.R. Spiegel, *Theory and Problems of Complex Variables with an Introduction to Conformal Mapping and its Applications*, Chap. 8. Schaum's Outline Series (McGraw-Hill, New York, 1964)
20. R.P. Feynman, R.B. Leighton, M. Sands, *The Feynman Lectures on Physics*, vol. 2 (Addison-Wesley, Redwood City, CA, 1989)
21. H. Lamb, *Hydrodynamics*, 6th edn., vol. 68 (Dover Publications, New York, 1945)
22. M.H. Sadd, *Elasticity: Theory, Applications and Numerics*, Chap. 10. 2nd edn. (Elsevier, New York, 2009)
23. L.L. Fischer, G.E. Beltz, Effect of crack geometry on dislocation nucleation and cleavage thresholds. Mater. Res. Soc. Symp. **539**, 57–62 (1999)
24. H. Cohn, *Conformal Mapping on Riemann Surfaces* (McGraw-Hill, New York, 1967)

Introduction to Fracture Mechanics

2

2.1 Introduction

The theory of elasticity used in Chap. 1 served the purpose of illustrating the close form of analytical procedures in order to develop constitutive equations for predicting failure of crack-free solids [1]. However, when solids contain flaws or cracks, the field equations are not completely defined by the theory of elasticity since it does not consider the stress singularity phenomenon near a crack tip. It only provides the means to predict general yielding as a failure criterion. Despite the usefulness of predicting yielding, it is necessary to use the principles of fracture mechanics to predict failure of solid components containing cracks.

Fracture mechanics is the study of mechanical behavior of cracked materials subjected to an applied load. In fact, Irwin [2] developed the field of fracture mechanics using the early work of Inglis [3], Griffith [4], and Westergaard [5]. Essentially, fracture mechanics deals with the irreversible process of rupture due to nucleation and growth of cracks. The formation of cracks may be a complex fracture process, which strongly depends on the microstructure of a particular crystalline or amorphous solid, applied loading, and environment. The microstructure plays a very important role in a fracture process due to dislocation motion, precipitates, inclusions, grain size, and type of phases making up the microstructure. All these microstructural features are imperfections and can act as fracture nuclei under unfavorable conditions. For instance, *brittle fracture* is a low-energy process (low-energy dissipation), which may lead to catastrophic failure without warning since the crack velocity is normally high. Therefore, little or no plastic deformation may be involved before separation of the solid. On the other hand, *ductile fracture* is a high-energy process in which a large amount of energy dissipation is associated with a large plastic deformation before crack instability occurs. Consequently, slow crack growth occurs due to strain hardening at the crack tip region.

© Springer International Publishing Switzerland 2017
N. Perez, *Fracture Mechanics*, DOI 10.1007/978-3-319-24999-5_2

Fig. 2.1 Sinusoidal stress vs.
interatomic displacement

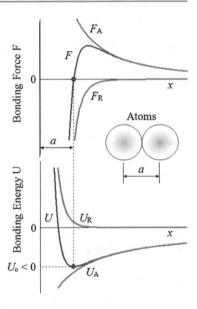

2.2   Theoretical Strength

Consider the predicament of how strong a perfect (ideal) crystal lattice should be
under an applied state of stress and the comparison of the actual and theoretical
strength of metals. This is a very laborious work to perform, but theoretical
approximations can be made in order to determine or calculate the stress required
for fracture of atomic bonding in crystalline or amorphous crystals.

Assume a simple sinusoidal stress-displacement law with a half period of $\lambda/2$
shown in Fig. 2.1 which predicts the simultaneous separation of atoms when the
atomic separation reaches a critical value.

For an ideal crystal subjected to a tensile load and a shear load, which generates
small displacements, the sinusoidal stress functions are

$$\sigma = \sigma_{max} \sin\left(\frac{2\pi x}{\lambda}\right) \simeq \left(\frac{2\pi x}{\lambda}\right)\sigma_{max} \tag{2.1a}$$

$$\tau = \tau_{max} \sin\left(\frac{2\pi x}{a_o}\right) \simeq \left(\frac{2\pi x}{a_o}\right)\tau_{max} \tag{2.1b}$$

Thus, the maximum theoretical tensile and shear stresses become

$$\sigma_{max} = \left(\frac{\lambda}{2\pi x}\right)\sigma \tag{2.2a}$$

$$\tau_{max} = \left(\frac{a_o}{2\pi x}\right)\tau \tag{2.2b}$$

Table 2.1 Theoretical and experimental fracture strength [6]

Material	E (MPa)	σ_{max}-Eq. (2.4)	σ_f-Exp.	σ_{max}/σ_f
Silica fibers	97.10	30.90	24.10	1.28
Iron whisker	295.20	94.00	13.10	7.18
Silicon whisker	165.70	52.70	6.50	8.11
Alumina whisker	496.20	158.00	15.20	10.39
Ausformed steel	200.10	63.70	3.10	20.55
Piano wire	200.10	63.70	2.80	22.75

The interpretation of Fig. 2.1 is that the strength to pull atoms apart increases with increasing atomic distance, reaches a maximum strength (peak strength) equals to the theoretical (cohesive) tensile strength $\sigma_{max} = \sigma_c$, and then decreases as atoms are further apart in the direction perpendicular to the applied stress. Consequently, atomic planes separate and the material cleaves perpendicularly to the tensile stress.

Assuming an elastic deformation process, Hooke's law gives the tensile modulus and the shear modulus of elasticity defined by

$$E = \frac{Tensile\ Stress}{Strain} = \frac{\sigma}{x/a_o} \tag{2.3a}$$

$$G = \frac{Shear\ Stress}{Strain} = \frac{\tau}{x/a_o} \tag{2.3b}$$

where $a_o = \lambda/2 =$ Equilibrium atomic distance (Fig. 2.1)

Combining Eqs. (2.2) and (2.3) yields the theoretical fracture strength of solid materials

$$\sigma_{max} = \frac{E}{\pi} \tag{2.4}$$

$$\tau_{max} = \frac{G}{2\pi} \tag{2.5}$$

Table 2.1 contains theoretical and experimental data for some elastic materials tested in tension.

The discrepancy between σ_{max} and σ_f values is due to the fact that the sinusoidal model assumes a concurrent fracture of atomic bonding until the atomic planes separate and σ_f is associated with plastic flow and dislocation motion. Physically, the discrepancy is due to the presence of small flaws or cracks on the surface or within the material.

Using the energy at fracture for a tension test, the fracture work per unit area can be defined by a simple integral

$$W' = \int_o^{\lambda/2} \sigma_{max} \sin\left(\frac{\pi x}{\lambda/2}\right) dx = \left(\frac{\lambda}{\pi}\right) \sigma_{max} \tag{2.6}$$

Fig. 2.2 Schematic variation of bonding force and bonding energy as functions of interatomic spacing x

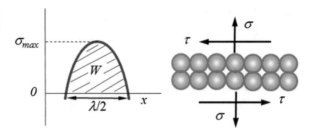

Letting $2\gamma = W'$ be the total surface energy required to form two new fracture surfaces and combining Eqs. (2.4) and (2.6) yields the theoretical tensile strength in terms of surface energy and equilibrium spacing

$$\sigma_c = \sqrt{\frac{E\gamma}{a_o}} \tag{2.7}$$

In addition, the atomic bonding in solids is related to bonding forces and energies. In fact, the atomic bonding is due to repulsive and attractive forces that keep the atoms together to form symmetrical arrays. These forces as well as the potential energies depend on the interatomic spacing or distance between adjacent atoms. Figure 2.2 schematically shows the forces and the energies as functions of interatomic spacing (separation distance between centers of two atoms) for two ideal atoms. In general, atoms are considered spherical electric structures having diameters in the order of 0.1 nm. According to the theoretical plot depicted in Fig. 2.2, both attractive and repulsive forces act together to keep the atoms at their equilibrium spacing. These forces depend on temperature and pressure. The general form of the potential or bonding energy (U) and the net force (F) are defined by

$$U = \frac{C_R}{x^n} - \frac{C_A}{x^m} = U_R + U_A \tag{2.8}$$

$$F = \frac{dU}{dx} = F_R + F_A \tag{2.9}$$

where $x =$ Interatomic distance
       $U_R =$ Repulsive energy
       $U_A =$ Attractive energy
       $F_R =$ Repulsive forces
       $F_A =$ Attractive force
       $C_R, C_A =$ Constants
       $n, m =$ Exponents

The curves in Fig. 2.2 are known as Condon-Morse curves and are used to explain the physical events of atomic displacement at a nanoscale. At equilibrium, the minimum potential energy and the net force are dependent of the interatomic spacing; that is,

$$U_o = f(a_o) < 0 \quad \text{and} \quad F = g(a_o) = 0 \tag{a}$$

However, if the interatomic spacing (a_o) is slightly perturbed elastically by the action of an applied load, a repulsive force builds up if $x < a_o$ or an attractive force builds up if $x > a_o$. Once the applied load is removed, the two atoms have the tendency to return to their equilibrium position at $x = a_o$

Conclusively, an array of atoms form a definite atomic pattern with respect to their neighboring atoms, and as a result, all atoms form a specific space lattice consisting of unit cells, such as body-centered cubic (BCC), hexagonal, monoclinic, and the like.

The reason atoms form a space lattice consisting of a unique atomic structure is due to the attractive and repulsive atomic forces being equal, but opposite in sense. Hence, the atoms are considered to be in their equilibrium state forming a particular structure. Thus, atoms are then bonded in a sea of electrons, forming metallic bonding. X-ray diffraction technique is used to reveal the type metallic structures. There are 7 types of crystal structures and 14 possible lattice geometries called Bravais lattices.

Any elastic perturbation of the lattice structure due to an external loading mode induces atomic deformation defined as the deformation strain (ϵ_x), which can be defined as a fractional change in the atomic spacing x (Fig. 2.2). Hence, $\epsilon_x = (x - a)/a$, x is the strained spacing.

Furthermore, Eq. (2.8) resembles the Lennard-Jones potentials [7] used to treat gases, liquids, and solids. There are other interatomic potential functions based on the quantum mechanical treatment of many particles. Among many references available in the literature, the book written by Michael Rieth [7] includes significant theoretical details for determining the potential energy for atomic interactions. This book includes the Buckingham, the Morse, the Lennard-Jones potentials and the Schommers potentials for aluminum, and so forth.

The most common Lennard-Jones potentials for materials 1 and 2 are of the form

$$U = 4\epsilon \left[\left(\frac{C}{x} \right)^{12} - \left(\frac{C}{x} \right)^{6} \right] \qquad (2.9a)$$

where the potential ϵ and the constant C are defined by the Lorentz-Berthelot mixing rules [7]

$$\epsilon = \epsilon_{12} = \sqrt{\epsilon_1 \epsilon_2} \qquad (2.9b)$$

$$C = C_{12} = \frac{C_1 + C_2}{2} \qquad (2.9c)$$

These variables are just empirical correction factors.

2.3   Stress Concentration Factor

Generally, structural components subjected to external loads should be analyzed or examined by determining the stress distribution on the loaded area and the theoretical stress concentration factor (K_t) at particular point about a notch with a radius of curvature (ρ). This can be accomplished by using a suitable Airy stress function ϕ or any other appropriate function. This section partially illustrates the methodology for determining the circumferential stress component in elliptical coordinates and K_t in infinite flat plates containing circular and elliptical holes. This is purposely done prior to the introduction of fracture mechanics in order for the reader to have a basic understanding of singular stress fields in components having notches instead of cracks.

Symmetric Elliptical Hole in an Infinite Plate Consider an infinite plate containing an elliptical hole with major axis 2a and minor axis 2b as shown in Fig. 2.3, where the elliptical and Cartesian coordinates are (ξ, ψ) and (x, y), respectively. It is assumed that the flat plate has uniform dimensions and contains a through-thickness smooth elliptical hole, which is symmetric about its center. The equation of the elliptical curve can be represented by $f_\xi (x, y) = \xi$ and the one for a hyperbola is $f_\psi (x, y) = \psi$, where ξ and ψ are constants [8].

   The equation of an ellipse in Cartesian coordinates is given by

$$\frac{x^2}{a^2} + \frac{y^2}{b^2} = 1 \tag{2.10}$$

Fig. 2.3 Elliptic coordinates in an infinite plate

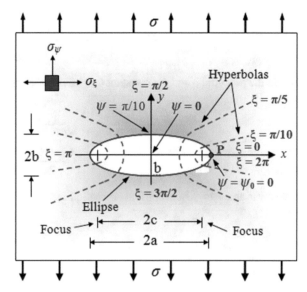

where

$$x = c \cosh \xi \cos \psi \tag{2.11a}$$

$$y = c \sinh \xi \sin \psi \tag{2.11b}$$

$$x + iy = c \cosh (\xi + i\psi) \tag{2.11c}$$

and

$$a = c \cosh (\xi_o) \tag{2.12a}$$

$$b = c \sinh (\xi_o) \tag{2.12b}$$

Here, the foci of the ellipse is at $x = \pm 2c$ which becomes the crack length when $\xi_o \to 0$ and $b \to 0$. In fact, the foci always lie on the major (longest) axis, spaced equally on each side of the center of the ellipse. The radius of the ellipse at the end of the major x-axis (point "P" in Fig. 2.3) is

$$\rho = \frac{b^2}{a} \tag{2.13}$$

Now, it is desirable to derive the maximum circumferential or tangential elastic stress component at an elliptical hole tip along the major axis $2a$ and the minor axis $2b$ [9]. Denote that and elliptical home becomes an elliptical crack if $2b \to 0$. Nevertheless, Inglis [3] derived the elastic stress distribution in an infinite plate subjected to a remote tension stress perpendicular to the major axis $2a$ of a flat plate. It was assumed that the plate width and height were $B >> 2a$ and $h >> 2b$, respectively, in order to avoid the effect of the plate boundary and to assure that the applied tension stress is remotely located from the elliptical hole surfaces. The resultant circumferential stress component for an elliptical hole (Fig. 2.3) is [3]

$$\sigma_\xi = \sigma e^{2\xi_o} \left[\frac{(1 + e^{-2\xi_o}) \sinh (2\xi_o)}{\cosh (2\xi_o) - \cos (2\psi)} - 1 \right] \tag{2.14}$$

Using Eq. (2.12) yields

$$\tanh \xi_o = \frac{b}{a} \tag{a}$$

The maximum local stress equation at the tip of the elliptic hole (point a in Fig. 2.3 where $\psi = 0$ and $\cos (2\psi) = 1$) can be derived by manipulating Eq. (2.14) along with the following hyperbolic relationships

$$e^{2\xi_o} = \sinh (2\xi_o) + \cosh (2\xi_o) \tag{b}$$

$$e^{-2\xi_o} = \sinh (2\xi_o) - \cosh (2\xi_o) \tag{c}$$

$$2 \sinh^2 (\xi_o) = \cosh (2\xi_o) - 1 \tag{d}$$

$$\cosh^2(2\xi_o) = \sinh^2(\xi_o) + \cosh^2(\xi_o) \tag{e}$$

$$\sinh(2\xi_o) = 2\sinh(\xi_o)\cosh(\xi_o) \tag{f}$$

The resultant equation is

$$\sigma_{\xi-\max} = \sigma_{\max} = \sigma_y = \sigma\left[1 + \frac{2\cosh(\xi_o)}{\sinh(\xi_o)}\right] \tag{2.15}$$

Combining Eqs. (2.12) and (2.15) yields the well-known expression in the literature for the maximum local stress in a plate containing an elliptic hole. Thus,

$$\sigma_{\max} = \sigma\left(1 + \frac{2a}{b}\right) \tag{2.16}$$

The reader should consult other references [10–12] in the field of theory of elasticity applied to components containing notches and specific stress concentration factors. With respect to Eq. (2.16), the expression $(1 + 2a/b)$ is known as the theoretical stress concentration factor for an ellipse. Thus,

$$K_t = 1 + \frac{2a}{b} = \frac{\sigma_{\max}}{\sigma} \tag{2.17}$$

Combining Eqs. (2.13) and (2.16) yields the axial stress equation as

$$\sigma_{\max} = \left(1 + 2\sqrt{\frac{a}{\rho}}\right)\sigma \tag{2.18}$$

For a sharp crack, $a \gg \rho$, $\sqrt{a/\rho} \gg 1$ and Eq. (2.18) becomes

$$\sigma_{\max} = \left(2\sqrt{\frac{a}{\rho}}\right)\sigma \tag{2.19}$$

Thus, the theoretical stress concentration factor becomes

$$K_t = 2\sqrt{\frac{a}{\rho}} = \frac{\sigma_{\max}}{\sigma} \tag{2.20}$$

In fact, the use of the stress concentration approach is meaningless for characterizing the behavior of sharp cracks because the theoretical axial stress concentration factor is $K_t \to \infty$ as $\rho \to 0$. Therefore, the elliptic hole becomes a sharp crack, and the stress intensity factor K_I is the most useful approach for analyzing structural and machine components containing sharp cracks.

Here, σ is the nominal stress or the driving force. If $a = b$, then Eq. (2.16) gives $K_t = 3$ and $\sigma_{\max} = 3\sigma$ for a circular hole. On the other hand, if $b \to 0$

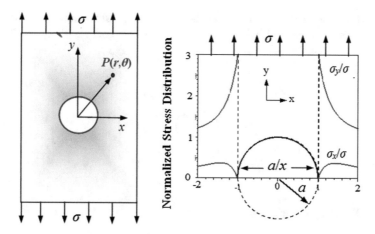

Fig. 2.4 Normalized axial and transverse stresses along the x-axis near a circular hole in a thin plate under tension loading

and $\rho \rightarrow 0$, then a sharp crack is formed and $\sigma_{max} \rightarrow \infty$, which is singular and meaningless. In addition, K_t is used to analyze the stress field at a point in the vicinity of a notch having a radius $\rho \gg 0$. However, if a crack is formed having $\rho \simeq 0$ at a microscopic level, the stress field at the crack tip is defined in terms of the stress intensity factor (K_I) instead of the stress concentration factor (K_t). In fact, microstructural discontinuities and geometrical discontinuities, such as notches, holes, grooves, and the like, are sources of crack initiation when the stress concentration factor is sufficiently high.

Symmetric Circular Hole in an Infinite Plate   Figure 2.4 shows the distribution of the axial and transverse normalized stresses along the x-axis near a circular hole in a wide, thin, and infinite plate loaded in tension.

Consider an infinite isotropic plate containing a circular hole as shown in Fig. 2.4. It is desirable to determine the elastic stresses in rectangular and polar coordinates when the plate is remotely loaded in tension.

The detailed analytical procedure for deriving the generalized stress equations in polar coordinates, based on the method of superposition, can be found in a book written by Dally and Riley [1]. Thus, these equations evaluated at point $P(r, \theta)$ are

$$\sigma_r = \frac{\sigma}{2} \left\{ \left(1 - \frac{a^2}{r^2} \right) \left[1 + \left(\frac{3a^2}{r^2} - 1 \right) \cos 2\theta \right] \right\}$$

$$\sigma_\theta = \frac{\sigma}{2} \left[\left(1 + \frac{a^2}{r^2} \right) + \left(1 + \frac{3a^4}{r^4} \right) \cos 2\theta \right] \qquad (2.21)$$

$$\tau_{r\theta} = \frac{\sigma}{2} \left[\left(1 + \frac{3a}{r^2} \right) \left(1 - \frac{a^2}{r^2} \right) \sin 2\theta \right]$$

Letting $\theta = 0$ and $r = x$ in Eq. (2.21) yields the stress distribution along the x-axis at point $P(x, 0)$ in Fig. 2.4, where $\sigma_r = \sigma_x$, $\sigma_\theta = \sigma_{yy}$ and $\tau_{r\theta} = \tau_{xy} = 0$. Thus, Eq. (2.21) becomes [1, 12]

$$\frac{\sigma_x}{\sigma} = \frac{3}{2} \left(\frac{a}{x}\right)^2 - \frac{3}{2} \left(\frac{a}{x}\right)^4$$
$$\frac{\sigma_y}{\sigma} = 1 + \frac{1}{2} \left(\frac{a}{x}\right)^2 + \frac{3}{2} \left(\frac{a}{x}\right)^4 \qquad (2.22)$$
$$\frac{\tau_{xy}}{\sigma} = 0$$

2.4   Griffith Crack Theory

The development of linear-elastic fracture mechanics (LEFM) started with Griffith work on glass [4]. Fundamentally, the Griffith theory considers the energy changes associated with incremental crack growth. He used an energy balance approach to predict the fracture stress of glass and noted in 1921 that when a stressed plate of an elastic material containing cracks, the potential energy per unit thickness (ΔU) decreased and the surface energy per unit thickness (U_s) increased during crack growth. Then, the total potential energy of the stressed solid body is related to the release of stored energy and the work done by the external loads. The "surface energy" arises from a nonequilibrium configuration of the nearest neighbor atoms at any surface in a solid [13–15].

Consider a large or an infinite brittle plate containing one center through-thickness crack of length $2a$ with two crack tips as depicted in Fig. 2.5. When the plate is subjected to a remote and uniform tensile load perpendicular to the crack plane along the x-axis, the stored elastic strain energy is released within a cylindrical volume of material of length B.

Fig. 2.5 A large plate containing one through-thickness central crack. Also shown are two idealized energy release areas ahead of the crack tips

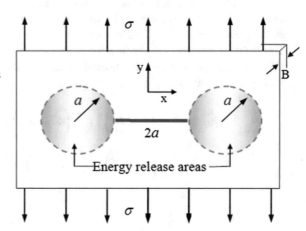

When the elastic or brittle solid body (specimen) is remotely loaded from the crack faces, the product of the released elastic strain energy density ($\int \sigma d\epsilon$) and the cylindrical volume element ($2\pi a^2 B$) about the crack (Fig. 2.5), where this energy is released, yields the elastic strain energy as

$$W_e = -2\left(\pi a^2 B\right) \int \sigma d\epsilon = -2\left(\pi a^2 B\right) \int E'\epsilon d\epsilon \tag{2.23}$$

$$W_e = -2\left(\pi a^2 B\right) \left(\frac{E'\epsilon^2}{2}\right) = -\left(\pi a^2 B\right)\left(\frac{\sigma^2}{E'}\right) \tag{2.24}$$

where $\sigma = E'\epsilon =$ Hooke's law

$E' = E$ for plane stress

$E' = E/\left(1 - v^2\right)$ for plane-strain conditions

$E =$ Modulus of elasticity (MPa)

$\epsilon =$ Elastic strain

$\sigma =$ Applied remote stress (MPa)

$a =$ One-half crack length (mm)

$v =$ Poisson's ratio

$4aB = 2(2aB) =$ Total surface crack area (mm^2)

$B =$ Thickness (mm)

Denote that the factor E' is introduced in Eq. (2.23) for controlling either plane stress or plane-strain condition. In addition, Eq. (2.24) can also be derived by inserting the Inglis displacement equation (μ_y) in the y-direction [3] into the following expression:

$$W_e = -4B \int_o^a \frac{1}{2}\sigma \mu_y dx = -4B \int_o^a \frac{1}{2}\sigma \left(\frac{2\sigma}{E'}\sqrt{a^2 - x^2}\right) dx \tag{a}$$

$$W_e = -\left(\frac{4B\sigma^2}{E'}\right) \int_o^a \sqrt{a^2 - x^2} dx = -\left(\frac{4B\sigma^2}{E'}\right)\left(\frac{\pi a^2}{4}\right) \tag{b}$$

$$W_e = -\left(\pi a^2 B\right)\left(\frac{\sigma^2}{E'}\right) \tag{2.25}$$

Now, the elastic surface energy for creating new crack surfaces during crack growth (from two crack tips) is [4]

$$W_s = 2\left(2aB\gamma_s\right) \tag{2.26}$$

where $\gamma_s =$ Specific surface energy for atomic bond breakage (J/mm^2)

For an elastically stressed solid body, Griffith energy balance takes into account the decrease in potential energy (due to the release of stored elastic energy and the work done by external loads) and the increase in surface energy resulting from the growing crack, which creates new surfaces. For the energy balance, the total elastic energy of the system, referred to as the total potential energy, takes the mathematical form

$$W = W_s + W_e = 2\left(2aB\gamma_s\right) - \left(\pi a^2 B\right)\left(\frac{\sigma^2}{E'}\right) \tag{2.27}$$

For convenience, divide Eq. (2.27) by the thickness B to get the total potential energy per unit thickness as

$$U = U_s + U_e \tag{2.28}$$

$$U = 2\left(2a\gamma_s\right) - \frac{\pi a^2 \sigma^2}{E'} \tag{2.29}$$

where U_s = Elastic surface energy per unit thickness (J/mm)
$\quad\ U_e$ = Released elastic energy per unit thickness (J/mm)

Thus, the Griffith's energy criterion for crack growth is $U_e \geq U_s$ when $dU/da = 0$. Then, the energy balance gives $4a\gamma_s E' = \beta\pi a^2\sigma^2$, from which the applied stress (σ), the crack length (a) or the strain energy release rate (G_I) for brittle solid materials are easily derivable. They are, respectively,

$$\sigma = \sqrt{\frac{(2\gamma_s)E'}{\pi a}} \tag{2.30}$$

$$a = \frac{(2\gamma_s)E'}{\pi\sigma^2} \tag{2.31}$$

$$G_I = 2\gamma_s = \frac{\pi a\sigma^2}{E'} \tag{2.32}$$

At fracture, Eqs. (2.30) through (2.32) give the critical entities. Rearranging Eq. (2.32) yields the elastic stress intensity factor

$$\sigma\sqrt{\pi a} = \sqrt{(2\gamma_s)E'} = \sqrt{G_I E'} \tag{2.33}$$

$$K_I = \sigma\sqrt{\pi a} \tag{2.34}$$

The parameter K_I is called the stress intensity factor which is the crack driving force, and its critical value is a material property known as fracture toughness, which, in turn, is the resistance force to crack extension [16]. The interpretation of Eq. (2.34) suggests that crack extension in brittle solids is completely governed by the critical value of the stress intensity factor.

Griffith observed that $\sigma_f\sqrt{a_c}$ was nearly a constant for six (6) cracked circular glass tubes under plane stress condition, Table III in ref. [4]. The average value is

$$\sigma_f\sqrt{a_c} = 0.2378 \pm 0.0062\,\text{ksi}\sqrt{\text{in}} \tag{a}$$

For convenience, using Eq. (2.34) yields the plane-strain fracture toughness

$$K_{IC} = \sigma_f \sqrt{\pi a_c} \simeq 0.42 \pm 0.01 \, \text{ksi} \sqrt{\text{in}} \qquad \text{(b)}$$

$$K_{IC} = 0.46 \pm 0.01 \, \text{MPa} \sqrt{\text{m}}$$

In addition, taking the second derivative of Eq. (2.29) with respect to the crack length yields

$$\frac{d^2 (U)}{da^2} = -\frac{2\pi \sigma^2 B}{E'} \qquad (2.35)$$

Denote that $d^2 U / da^2 < 0$ represents an unstable system. Consequently, the crack will always grow [15]. Fundamentally, linear-elastic fracture mechanics requires a stress analysis approach to predict nonphysical or conceptual infinite local stresses ($\sigma_{ij} \rightarrow \infty$) at a crack tip despite that yielding occurs, to an extent, in most engineering brittle solids. Glass and the pure brittle materials are an exception. The yielding process truncates the local stresses, specifically the stress perpendicular (σ_y) to the crack plane. Eventually, the maximum applied stress is a critical (σ_c) or fracture stress (σ_f) that causes fracture of the solid is less than the yield strength (σ_{ys}) of the solid body due to the existence of cracks or defects.

Example 2.1. *A large and wide brittle plate containing a single-edge crack (a) fractures at a tensile stress of 4 MPa. The critical strain energy release rate (G_c) and the modulus of elasticity (E) are $4 \, J/m^2$ and $65,000 \, MPa$, respectively. Assume plane stress condition and include the thickness $B = 3 \, mm$ in all calculations. (a) Plot the theoretical total surface energy (U_s), the released strain energy (U_e), and the total potential energy change (W). Interpret the energy profiles. Determine (b) the critical crack length and (c) the maximum potential energy change ($W_{\max}$); (d) will the crack grow unstably? (e) What is the critical stress intensity factor for this brittle plate?*

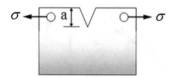

Solution. *Given data:*
$\sigma = 4 \, \text{MPa}; \, E' = E = 65,000 \, \text{MPa}$ *for plane stress condition*
$G_c = 4 \, \text{J/m}^2 = 4 \times 10^{-6} \, \text{J/mm}^2$
$\gamma_s = G_c/2 = 2 \, \text{J/m}^2 = 2 \times 10^{-6} \, \text{J/mm}^2$

(a) *For single-edge crack with one crack tip, Eq. (2.27) is divided by 2. Thus, the surface energy becomes*

$$W_s = 2aB\gamma_s = (2)\left(3 \times 10^{-3} \text{ m}\right)\left(2 \text{ J/m}^2\right)a = \left(12 \times 10^{-3} \text{ J/m}\right)a$$

$$W_s = \left(12 \times 10^{-6} \text{ J/mm}\right)a$$

On the other hand, the released strain energy is also divided by 2 so that

$$W_e = -\left(\pi a^2 B\right)\left(\frac{\sigma^2}{2E}\right)$$

$$W_e = -\frac{(\pi)\left(3 \times 10^{-3} \text{ m}\right)(4 \text{ MPa})^2 a^2}{(2)(65000 \text{ MPa})} = -\left(1.16 \times 10^{-6} \text{ MPa m}\right)a^2$$

$$W_e = -(1.16 \text{ Pa m})a^2 = \left(1.16 \text{ J/m}^2\right)a^2$$

$$W_e = -\left(1.16 \times 10^{-6} \text{ J/mm}^2\right)a^2$$

Thus, the potential energy change becomes

$$W = W_s + W_e$$

$$W = \left(12 \times 10^{-6} \text{ J/mm}\right)a - \left(1.16 \times 10^{-6} \text{ J/mm}^2\right)a^2$$

The energy profiles are given below. The total potential energy associated with crack growth is simply the sum of the surface energy and the released strain energy. The former energy is needed for creating new crack surfaces by allowing the breakage of atomic bonds. The latter energy is negative because it is released during crack growth and it is needed for unloading the regions near the crack flanks.

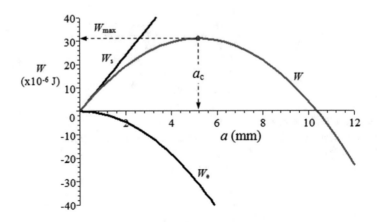

Moreover, stable crack growth occurs when $W < W_{max}$, but $W > 0$. At the maximum potential energy, the crack length reaches a critical value $a = a_c$ and the slope becomes $dW/da = 0$. Beyond this critical state, the potential energy change decreases very significantly because crack growth continues to occur at a very high velocity causing catastrophic failure.

(b) *The critical crack length can be determined by letting the first-order derivative of the total potential energy equals to zero. Thus,*

$$W = \left(12 \times 10^{-6} \text{ J/mm}\right) a - \left(1.16 \times 10^{-6} \text{ J/mm}^2\right) a^2$$

$$\left.\frac{dW}{da}\right|_{a=a_c} = \left(12 \times 10^{-6} \text{ J/mm}\right) - \left(2.32 \times 10^{-6} \text{ J/mm}^2\right) a_c = 0$$

$$a_c = \frac{12 \times 10^{-6}}{2.32 \times 10^{-6}} = 5.17 \text{ mm} = 0.20 \text{ in}$$

Therefore, $a_c = 5.17$ mm is not a very large crack, but it is the maximum allowable crack length in the system. If a similar structure is to be put in service, then a crack detection technique has to be implemented in order to avoid fracture when $a = a_c$.

(c) *The maximum potential energy is*

$$W_{max} = \left(12 \times 10^{-6} \text{ J/mm}\right) a_c - \left(1.16 \times 10^{-6} \text{ J/mm}^2\right) a_c^2$$

$$W_{max} = \left(12 \times 10^{-6} \text{ J/mm}\right) (5.17 \text{ mm}) - \left(1.16 \times 10^{-6} \text{ J/mm}^2\right) (5.17 \text{ mm})^2$$

$$W_{max} = 31.03 \times 10^{-6} \text{ J}$$

Thus, the point $(a_c, W_{max}) = \left(5.17 \text{ mm}, 31.03 \times 10^{-6} \text{ J}\right)$ is plotted on the figure above. At this critical point, the derivative of the total potential energy change is $d(W)/da = 0$ and $W = W_{max}$ at $a = a_c$.

(d) *From part* **(a)***,*

$$W = \left(12 \times 10^{-6} \text{ J/mm}\right) a - \left(1.16 \times 10^{-6} \text{ J/mm}^2\right) a^2$$

$$\frac{dW}{da} = \left(12 \times 10^{-6} \text{ J/mm}\right) - \left(2.32 \times 10^{-6} \text{ J/mm}^2\right) a$$

$$\frac{d^2 W}{da^2} = -2.32 \times 10^{-6} \text{ J/mm}^2$$

Therefore, the crack will grow unstable because $d^2 W/da^2 < 0$.

(e) *The critical stress intensity factor or the plane stress fracture toughness for the brittle plate is determined by using Eq. (2.34). Thus,*

$$K_C = \sigma \sqrt{\pi a_c}$$

$$K_C = (4 \text{ MPa}) \sqrt{\pi \, (5.17 \times 10^{-3} \text{ m})}$$

$$K_C = 0.51 \text{ MPa}\sqrt{\text{m}}$$

In addition, combining Eqs. (2.30) and (2.34) along $\beta = 1$, $K_I = K_C$ and $a = a_c$ gives

$$K_C = \sqrt{2\gamma_s E} = \sqrt{2\left(2 \text{ J/m}^2\right)(65000 \times 10^6 \text{ Pa})}$$

$$K_C = \sqrt{2 \, (2 \text{ Pa m}) \, (65000 \times 10^6 \text{ Pa})}$$

$$K_C = 5.1 \times 10^5 \text{ Pa}\sqrt{\text{m}}$$

$$K_C = 0.51 \text{ MPa}\sqrt{\text{m}}$$

Also, combining Eqs. (2.29) and (2.34) yields

$$K_C = \sqrt{\frac{2E}{a_c B} \left(2a_c B\gamma_s - W_{\max}\right)} = 0.51 \text{ MPa}\sqrt{\text{m}}$$

Therefore, the above mathematical equations give the same low result for K_C, which, in turn, indicates that the material is brittle and it is an important property of the material for design applications. In fact, K_C is the plane stress fracture toughness that defines the material resistance to brittle fracture when the crack length reaches a critical value a_c. Furthermore, a low K_C value implies that the material absorbs a small quantity of strain energy prior fracture. This indicates that the materials is elastic or brittle. Hence, brittle materials can be used in specific engineering applications.

On the other hand, a large K_C value specifies a larger consumption of strain energy and the material fractures in a tearing-like mode. This, then, defines a ductile material, which may have significant applications in industrial structures.

2.5   Strain Energy Release Rate

It is well known that plastic deformation occurs in engineering metal, alloys, and some polymers. Due to this fact, Irwin [2] and Orowan [17] modified Griffith's elastic surface energy expression, Eq. (2.32), by adding a plastic deformation energy or plastic strain work γ_p in the fracture process. For tension loading, the total elastic-plastic strain energy is known as the strain energy release rate G_I, which is the energy per unit crack surface area available for infinitesimal crack extension [16]. Thus,

$$G_I = 2\left(\gamma_s + \gamma_p\right) \tag{2.36}$$

$$G_I = \frac{\pi a \sigma^2}{E'} \tag{2.37}$$

Here, $E' = E/\beta$. Rearranging Eq. (2.37) gives the stress equation as

$$\sigma = \sqrt{\frac{E'G_I}{\pi a}} \tag{2.38}$$

Combining Eqs. (2.34) and (2.37) yields

$$G_I = \frac{K_I^2}{E'} \tag{2.39}$$

This is one of the most important relations in the field of linear fracture mechanics. Hence, Eq. (2.39) suggests that G_I represents the material's resistance (R) to crack extension, and it is known as the crack driving force. On the other hand, K_I is the intensity of the stress field at the crack tip.

The condition of Eq. (2.39) implies that $G_I = R$ before relatively slow crack growth occurs. However, rapid crack growth (propagation) takes place when $G_I \rightarrow G_{IC}$, which is the critical strain energy release rate known as the crack driving force or fracture toughness of a material under tension loading. Consequently, the fracture criterion by G_{IC} establishes crack propagation when $G_I \geq G_{IC}$. In this case, the critical stress or fracture stress σ_c and the critical crack driving force G_{IC} can be predicted using Eq. (2.39) when the crack is unstable. Hence, the critical or fracture stress is defined as

$$\sigma_f = \sigma_c = \sqrt{\frac{E'G_{IC}}{\pi a}} \tag{2.40}$$

Griffith assumed that the crack resistance R consisted of surface energy only for brittle materials. This implies that $R = 2\gamma_s$, but most engineering materials undergo, to an extent, plastic deformation so that $R = 2\left(\gamma_s + \gamma_p\right)$. Figure 2.6 shows a plastic zone at the crack tip representing plasticity or localized yielding, induced by an external nominal stress. This implies that the energy γ_p is manifested due to this small plastic zone in the vicinity of the crack tip.

It is clear that the internal stresses on an element of an elastic-plastic boundary are induced by plasticity and are temperature-dependent tensors. The stress in front of the crack tip or within the plastic zone exceeds the local microscopic yield stress,

Fig. 2.6 Single-edge crack configuration showing the plastic zone and stresses at the crack tip is rectangular and polar coordinates

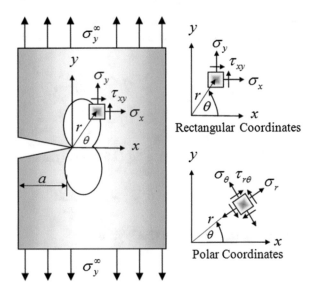

which may be defined as the theoretical or cohesive stress for breaking atomic bonds. If microscopic plasticity through activated slip systems does not occur, as in glasses, then a linear-elastic fracture is achieved as the controlling fracture process. In essence, the fracture process is associated with plasticity at a microscopic level.

If large plasticity occurs at the crack tip, then the crack blunts and its radius of curvature increase. This plastic deformation process is strongly dependent on the temperature and microstructure. Regardless of the shape of the plastic zone, the irreversible crack tip plasticity is an indication of a local strain hardening process during which slip systems are activated and dislocations pile up and dislocation interaction occurs.

Example 2.2. *This worked example illustrates the application of fracture mechanics to determine the behavior of a cracked solid body subjected to a static stress. A hypothetical (1-m) × (100-mm) × (3-mm) sheet of glass containing a 0.4-mm-long central crack (slit) is loaded in tension by a hanging person having a mass of 80 Kg. (a) Calculate K_I and G_I at fracture. (b) Determine whether or not the glass sheet will fracture. The fracture toughness and the modulus of elasticity of glass are $K_{IC} = 0.80\,MPa\sqrt{m}$ and $E = 60\,GPa$, respectively. Calculate (c) the critical crack length (a_c) which is the maximum allowable crack size before fast fracture occurs, (d) the fracture or critical stress, (e) the reduction in strength if the crack-free (sound) fracture stress is 165 MPa, and (f) the maximum allowable mass (m_c). Assume a Poisson's ratio of 0.3 and plane stress conditions because the plate is very thin.*

Solution.

(a) *The applied mass has to be converted to a load, which in turn is divided by the cross-sectional area of the plate as if the crack is not present. Also, the specimen configuration is given below.*

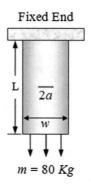

Fixed End

L $\overline{2a}$

w

$m = 80\ Kg$

The applied stress, the stress intensity factor, and the strain energy release rate are, respectively,

$$\sigma = P/A_o = \frac{(80\ Kg)\left(9.81\ m/s^2\right)}{(100 \times 10^{-3}\ m)\,(3 \times 10^{-3}\ m)} = 262\ MPa$$

$$K_I = \sigma\sqrt{\pi a} = (262\ MPa)\sqrt{\pi\,(0.2 \times 10^{-3}\ m)} = 0.07\ MPa\sqrt{m}$$

$$G_I = \frac{K_I^2}{E} = \frac{(0.07\ MPa\sqrt{m})^2}{11 \times 10^3\ MPa} = 4.45 \times 10^{-7}\ MPa = 0.445\ J/m^2$$

(b) *The sheet of glass will not fracture because $K_I < K_{IC}$.*

(c) *Let $K_I < K_{IC}$ and $a < a_c$ in Eq. (2.34) so that the maximum allowable crack length under the loading condition becomes*

$$K_{IC} = \sigma\sqrt{\pi a_c}$$

$$a_c = \frac{1}{\pi}\left(\frac{K_{IC}}{\sigma}\right)^2 = \frac{1}{\pi}\left(\frac{0.80\ MPa\sqrt{m}}{262\ MPa}\right)^2 = 29.68\ mm$$

(d) *From Eq. (2.39), $G_I = G_C$*

$$G_C = \frac{K_{IC}^2}{E} = \frac{(0.80\ MPa\sqrt{m})^2}{60 \times 10^3\ MPa}$$

$$G_C = 1.067 \times 10^{-5}\ MPa \cdot m = 10.67\ J/m^2$$

From Eq. (2.38), the applied stress at fracture is just the critical stress; that is, $\sigma = \sigma_c$

$$\sigma_c = \sqrt{\frac{EG_C}{\pi a}} = \sqrt{\frac{(60 \times 10^3 \text{ MPa}) (9.71 \times 10^{-6} \text{ MPa} \cdot \text{m})}{\pi (0.2 \times 10^{-3} \text{ m})}}$$

$$\sigma_c = 31.92 \text{ MPa}$$

The critical stress for fracture can also be calculated using the critical stress intensity factor K_{IC}. Thus,

$$\sigma_c = \frac{K_{IC}}{\sqrt{\pi a_c}} = \frac{0.80 \text{ MPa}\sqrt{\text{m}}}{\sqrt{\pi (0.2 \times 10^{-3} \text{ m})}}$$

$$\sigma_c = 31.92 \text{ MPa}$$

(e) *The reduction in strength due to a small crack is*

$$\frac{165 - 31.92}{165} \times 100\% = 80.65\%$$

This particular result indicates how critical is the presence of small defects in solid bodies.

(f) *The maximum allowable or critical mass can be determined using the following critical stress equation*

$$\sigma_c = \frac{m_c g}{A_o}$$

$$m_c = \frac{A_o \sigma_c}{g}$$

Hence, the critical mass is

$$m_c = \frac{A_o \sigma_c}{g} = \frac{(100 \times 10^{-3} \text{ m}) (3 \times 10^{-3} \text{ m}) (31.92 \times 10^6 \text{ N/m}^2)}{(9.81 \text{ m/s}^2)}$$

$$m_c = 976.15 \text{ Kg}$$

The critical mass can also be calculated using the critical stress intensity factor (plane-strain fracture toughness) as follows:

$$m_c = \frac{A_o K_{IC}}{g\sqrt{\pi a_c}} = \frac{(100 \times 10^{-3} \text{ m}) (3 \times 10^{-3} \text{ m}) (0.80 \times 10^6 \text{ Pa}\sqrt{\text{m}})}{(9.81 \text{ m/s}^2) \sqrt{\pi (0.2 \times 10^{-3} \text{ m})}}$$

$$m_c = 976.01 \text{ Kg}$$

The slight difference in the critical mass is due to the truncation error involved in all calculations.

It is obvious, in general, that the existence of different crack configurations is quite detrimental or fatal to structures under a loading mode. The current example is very simple, but exhibits an important analysis in determining the critical crack length a structural component can tolerate before crack propagation toward fracture takes place. This example also reflects the importance of fracture mechanics in designing applications to assure structural integrity during a design lifetime.

2.6 Grain-Boundary Strengthening

This technique is used to enhance the strength and the fracture toughness of polycrystalline materials, such as low-carbon steels. The classical Hall–Petch model [18, 19] is commonly used to predict the mechanical behavior of polycrystalline materials since it correlates the yield strength with the grain size (d) to an extent. The Hall–Petch model is based on a planar dislocation pileup mechanism in an infinite and homogeneous medium as schematically shown in Fig. 2.7. The dislocation pileup is modeled as a series of edge dislocations ($\perp$) emanating from a source toward the grain boundary on a particular slip plane.

The mathematical form of the Hall–Petch model for the yield strength is

$$\sigma_{ys} = \sigma_o + k_y d^{-1/2} \tag{2.41}$$

where σ_o = Lattice friction stress (MPa)

k_y = Dislocation locking term (MPa$\sqrt{m}$)

Briefly, if a dislocation source is activated, then it causes dislocation motion to occur toward the grain boundary, which is the obstacle suitable for dislocation pileup. Thus, dislocation motion encounters the lattice friction stress σ_o as dislocations move on a slip plane toward a grain boundary. This dislocation-based model assumes that dislocation motion is the primary mechanism for plastic flow in crystalline materials and it is the basis for the Hall–Petch equation. In addition,

Fig. 2.7 Edge dislocation pileup model

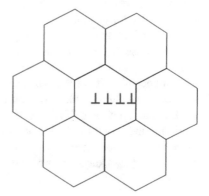

the pileup causes a stress concentration at the grain boundary, which eventually fractures when the local stress (shear stress) reaches a critical value. Therefore, other dislocation sources are generated. This is a possible mechanism for explaining the yielding phenomenon from one grain to the next. However, the grain size dictates the size of dislocation pileup, the distance dislocations must travel, and the dislocation density associated with yielding. This implies that the finer the grains, the higher the yield strength.

If a suitable volume of hard particles exists in a fine-grain material, the yield stress is enhanced further since three possible strengthening mechanisms are present, that is, solution strengthening, fine-grain strengthening, and particle (dispersion) strengthening. If these three strengthening mechanisms are activated, then the Hall–Petch model is not a suitable model for explaining the mechanical behavior of polycrystalline materials. This suitable explanation is vital in understanding that the grain size plays a major role in determining material properties such as the yield strength and fracture toughness.

Further, the Hall–Petch model can also be used to correlate hardness and grain size as indicated below:

$$H = H_o + k_h d^{-1/2} \tag{2.42}$$

Here, H_o and k_h are constants which are determined through a curve fitting procedure. Nonetheless, the physical foundation of this empirical equation is assumed to be associated with dislocation pileup within grains.

The understanding of the mechanical behavior of polycrystalline materials from nano- to microscales is very important scientifically and technologically because the grain size (d) plays a critical role in designing materials having desired mechanical properties.

The crossover from normal to inverse Hall–Petch model is depicted in Fig. 2.8 [20, 21]. The inverse Hall–Petch model is found in the new generation of advanced materials with nanostructures or nano-grains [7, 22–25].

Figure 2.8 illustrates three different regions: (1) the micro-grain region in which the conventional Hall–Petch can be used, (2) the transitional region where the maximum hardness corresponds to a critical grain size (d_c) and beyond this point the Hall–Petch model is reversed having a negative slope, and (3) the nano-grain region in which the grains are very small leading to large values of $d^{-1/2}$, which in turn cause a decrease in hardness. The specific hardness or yield strength profile for materials having very small grains at a nanoscale depends on the inherent peculiarities of the nanostructures, which have large grain-boundary areas. Currently, excellent publications on the reverse Hall–Petch model are available in the literature; however, this is a subject of so much controversy [21].

Furthermore, letting the crack length be in the order of the average grain size ($d = a$), it can be shown that both yield strength (σ_{ys}) and the critical strain energy release rate (G_{IC}) known as fracture toughness depend on the grain size. Using the

Fig. 2.8 Schematic hardness profile showing the normal to inverse Hall–Petch model

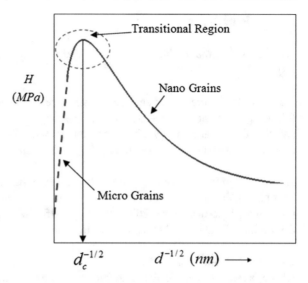

Hall–Petch equation, Eq. (2.41), for σ_{ys} and Eq. (2.40) for σ_f, it is clear that these stress entities depend on the grain size. Hence,

$$\sigma_f = \left(\frac{E'G_{IC}}{\pi}\right)^{1/2} d^{-1/2} = k_f d^{-1/2} \tag{2.43}$$

where $k_f = $ Constant (MPa$\sqrt{\text{m}}$)

Denote that Eqs. (2.41) and (2.43) predict that $\sigma_{ys} = f(d)$ and $\sigma_f = f(d)$. The slopes of these functions, k_y and k_f, respectively, have the same units, and they may be assumed to be related to fracture toughness. The slope k_y is referred to as the dislocation locking term that restricts yielding from a grain to the adjacent one. Mathematically, the analysis of Eq. (2.40) through (2.43) for materials having temperature and grain size dependency indicate that $\sigma_{ys} \to \infty$, $\sigma_f \to \infty$ and $H \to \infty$ as $d \to 0$. Physically, these entities have limited values and $\sigma_f \geq \sigma_{ys}$ due to the inherent friction stress (σ_o) at a temperature T_1. At a temperature $T_2 > T_1$, σ_{ys} decreases since σ_o and k_y also decrease, and G_{IC} increases and, therefore, σ_f must decrease.

One can observe that $\sigma_{ys} \to \sigma_o$ for $d^{-1/2} = 0$, which means that σ_o is regarded as the yield stress of a single crystal. However, $\sigma_{ys} \to \sigma_o$ and $\sigma_f \to \sigma_o$ as $d \to \infty$ is an unrealistic case. Therefore, grain size refinement is a useful strengthening mechanism for increasing both σ_{ys} and σ_f.

Finally, if one combines Eqs. (2.34) and (2.42) along with $d = a$, one can determine that the K_I is inversely proportional to hardness. Thus,

$$K_I = \frac{k_h \sigma \sqrt{\pi}}{H - H_o} \approx \frac{k_h \sigma \sqrt{\pi}}{H} \tag{2.44}$$

Here, it can be assumed that k_h is a correction factor and $H \gg H_o$.

2.7   Problems

2.1. Show that the applied stress is $\sigma \rightarrow 0$ when the crack tip radius is $\rho \rightarrow 0$. Explain.

2.2. In order for crack propagation to take place, the strain energy is defined by the following inequality $U(a) - U(a + \Delta a) \geq 2\Delta a \gamma$, where Δa is the crack extension and γ is the surface energy. Show that the crack driving force or the strain energy release rate at instability is defined by $G \geq dU(a)/da$.

2.3. One (1-mm) $\times$ (15-mm) $\times$ 100-mm steel strap has a 3-mm-long central crack. This strap is loaded in tension to failure. Assume that the steel is brittle having the following properties: $E = 207\,\text{GPa}$, $\sigma_{ys} = 1500\,\text{MPa}$, and $K_{IC} = 70\,\text{MPa}\sqrt{\text{m}}$. Determine (a) the critical stress and (b) the critical strain energy release rate.

2.4. Suppose that a structure made of plates has one cracked plate. If the crack reaches a critical size, will the plate fracture or the entire structure collapse? Explain.

2.5. What is crack instability according to Griffith criterion?

2.6. Assume that a quenched 1.2%C-steel plate has a penny-shaped crack. Will the Griffith theory be applicable to this plate?

2.7. Will the Irwin theory be valid for a changing plastic zone size during crack growth?

2.8. What are the major roles of the surface energy and the stored elastic energy during crack growth?

2.9. What does happen to the elastic energy during crack growth?

2.10. What does $dU/da = 0$ mean?

2.11. Derive Eq. (2.29) starting with Eq. (2.20).

References

1. J.W. Dally, W.F. Riley, *Experimental Stress Analysis*, 3rd edn. (McGraw-Hill, New York, 1991)
2. G.R. Irwin, Fracture I, in *Handbuch der Physik VI*, ed. by S. Flugge (Springer, New York, 1958) 558–590
3. C.E. Inglis, Stresses in a plate due to the presence of cracks and sharp corners. Trans. Inst. Nav. Archit. **55**, 219–241 (1913)

4. A.A. Griffith, The phenomena of rupture and flows in solids. Philos. Trans. R. Soc. **221**, 163–198 (1921)
5. H.M. Westergaard, Bearing pressures and cracks. J. Appl. Mech. **G1**, A49–A53 (1939)
6. W.J. McGregor Tegart, *Elements of Mechanical Metallurgy* (Macmillan, New York, 1966)
7. M. Rieth, *Nano-Engineering in Science and Technology : An Introduction to the World of Nano-Design*. Series on the Foundations of Natural Science and Technology, vol. 6 (World Scientific, Singapore, 2003)
8. K. Ashbee, *Fundamental Principles of Fiber Reinforced Composites*, 2nd edn. (CRC Press, West Palm Beach, 1993), pp. 211–218
9. A.P. Boresi, O.M. Sidebottom, *Advanced Mechanics of Materials* (Wiley, New York, 1985), pp. 534–538
10. J.A. Collins, *Failure of Materials in Mechanical Design: Analysis, Prediction, Prevention*, 2nd edn. (Wiley, New York, 1993), pp. 422–427
11. S.P. Timoshenko, J.N. Goodier, *Theory of Elasticity*, 3rd edn. (McGraw-Hill, New York, 1970), p. 193
12. H.J. Grover, *Fatigue Aircraft Structures* (NAVIR 01-1A-13 Department of the Navy, Washington, DC, 1966)
13. D. Broek, *Elementary Engineering Fracture Mechanics*, 4th edn. (Kluwer Academic, Boston, 1986)
14. K. Hellan, *Introduction to Fracture Mechanics* (McGraw-Hill, New York, 1984)
15. R.W. Hertzberg, *Deformation and Fracture Mechanics of Engineering Materials*, 3rd edn. (Wiley, New York, 1989)
16. J.M. Barsom, S.T. Rolfe, *Fracture and Fatigue Control in Structures: Applications of Fracture Mechanics*, Chap. 2, 3rd edn. (American Society For Testing and Materials, West Conshohocken, PA, 1999)
17. E. Orowan, *Fatigue and Fracture of Metals* (MIT Press, Cambridge, MA, 1950), p. 139
18. E.O. Hall, The deformation and ageing of mild steel: III discussion of results. Proc. Phys. Soc. B Lond. **64**(9), 747–753 (1951)
19. N.J. Petch, The cleavage strength of polycrystals. J. Iron Steel Inst. **174**, 25–28 (1953)
20. T.G. Nieh, J. Wadsworth, Hall-petch relation in nanocrystalline solids. Scr. Mater. **25**, 955–958 (1991)
21. M. Yu Gutkin, I.A. Ovid'ko, C.S. Pande, Theoretical models of plastic deformation processes in nanocrystalline matertials. Rev. Adv. Mater. Sci. **2**(1), 80–102 (2001)
22. A.S. Edelstein, R.C. Cammaratra, *Nanomaterials: Synthesis, Properties and Applications* (Taylor & Francis, London, 1998)
23. H.S. Nalwa (ed.), *Handbook of Nanostructured Materials and Nanotechnology*, vol. 1–5 (Academic Press, San Diego, 1999)
24. C.P. Poole Jr., F.J. Owens, *Introduction to Nanotechnology* (Wiley, Hoboken, 2003)
25. G.M. Chow, I.A. Ovid'ko, T. Tsakalakos (eds.), *Nanostructured Films and Coatings*. NATO Science Series (Kluwer Academic, Dordrecht, 2000)

Linear-Elastic Fracture Mechanics

3

3.1 Introduction

Solid bodies containing cracks can be characterized by defining a state of stress near a crack tip and the energy balance coupled with fracture. Introducing the Westergaard stress function and will allow the development of a significant stress analysis at the crack tip. These particular functions can be found elsewhere [12]. For instance, Irwin [20] treated the singular stress field by introducing a quantity known as the elastic **stress intensity factor**, which is used as the controlling parameter for evaluating the critical state of a crack.

The theory of linear-elastic fracture mechanics (LEFM) is integrated in this chapter using an analytical approach that will provide the reader useful analytical steps. Thus, the reader will have a clear understanding of the concepts involved in this particular engineering field and will develop the skills for a mathematical background in determining the elastic stress field equations around a crack tip. The field equations are assumed to be within a small plastic zone ahead of the crack tip. If the plastic zone is sufficiently small, the small-scale yielding (SSY) approach is used for characterizing brittle solids and for determining the stress and strain fields when the size of the plastic zone is sufficiently smaller than the crack length; that is, $r \ll a$. In contrast, a large-scale yielding (LSY) is for ductile solids, in which $r \geq a$.

Most static failure theories assume that the solid material to be analyzed is perfectly homogeneous, isotropic, and free of stress risers or defects, such as voids, cracks, inclusions, and mechanical discontinuities (indentations, scratches, or gouges). Actually, fracture mechanics considers structural components having small flaws or cracks which are introduced during solidification, quenching, welding, machining, or handling process. However, cracks that develop in service are difficult to predict and account for preventing crack growth.

© Springer International Publishing Switzerland 2017
N. Perez, *Fracture Mechanics*, DOI 10.1007/978-3-319-24999-5_3

3.2   Modes of Loading

Assume that a solid body (material) containing a crack is subjected to a continuously and gradually increasing loading mode. Initially the crack grows stably until the applied stress level reaches a critical level ($\sigma \rightarrow \sigma_c$). Once $\sigma > \sigma_c$, the crack propagates (rapid growth) and, as a result, the solid body fractures (breaks). In order to understand crack growth, it is relevant to firstly study the mechanical behavior of a crack-free body based on its microstructure and the related mode of elastic or elastic-plastic deformation.

Consider a crack in a homogeneous and linear-elastic body being subjected to a particular loading mode of crack displacement, inducing a stress distribution ahead of the crack tip responsible for stable growth. In this case, a load-displacement curve is generated for determining the critical stress level and related critical stress intensity factor, which depends on the applied stress, specimen size, geometry, and crack size. Hence, the mechanical behavior of a solid containing a crack of a specific geometry and size can be predicted by evaluating the elastic stress intensity factors K_I, K_{II}, and K_{III} for the specimen geometries shown in Fig. 3.1a.

On the other hand, Fig. 3.1b depicts the elastic stresses ahead of the crack tip that must be derived in Cartesian coordinates using theory of elasticity. Polar coordinates will be used for this purpose in a later chapter.

The reader should denote the meaning of each subscript assigned to each elastic stress intensity factor. Although a combined loading can be encountered in structural

Fig. 3.1 Stress loading modes and crack coordinate system. (**a**) Stress modes and (**b**) stress components ahead of the crack tip

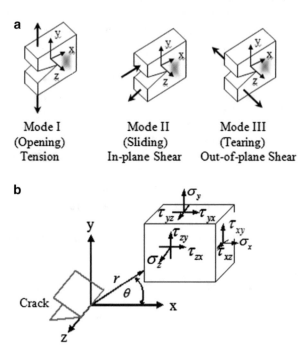

components, K_I is the most studied and evaluated experimentally for determining its critical value called fracture toughness, which is a material property.

In addition, specimens and structural components having flaws or cracks can be loaded to various levels of the applied elastic stress intensity factor for a particular stress mode as shown in Fig. 3.1a. This is analogous to sound components being loaded to various levels of the applied stress σ.

If crack growth occurs along the crack plane perpendicular to the direction of the applied external loading mode, then the elastic stress intensity factors (K_I, K_{II} and K_{III}) are defined according to the American Society for Testing Materials (ASTM) E399 Standard Test Method. Hence,

$$K_I = \lim_{r \to 0} \left(\sigma_{yy} \sqrt{2\pi r} \right) f_I(\theta) = \sigma_y \sqrt{2\pi r} \quad @ \; \sigma_y = \sigma_y(r, \theta = 0) \qquad (3.1)$$

$$K_{II} = \lim_{r \to 0} \left(\tau_{xy} \sqrt{2\pi r} \right) f_{II}(\theta) = \tau_{xy} \sqrt{2\pi r} \quad @ \; \tau_{xy} = \tau_{xy}(r, \theta = 0) \qquad (3.2)$$

$$K_{III} = \lim_{r \to 0} \left(\tau_{yz} \sqrt{2\pi r} \right) f_{III}(\theta) = \tau_{yz} \sqrt{2\pi r} \quad @ \; \tau_{yz} = \tau_{yz}(r, \theta = 0) \qquad (3.3)$$

where $f_I(\theta)$, $f_{II}(\theta)$, and $f_{III}(\theta)$ are trigonometric functions to be derived analytically, r is the plastic zone size, and K_I is the stress intensity factor developed by Irwin [20].

The stress intensity factor for a particular crack configuration and specimen geometry can be defined as a general function

$$K_i = f(\sigma, \; Crack \; Configuration, \; Specimen \; Geometry, \; Temp.) \qquad (3.4)$$

where $i = I, II, III$.

The parameter K_i can be used to determine the static or dynamic fracture stress, the fatigue crack growth rate, and corrosion crack growth rate. For elastic materials, the strain energy release rate G_i, known as the crack driving force, is related to the stress intensity factor and the modulus of elasticity. Thus, G_i also depends on several variables as K_i does. The relationship between these two fracture mechanics parameters is

$$G_i = \frac{K_i^2}{E'} \qquad (3.5)$$

where $E' = E$ for plane stress (MPa)
$\qquad E' = E/\left(1 - v^2\right)$ for plane strain (MPa)
$\qquad E =$ Elastic modulus of elasticity (MPa)
$\qquad v =$ Poisson's ratio

This expression, Eq. (3.5), is a fundamental mathematical model in the field of fracture mechanics, specifically for mode I loading since Griffith's work on sheets of glass containing central cracks was published. Denote that both parameters K_i and G_i can become material properties at a specific stress level for crack instability to occur.

3.3   Westergaard Stress Function

The Westergaard single stress function (complex function) gives the general solution for stresses and displacements near the crack tip subjected to a particular loading mode.

Consider mode I as a basic loading for crack growth. The coordinate system defining a double-ended crack in a complex z-plane is shown in Fig. 3.2. Consider the coordinates origin at the center of the elliptical crack so that the complex variable z is defined as $z = x + iy$. Thus, z represents a point in the z-plane where the elastic stresses, σ_x, σ_y, and τ_{xy}, are determined at $\alpha = \pi/2 - \theta$. However, the origin may be moved to the right crack tip as illustrated in Fig. 3.2. In this case, $z = r$ and if $y = o$, then $-x < z < x$, $\theta = 0$ and σ_y is the only nonzero elastic stress. This means that the crack occupies a straight segment treated as a straight line along the x-axis in Cartesian coordinates subjected to a remote and perpendicular force (F_y^∞) or stress (σ_y^∞).

This is a classical representation of Westergaard's approach for developing stress functions near an elliptical crack of total length $2a$. The primary goal using the approach is to derive stresses and strains ahead of a crack tip. Recall that strains are proportional to stresses, while displacements are derived from integration of the strains.

The Airy and Westergaard stress functions are defined in terms of the complex function Z, respectively [12, 23]:

$$\phi = \mathrm{Re}\,\overline{Z} + y\,\mathrm{Im}\,\overline{Z} \qquad \text{(Airy)} \qquad (3.6)$$

$$Z(z) = \mathrm{Re}\,Z + i\,\mathrm{Im}\,Z \qquad \text{(Westergaard)} \qquad (3.7)$$

where $\mathrm{Re}\,\overline{Z}$ = Real part
$\mathrm{Im}\,\overline{Z}$ = Imaginary part
Z = Analytic stress function
$i = \sqrt{-1}$
$i^2 = -1$

Fig. 3.2 The crack tip stress field in complex coordinates

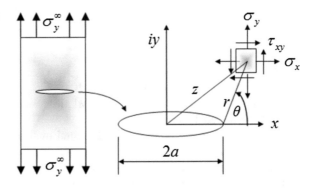

Below are some supportive and useful expressions for developing the elastic stress functions. Hence,

$$Z = \int Z'dz \quad \text{or} \quad Z' = \frac{dZ}{dz} \tag{3.9a}$$

$$\overline{Z} = \int Zdz \quad \text{or} \quad Z = \frac{d\overline{Z}}{dz} \tag{3.9b}$$

$$\overline{\overline{Z}} = \int \overline{Z}dz \quad \text{or} \quad \overline{Z} = \frac{d\overline{\overline{Z}}}{dz} \tag{3.9c}$$

$$\frac{\partial (\operatorname{Re} Z)}{\partial x} = \operatorname{Re} Z' \tag{3.9d}$$

$$\frac{\partial (\operatorname{Re} Z)}{\partial y} = -\operatorname{Im} Z' \tag{3.9e}$$

$$\frac{\partial (\operatorname{Im} Z)}{\partial x} = \operatorname{Im} Z' \tag{3.9f}$$

$$\frac{\partial (\operatorname{Im} Z)}{\partial y} = \operatorname{Re} Z' \tag{3.9g}$$

$$\frac{\partial \left(-y \operatorname{Re} Z + \operatorname{Im} \overline{Z}\right)}{\partial x} = y \operatorname{Im} Z' - \operatorname{Re} Z \tag{3.9h}$$

For the ellipse in Fig. 3.2, the complex function and its conjugate are, respectively,

$$z = x + iy = r \exp(i\theta) \tag{3.10}$$

$$\bar{z} = x - iy = r \exp(-i\theta) \tag{3.11}$$

and

$$\exp(\pm i\theta) = \cos \theta \pm i \sin \theta \tag{3.12}$$

$$y = r \sin \theta \tag{3.13}$$

Furthermore, the **Cauchy–Riemann equations** in some domain D are considered fundamental and sufficient for a complex function to be analytic in region D. Letting $z = u(x, y) + iv(x, y)$, then the Cauchy–Riemann important theorem states that $\partial u / \partial x = \partial v / \partial y$ and $\partial u / \partial y = -\partial v / \partial x$ so that the function z be analytic. The Cauchy–Riemann condition is of great practical importance for determining elastic stresses. Hence,

$$\nabla^2 \operatorname{Re} Z = \nabla^2 \operatorname{Im} Z = 0 \tag{3.14}$$

which must be satisfied by an Airy stress function. The harmonic operator ∇^2 in rectangular and polar coordinates takes the form

$$\nabla^2 = \frac{\partial^2}{\partial x^2} + \frac{\partial^2}{\partial y^2} + \frac{\partial^2}{\partial z^2} \tag{3.15a}$$

$$\nabla^2 = \frac{\partial^2}{\partial r^2} + \frac{1}{r}\frac{\partial}{\partial r} + \frac{1}{r^2}\frac{\partial^2}{\partial \theta^2} \tag{3.15b}$$

Assume that the Airy partial derivatives given by Eq. (1.40) are applicable to homogeneous and elastic materials without body forces. Hence,

$$\sigma_x = \frac{\partial^2 \phi}{\partial y^2} \tag{3.16}$$

$$\sigma_y = \frac{\partial^2 \phi}{\partial x^2} \quad \text{(Airy)} \tag{3.17}$$

$$\tau_{xy} = -\frac{\partial^2 \phi}{\partial x \partial y} \tag{3.18}$$

Thus,

$$\frac{\partial \phi}{\partial x} = \frac{\partial \left(\operatorname{Re} \overline{Z} \right)}{\partial x} + \frac{\partial \left(y \operatorname{Im} \overline{Z} \right)}{\partial x} = \operatorname{Re} \overline{Z} + y \operatorname{Im} Z \tag{3.19}$$

$$\frac{\partial^2 \phi}{\partial x^2} = \frac{\partial \left(\operatorname{Re} \overline{Z} \right)}{\partial x} + \frac{\partial \left(y \operatorname{Im} Z \right)}{\partial x} = \operatorname{Re} Z + y \operatorname{Im} Z'$$

$$\frac{\partial \phi}{\partial y} = \frac{\partial \left(\operatorname{Re} \overline{Z} \right)}{\partial y} + \frac{\partial \left(y \operatorname{Im} \overline{Z} \right)}{\partial y} = -\operatorname{Im} \overline{Z} + y \operatorname{Re} Z + \operatorname{Im} \overline{Z} = y \operatorname{Re} Z$$

$$\frac{\partial^2 \phi}{\partial y^2} = \frac{\partial \left(y \operatorname{Re} Z \right)}{\partial y} = \operatorname{Re} Z \frac{\partial \left(y \right)}{\partial y} + y \frac{\partial \left(\operatorname{Re} Z \right)}{\partial y} = \operatorname{Re} Z - y \operatorname{Im} Z'$$

$$\frac{\partial^2 \phi}{\partial x \partial y} = \frac{\partial \left(\operatorname{Re} \overline{Z} \right)}{\partial y} + \frac{\partial \left(y \operatorname{Im} Z \right)}{\partial y} = -\operatorname{Im} Z + y \operatorname{Re} Z' + \operatorname{Im} Z = y \operatorname{Re} Z'$$

Substituting Eq. (3.19) into (3.18) along with $\Omega = 0$ yields the Westergaard stress function in two dimensions:

$$\sigma_x = \operatorname{Re} Z - y \operatorname{Im} Z'$$

$$\sigma_y = \operatorname{Re} Z + y \operatorname{Im} Z' \quad \text{(Westergaard)} \tag{3.20}$$

$$\tau_{xy} = -y \operatorname{Re} Z'$$

These are the stresses, Eq. (3.20), that were proposed by Westergaard as the stress singularity field at the crack tip. However, additional terms must be added to the stress functions, Eq. (3.20), analytic over an entire region, for an adequate representation of a stress field adjacent to the crack tip. This implies that when solving practical problems, additional boundary conditions must be imposed on the stresses.

Consequently, this leads to the well-known boundary value problem in which the boundary value method (BVM), which is an alternative technique to the most commonly finite element method (FEM) and finite difference method (FDM). Therefore, the original Westergaard stress function no longer gives a unique solution, despite that the analysis of the stress field in a region near the crack tip is of extreme importance. Thus, the stress intensity factor represents the intensity of these stresses.

3.3.1 Far-Field Boundary Conditions

Consider the **classical problem** in fracture mechanics in which a homogeneous or isotropic infinite plate containing a central elliptical crack, shown in Fig. 3.2, is subjected to uniaxial tension. In this case, the Westergaard stress function Z is used as an analytic function because its derivative dZ/dz is defined unambiguously. If the coordinate origin is located at the center of the ellipse, then the Westergaard stress function is

$$Z = \frac{z\sigma}{\sqrt{z^2 - a^2}} = \frac{\sigma}{\sqrt{1 - (a/z)^2}} \tag{3.21}$$

However, the far-field boundary conditions require that $z \to \infty$ and $z \gg a$. Consequently, Eq. (3.21) becomes

$$Z = \frac{z\sigma}{\sqrt{z^2}} = \sigma \tag{3.22}$$

from which $\operatorname{Re} Z = \sigma$ and $\operatorname{Im} Z' = 0$. Substituting Eq. (3.22) into (3.20) yields

$$\sigma_x = \operatorname{Re} Z - y \operatorname{Im} Z' = \sigma$$
$$\sigma_y = \operatorname{Re} Z + y \operatorname{Im} Z' = \sigma \tag{3.23}$$
$$\tau_{xy} = -y \operatorname{Re} Z' = 0$$

On the crack surface: If $y = 0$ and $z = x$ for $-a \le x \le +a$, then $\operatorname{Re} Z = 0$ and $\sigma_x = \sigma_y = \tau_{xy} = 0$. Therefore, Z is not an analytical function because it does not have a unique derivative at a point z.

3.3.2   Near-Field Boundary Conditions

Locate the origin at the crack tip in Fig. 3.2 so that the Westergaard stress function becomes [12]

$$Z = \frac{(z+a)\,\sigma}{\sqrt{(z+a)^2 - a^2}} = \frac{(z+a)\,\sigma}{\sqrt{z^2 + 2az}} \tag{3.24}$$

Here, the near-field boundary conditions are $z \ll a$ near the crack tip and $z^n = r^n \exp(in\theta)$, where n is a real number. Hence, Eq. (3.24) becomes

$$Z = \frac{a\sigma}{\sqrt{2az}} \tag{3.25a}$$

$$Z = \sigma\sqrt{\frac{a}{2}}z^{-1/2} = \sigma\sqrt{\frac{a}{2r}}\exp\left(-i\theta/2\right) \tag{3.25b}$$

$$Z = \sigma\sqrt{\frac{a}{2r}}\left(\cos\frac{\theta}{2} - i\sin\frac{\theta}{2}\right) \tag{3.25c}$$

from which the real and imaginary parts are extracted as

$$\operatorname{Re} Z = \sigma\sqrt{\frac{a}{2r}}\cos\frac{\theta}{2} \tag{3.26a}$$

$$\operatorname{Im} Z = -\sigma\sqrt{\frac{a}{2r}}\sin\frac{\theta}{2} \tag{3.26b}$$

The Westergaard stress function, Eq. (3.20), along the crack line, where $y = 0$ and $\theta = 0$, takes the form

$$\sigma_x = \operatorname{Re} Z - y\operatorname{Im} Z' = \sigma\sqrt{\frac{a}{2r}} = \sigma\sqrt{\frac{a}{2r}}$$

$$\sigma_y = \operatorname{Re} Z + y\operatorname{Im} Z' = \sigma\sqrt{\frac{a}{2r}} \tag{3.27}$$

$$\tau_{xy} = -y\operatorname{Re} Z' = 0$$

If the plastic zone ahead of the crack tip is $r \to 0$, then the stress becomes $\sigma \to \infty$. This stress defines what is referred to as a "singularity" state of stress which is in the order of $r^{-1/2}$ along the x-axis.

Substituting Eq. (3.27) into (3.1) yields the stress intensity factor for infinite specimen dimensions (uncorrected) and finite specimen dimensions (corrected) under mode I, respectively:

$$K_I = \sigma\sqrt{\pi a} \qquad \text{(uncorrected)} \tag{3.28}$$

$$K_I = \alpha\sigma\sqrt{\pi a} \qquad \text{(corrected)} \tag{3.29}$$

Fig. 3.3 Linear behavior of
the stress intensity factor

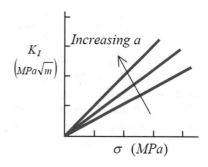

where α = Specimen finite geometry correction factor

$\alpha = f\,(a/w)$ for plates and $\alpha = f\,(d_i/d_o)$ for cylindrical components

a = Crack length (Also called crack size)

w = Specimen width

d_i, d_o = Inside and outside diameters

The equation $\alpha = f\,(a/w)$ is a function of normalized crack length a/w and it makes the surface traction stresses vanish [27]. Basically, α is a geometry function that considers the effect of crack configuration and boundary conditions. In fact, α is just a multiplier. If $a/w \rightarrow 0$, then $\alpha - f\,(0) \geq 1$. Hence, Eq. (3.29) reduces to (3.28). In addition, K_I has units in $MPa\sqrt{m}$ or $ksi\sqrt{in}$.

The K_I expression, Eq. (3.28), is for an infinite (large) plate and it is a linear function of the applied stress (σ), and it increases with initial crack size (a). This is shown in Fig. 3.3. On the other hand, Eq. (3.29) is for a finite plate. Thus, the stress intensity factor is corrected due to specimen geometry effects. In certain cases, crack shape and plasticity effects can also be included in the K_I equation. The onset of crack propagation is a **critical condition** so that the crack a) extends suddenly by tearing in a shear-rupture failure or b) extends suddenly at high velocity for cleavage fracture. All this means that the crack is unstable when a critical condition exists due to an applied load. In this case, the stress intensity factor K_I reaches a critical magnitude, and it is treated as a material property called fracture toughness, which represents the resistance of a elastic material to fracture. For sufficiently thick specimens, $K_I = K_{IC}$ is called plane-strain fracture toughness, and for thin specimens, $K_I = K_C$ is known as plane stress fracture toughness.

Moreover, K_{IC} and K_C are material properties that measure crack resistance. Since Mode I testing the most common loading condition, K_{IC} is the most used in designing applications. Nonetheless, the fracture criterion by K_{IC} states that crack propagation occurs when $K_I \geq K_{IC}$ for brittle materials. This simply implies that the crack extends to reach a critical crack length ($a = a_c$), defining a critical state in which the crack velocity may reach the magnitude of the speed of sound for most brittle materials. In fact, the crack velocity is referred to as the subcritical crack growth rate, da/dt, for time-dependent cracking. Particularly, the crack velocity in stress corrosion cracking (SCC) studies takes the form [28, 29]

$$\frac{da}{dt} = c_1 K_I^n \quad \text{(for mode I)} \tag{3.29a}$$

where c_1 and n are material constants that depend on the environment.

Furthermore, the LEFM theory is well documented, and the ASTM E399 Standard Test Method, Vol. 03.01, validates the K_{IC} data and assures the starting crack length and the minimum plate thickness through Brown and Strawley [9] empirical equation. Thus,

$$a, B \geq 2.5 \left(\frac{K_{IC}}{\sigma_{ys}}\right)^2 \quad \text{for } a >> r \tag{3.30}$$

This is a maximum constraint set forth by requiring large specimen dimensions, a deep crack, and a very small plastic zone.

3.4   Specimen Geometries

3.4.1 Through-Thickness Cracks

In general, the successful application of linear elastic fracture mechanics to structural analysis, fatigue, and stress corrosion cracking requires a known stress intensity factor equation for a particular specimen configuration. Cracks in bodies of finite size are important since cracks pose a threat to the instability and safety of an entire structure. Ordinarily, structural integrity can be assessed by using proper design methodology and by periodic nondestructive evaluation (NDE) during service. However, the presence of cracks in stressed structural components can reduce the material strength and its design lifetime or cause fracture [31, 41, 42].

Accordingly, fracture mechanics provides a methodology for evaluating the behavior of cracked solid bodies subjected to stresses and strains. In fact, mode I (opening) loading system is the most-studied and evaluated mode for determining the mechanical behavior of solids having specific geometries exposed to a particular environment. Some selected specimen geometries and related through-thickness crack configurations are shown in Table 3.1 along with the related geometry correction factors, $\alpha = f(a/w)$.

These specimens are subjected to a remote external load. An exception is the K_I equation for an embedded (internal) pressurized crack in Table 3.1 is not corrected because the crack is only subjected to an internal pressure on the crack surfaces.

The stress intensity factor (K_I) for mode I is normally corrected through a finite geometry correction factor α expression, and its critical value is referred to as plane-strain fracture toughness K_{IC} under plane-strain conditions.

In addition, there are several finite geometry correction factor α for a plate containing a through-thickness center crack. For convenience, some expressions for $\alpha = f(a/w)$ are listed in Table 3.1, and others are given below as per individual authors.

Table 3.1 Through-thickness crack configurations for solid specimens [37]

Center Cracked Specimen 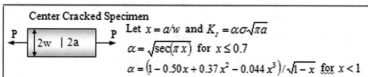	Let $x = a/w$ and $K_I = \alpha\sigma\sqrt{\pi a}$ $\alpha = \sqrt{\sec(\pi x)}$ for $x \leq 0.7$ $\alpha = (1 - 0.50x + 0.37x^2 - 0.044x^3)/\sqrt{1-x}$ for $x < 1$
Single-Edge Cracked Specimen 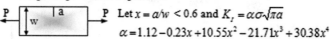	Let $x = a/w < 0.6$ and $K_I = \alpha\sigma\sqrt{\pi a}$ $\alpha = 1.12 - 0.23x + 10.55x^2 - 21.71x^3 + 30.38x^4$
Double-Edge Cracked Specimen 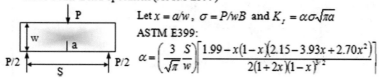	Let $x = a/w < 0.7$ and $K_I = \alpha\sigma\sqrt{\pi a}$ for $x < 1$ (Tada 1975) $\alpha = (1.12 - 0.561x - 0.205x^2 + 0.471x^3 - 0.190x^4)/\sqrt{1-x}$
Three-Point Bend Specimen (ASTM E399) 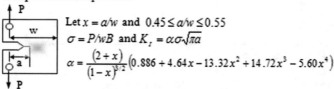	Let $x = a/w$, $\sigma = P/wB$ and $K_I = \alpha\sigma\sqrt{\pi a}$ ASTM E399: $\alpha = \left(\dfrac{3}{\sqrt{\pi}}\dfrac{S}{w}\right)\left[\dfrac{1.99 - x(1-x)(2.15 - 3.93x + 2.70x^2)}{2(1+2x)(1-x)^{3/2}}\right]$ Ref. [41]: For $S/w = 4$, $\alpha = \left(\dfrac{3}{\sqrt{\pi}}\dfrac{S}{w}\right)\left[1.93 - 3.07x + 14.53x^2 - 25.11x^3 + 25.80x^4\right]$ Ref. [41]: For $S/w = 8$, $\alpha = \left(\dfrac{3}{\sqrt{\pi}}\dfrac{S}{w}\right)\left[1.96 - 2.75x + 13.66x^2 - 23.98x^3 + 25.22x^4\right]$
Compact Tension Specimen 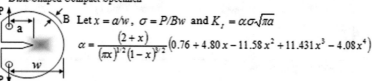	Let $x = a/w$ and $0.45 \leq a/w \leq 0.55$ $\sigma = P/wB$ and $K_I = \alpha\sigma\sqrt{\pi a}$ $\alpha = \dfrac{(2+x)}{(1-x)^{3/2}}(0.886 + 4.64x - 13.32x^2 + 14.72x^3 - 5.60x^4)$
Disk-Shaped Compact Specimen	Let $x = a/w$, $\sigma = P/Bw$ and $K_I = \alpha\sigma\sqrt{\pi a}$ $\alpha = \dfrac{(2+x)}{(\pi x)^{1/2}(1-x)^{3/2}}(0.76 + 4.80x - 11.58x^2 + 11.431x^3 - 4.08x^4)$
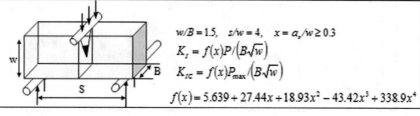	$w/B = 1.5$, $s/w = 4$, $x = a_s/w \geq 0.3$ $K_I = f(x)P/(B\sqrt{w})$ $K_{Ic} = f(x)P_{max}/(B\sqrt{w})$ $f(x) = 5.639 + 27.44x + 18.93x^2 - 43.42x^3 + 338.9x^4$

(continued)

Table 3.1 (Continued)

Arc-Shaped Tension Specimens	

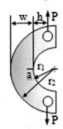

Let $x = a/w$, $\sigma = P/Bw$ and $K_I = \alpha\sigma\sqrt{\pi a}$

$$g(x) = (2h/w + 1.9 + 1.1x)\left[1 + 0.25(1-x)^2\left(1 - \frac{r_1}{r_2}\right)\right]$$

$$f(x) = \frac{\sqrt{x}}{(1-x)^{3/2}}\left[3.74 - 6.30x + 6.32x^2 - 2.43x^3\right]$$

$$\alpha = \frac{g(x)f(x)}{\sqrt{\pi x}}$$

Radial Cracked Specimen

Let $x = a/R$, $\sigma = P/wB$ and $K_I = \alpha\sigma\sqrt{\pi(a+2R)/2}$

For one crack,

$\alpha = 1.03 + 0.35x$ for $0 < x \le 10$

$\alpha = 3.36$ for $a << R$

For two cracks,

$\alpha = 0.75 + 0.59x$ for $0 < x \le 10$

Elliptically Cracked Specimen

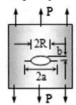

Let $x = a/R$, $b/R = 0.25$, $\sigma = P/wB$ and $K_I = \alpha\sigma\sqrt{\pi a}$

$\alpha = -84.375 + 225.47x - 199.25x^2 + 58.63x^3$

Incline Center cracked Plate

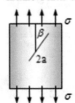

$K_I = \alpha_I\sigma\sqrt{\pi a}$ & $K_{II} = \alpha_{II}\sigma\sqrt{\pi a}$

$\alpha_I = \sin^2(\beta)$ $\alpha_{II} = \sin(\beta)\cos(\beta)$

Embedded Cracked Specimen

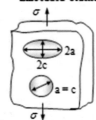

Let $x = a/c$ and $K_I = \alpha\sigma\sqrt{\pi a}$

For an elliptical crack, $\alpha = \dfrac{8\sqrt{x}}{\pi(3 + x^2)}$

For a circular crack, $x = a/c = 1$ and $\alpha = 2/\pi$

$$\alpha = \sqrt{\sec\left(\frac{\pi a}{w}\right)} \quad \text{for } 0 < a \leq 0.5 \ [16] \tag{3.31a}$$

$$\alpha = \sqrt{\frac{w}{\pi a} \tan\left(\frac{\pi a}{w}\right)} \quad \text{for } 0 < a \leq 0.5 \ [21] \tag{3.31b}$$

$$\alpha = \frac{1 - (a/w) + 1.304\,(a/w)^2}{\sqrt{1 - 2a/w}} \quad \text{for } 0 < a \leq 0.5 \ [25] \tag{3.32}$$

These α equations yield similar values. Therefore, the reader is bound to use one particular α equation based on a smart judgment. The reason for using α is that most current crack growth models rely on an assessment of geometry correction factors for the stress intensity factor to provide transferability between different crack configurations. This then ensures that calculations of α and K_I are sufficiently accurate to produce reliable crack growth modeling. Also, α accounts for a finite plate width.

Example 3.1. *Consider a 4340 steel plate with $K_{IC} = 41\,MPa\sqrt{m}$ and $\sigma_{ys} = 1900\,MPa$. Use some of the single-edge specimen configurations from Table 3.1 to determine the least allowable applied tension stress (σ) for common specimen dimensions $w = 10\,mm$, $B = w/2$, $S = 2.4w$, and $a = 5\,mm$. Start with the center cracked specimen.*

Solution. $K_I = \alpha\sigma\sqrt{\pi a}$; $\sigma = K_{IC}/\left(\alpha\sqrt{\pi a}\right)$; $a = 5\,\text{mm}$; $x = a/w = 0.5$
  Center cracked specimen

$$\alpha = \left(1 - 0.5x + 0.37x^2 - 0.044x^3\right)/\sqrt{1 - x} = 1.1837$$

$$\sigma = \left(41\,\text{MPa}\sqrt{m}\right) / \left((1.1837)\sqrt{\pi\,(5 \times 10^{-3}\,\text{m})}\right) = 276.36\,\text{MPa}$$

Single-edge cracked specimen

$$\alpha = 1.12 - 0.23x + 10.55x^2 - 21.70x^3 + 30.38x^4 = 2.8288$$

$$\sigma = \left(41\,\text{MPa}\sqrt{m}\right) / \left((2.8288)\sqrt{\pi\,(5 \times 10^{-3}\,\text{m})}\right) = 115.64\,\text{MPa}$$

Single-edge bend specimen

$$\alpha = \left(\frac{3S}{w\sqrt{\pi}}\right)\frac{1.99 - x\,(1 - x)\,\left(2.15 - 3.93x + 2.70x^2\right)}{2\,(1 + 2x)\,(1 - x)^{3/2}} = 5.0985$$

$$\sigma = \left(41\,\text{MPa}\sqrt{m}\right) / \left[(5.0985)\sqrt{\pi\,(5 \times 10^{-3}\,\text{m})}\right] = 64.16\,\text{MPa}$$

Disk-shaped compact specimen

$$\alpha = \frac{(2+x)\left(0.76 + 4.8x - 11.58x^2 + 11.43x^3 - 4.08x^4\right)}{(\pi x)^{1/2}(1-x)^{3/2}} = 8.1173$$

$$\sigma = \left(41\ \text{MPa}\sqrt{m}\right) / \left((8.1173)\ \sqrt{\pi\ (5 \times 10^{-3}\ \text{m})}\right) = 40.30\ \text{MPa}$$

Compact tension specimen

$$\alpha = \frac{(2+x)}{(1-x)^{3/2}}\left(0.886 + 4.64x - 13.32x^2 + 14.72x^3 - 5.60x^4\right)$$

$$\alpha = 9.6591$$

$$\sigma = \left(41\ \text{MPa}\sqrt{m}\right) / \left((9.6591)\ \sqrt{\pi\ (5 \times 10^{-3}\ \text{m})}\right) = 33.87\ \text{MPa}$$

Therefore, the fracture stress as well as fracture toughness depends on the specimen configuration and geometry. This exercise induces to conclude that normalized crack depth (a/w), specimen width (w), and thickness (B) affect the fracture stress as well as fracture toughness. It should also be mentioned that aggressive environments affect, in general, material properties.

Example 3.2. *A large plate (2a << w) containing a through-thickness center crack 40-mm long is subjected to a tension stress as shown below and illustrated in Table 3.1. If the crack growth rate is 10 mm/month and fracture is expected at 10 months from now, calculate the fracture stress. Use the plane-strain fracture toughness $K_{IC} = 30\ MPa\sqrt{m}$.*

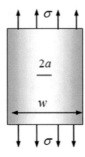

Solution. *If the crack growth rate is defined by*

$$\frac{da}{dt} = \frac{2a_c - 2a}{t}$$

then

$$2a_c = 2a + t\left(\frac{da}{dt}\right)$$

$$2a_c = 40\,\text{mm} + (10\,\text{month})\,(10\,\text{mm/month})$$

$$2a_c = 140\ \text{mm}$$

$$a_c = 70\ \text{mm}$$

Therefore, the total critical crack length is 100 mm longer than the original or
$a_c = 3.5a$.

   *Now, the fracture stress can be calculated using Eq. (3.29). A large plate implies
that $a/w \to 0$ and from Table 3.1 the geometric correction factor becomes*

$$\alpha = \sqrt{\sec\left[\pi\,(a/w)\right]} \to 1$$

Thus, the fracture stress becomes

$$\sigma_f = \frac{K_{IC}}{\alpha\sqrt{\pi a_c}}$$

$$\sigma_f = \frac{30\,\text{MPa}\sqrt{\text{m}}}{(1)\sqrt{\pi\,(70 \times 10^{-3}\ \text{m})}} = 64\ \text{MPa}$$

*This result indirectly implies that $\sigma_f \ll \sigma_{ys}$ because σ_f has a significantly low
value (magnitude).*

Example 3.3. *A large and thick brittle plate containing a 4-mm long through-
thickness center crack fractures when it is subjected to an external tensile stress
of 7 MPa. Calculate the strain energy release rate using (a) the Griffith's theory and
(b) the LEFM approach. Should there be a significant difference between results?
Explain. Assume stable crack growth and use the following material's properties
$E = 62,000\,MPa$ and $v = 0.20$.*

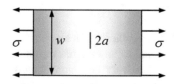

Solution.

(a) *For a total crack size of $2a = 4\,mm$, Eq. (2.35) yields the critical strain energy
release rate (fracture toughness in terms of strain energy) at fracture*

$$\sigma_f = \sqrt{\frac{E'G_{IC}}{\pi a}}$$

$$G_{IC} = \frac{\pi a \sigma_f^2}{E'} = \frac{\pi \left(1 - v^2\right) a \sigma_f^2}{E}$$

$$G_{IC} = \frac{\pi \left(1 - 0.2^2\right) \left(2 \times 10^{-3} \text{ m}\right) (7 \text{ MPa})^2}{62{,}000 \text{ MPa}}$$

$$G_{IC} = 4.77 \times 10^{-6} \text{ MPa m} = 4.77 \text{ J/m}^2$$

(b) *Using Eq. (3.29) along with* $\alpha = \sqrt{\sec\left(\pi a/w\right)} = 1$ *(see Table 3.1) and letting the stress intensity factor reach its critical value under plane-strain condition,* $K_I = K_{IC}$, *give*

$$K_{IC} = \alpha \sigma \sqrt{\pi a}$$

$$K_{IC} = (1) (7 \text{ MPa}) \sqrt{\pi \left(2 \times 10^{-3} \text{ m}\right)}$$

$$K_{IC} = 0.555 \text{ MPa}\sqrt{m}$$

This is a very small value, implying that the material is very brittle such as glass. From Eq. (2.35),

$$G_{IC} = \frac{K_{IC}^2}{E'} = \frac{\left(1 - v^2\right) K_{IC}^2}{E}$$

$$G_{IC} = \frac{\left(1 - 0.2^2\right) \left(0.555 \text{ MPa}\sqrt{m}\right)^2}{62{,}000 \text{ MPa}}$$

$$G_{IC} = 4.77 \times 10^{-6} \text{ MPa m} = 4.77 \text{ J/m}^2$$

These results indicate that there should not be any difference in the strain energy release rate because either approach gives the same result.

3.4.2 Elliptical Cracks

This section deals with elliptical, semielliptical, circular, and semicircular cracks in structural components subjected to an external remote stress. In fact, small cracks developing from rivet holes are very common defects encountered in lap joints of plates. A particular specimen containing an embedded elliptical crack can be analyzed using a semielliptical crack geometry due to symmetry. This is depicted in Fig. 3.4. It can be assumed that the semielliptical crack (solid curve) can grow to a semicircle (dash curve). The goal in this type of problem is to determine the stress intensity factor (K_I) on the perimeter of the crack at angle λ when the crack length (a) is the depth of the semi-ellipse.

Fig. 3.4 Embedded
semielliptical crack [19]

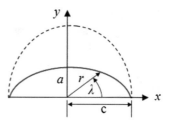

It is common to encounter embedded and surface elliptical cracks in certain
structural components. In such a case, the stress intensity factor K_I has received
ample consideration by the engineering and the scientific community to predict
fracture susceptibility.

According to Irwin's analysis [21], which is also cited in Broek's book [8], on
an infinite plate containing an embedded elliptical or semielliptical crack remotely
loaded in tension, the stress intensity factor is defined by

$$K_I = \frac{\sigma \sqrt{\pi a}}{\Phi} \left[\sin^2 \lambda + \left(\frac{a}{c} \right)^2 \cos^2 \lambda \right]^{1/4} \tag{3.33}$$

which can be evaluated at a point on the perimeter of the crack. This point is located
at an angle λ with respect to the direction of the applied tensile stress. Also shown
in Fig. 3.4 is a semielliptical surface crack that grows to a semicircle as in the case
of some pressure vessels.

The term Φ in Eq. (3.33) accounts for the crack shape effects, and it takes the
form of an elliptic integral of the second kind defined as [21]

$$\Phi = \int_o^{\pi/2} \left[1 - \left(\frac{c^2 - a^2}{c^2} \right) \sin^2 \lambda \right]^{1/2} d\lambda \tag{3.34}$$

where a and c are as defined in Fig. 3.4. This mathematical expression can be
simplified in order to obtain the geometry correction factor for a circular crack.
This can be accomplished by letting the crack depth become equals to the radius of
the circle; that is, if $a = c$, the ellipse becomes a circle and the integral in Eq. (3.34)
gives

$$\Phi = \int_o^{\pi/2} d\lambda = \frac{\pi}{2} \tag{3.34a}$$

Substituting $\Phi = \pi/2$ in Eq. (3.33) yields the stress intensity factor for a circular
crack of radius a, known as a penny-shaped crack, embedded in an infinite solid
subjected to a uniform tension stress developed by Sneddon [34]

$$K_I = \frac{2}{\pi} \sigma \sqrt{\pi a} \tag{3.35}$$

In this specific case, the finite geometry correction factor for a circular crack is simply given by $\alpha = 1/\Phi = 2/\pi$. Actually, Eq. (3.35) can also be used for a specimen containing semicircular surface flaw.

Furthermore, values of Φ can be found in mathematical tables, or it can be approximated by expanding Eq. (3.34) in a Taylor's series form [8]. Hence, Φ becomes

$$\Phi = \frac{\pi}{2}\left[1 - \frac{1}{4}\left(\frac{c^2 - a^2}{c^2}\right) - \frac{3}{64}\left(\frac{c^2 - a^2}{c^2}\right)^2 - \ldots\ldots\right] \tag{3.36}$$

Neglecting higher terms in the series, the margin of error is not significant; therefore, the correction factor Φ can further be approximated using the first two terms in the series given by Eq. (3.36). Thus,

$$\Phi = \frac{\pi}{2}\left[1 - \frac{1}{4}\left(\frac{c^2 - a^2}{c^2}\right)\right] = \frac{\pi}{2}\left[\frac{3}{4} + \left(\frac{a}{2c}\right)^2\right] \tag{3.37}$$

Inserting Eq. (3.37) into (3.33) gives

$$K_I = \frac{2\sigma\sqrt{\pi a}}{\pi\left[3/4 + (a/2c)^2\right]}\left[\sin^2\lambda + \left(\frac{a}{c}\right)^2\cos^2\lambda\right]^{1/4} \tag{3.38}$$

Evaluating Eq. (3.38) on the perimeter of the ellipse yields the stress intensity factor for two extreme cases. Thus,

$$K_I = \frac{2\sigma\sqrt{\pi a}}{\pi\left[3/4 + (a/2c)^2\right]} \qquad @ \ \lambda = \pi/2 \ \text{(maximum)} \tag{3.39}$$

$$K_I = \frac{2\sigma\sqrt{\pi a}}{\pi\left[3/4 + (a/2c)^2\right]}\sqrt{\frac{a}{c}} \quad @ \ \lambda = 0 \ \text{(minimum)} \tag{3.40}$$

The condition $\lambda = \pi/2$ is vital in evaluating an elliptical crack behavior because K_I is computed as a maximum applied entity for predicting crack instability at fracture.

3.4.3 Cylindrical Pressure Vessels

The stress intensity factor K_I at a point on the perimeter of embedded elliptical or circular cracks is located at an angle λ which is described below. In this section, the influence of internal surface cracks are analyzed in a cylindrical pressure vessel as schematically depicted in Fig. 3.5.

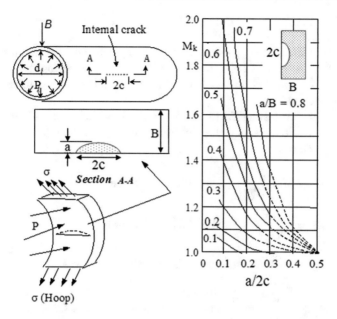

σ (Hoop)

Fig. 3.5 Front face correction factor MK for a semielliptical surface flaw in thin-wall pressure vessel [6, 24]

The general mathematical definition of K_I in an infinite body subjected to a remotely uniform tensile stress is defined by Irwin [21]

$$K_I = \sigma \sqrt{\frac{\pi a}{Q}} \left[\sin^2 \lambda + \left(\frac{a}{c}\right)^2 \cos^2 \lambda \right]^{1/4} \tag{3.41}$$

where Q is the shape factor given by

$$Q = \Phi^2 - \frac{2}{3\pi} \left(\frac{\sigma}{\sigma_{ys}}\right)^2 \tag{3.42}$$

It is very common in machine parts to have a combination of stresses on the same part leading to different values of K_I due to different applied modes of type I loading.

Of particular interest in this section is the pressure vessel schematically depicted in Fig. 3.5 in which the hoop stress and the internal pressure act as a combined loading system. Consequently, the principle of superposition requires that the total stress intensity factor K_I be the sum of each stress intensity factor components [5,8].

For a **cylindrical pressure vessel** (Fig. 3.5), the principle of superposition assumes that K_I arises from the hoop stress (σ_h) and the internal pressure (P_i), which loads the crack surfaces. In this case, the total stress intensity factor due to the hoop stress and the internal pressure is

$$K_I(\sigma) = K_I(\sigma_h) + K_I(P_i) \tag{3.42a}$$

Here, the total stress $\sigma = \sigma_h + P_i$ and $K_I(\sigma) = K_I$ will cause fracture if the critical crack depth is $a_c < B$. Conversely, if $a = B$, then the criterion that governs the failure mode is known as the leak before break, which is a concept to be dealt with in the next section.

Accordingly, evaluating Eq. (3.41) at $\lambda = \pi/2 = 90°$ (Fig. 3.4) and including additional correction factors yields the stress intensity factor for thin-wall and thick-wall vessels. Thus,

$$K_I = MM_k\sigma\sqrt{\frac{\pi a}{Q}} \quad \text{For thin-wall vessels} \tag{3.43a}$$

$$K_I = \sigma\sqrt{\frac{\pi a}{Q}} \quad \text{For thick-wall vessels} \tag{3.43b}$$

$$\sigma_h = \frac{P_i d_i}{2B} \quad \text{For thin-wall vessels, } d_o/d_i < 1.1 \ \& \ B \le \frac{d_i}{20} \tag{3.43c}$$

$$\sigma_h = \left[\frac{(d_o/d_i)^2 + 1}{(d_o/d_i)^2 - 1}\right] P_i \quad \text{For thick-wall vessels, } d_o/d_i > 1.1 \tag{3.43d}$$

where M = Magnification correction factor [24]
M_k = Front face correction factor [5, 24] shown in Fig. 3.5
σ_h = Hoop stress, d_i = Internal diameter and d_o = Outer diameter
The magnification correction factor takes the form [5, 6, 24]

$$M = 0.4 + 1.2\left(\frac{a}{B}\right) \quad \text{For } 0.5 \le \frac{a}{B} \le 1.0 \tag{3.44}$$

The hoop stress (σ_h) defined above will be derived later for cylindrical pressure vessels only. Combining Eqs. (3.37) and (3.42) yields a convenient mathematical expression for predicting the shape factor

$$Q = \left(\frac{\pi}{2}\right)^2\left[\frac{3}{4} + \left(\frac{a}{2c}\right)^2\right]^2 - \frac{7}{33}\left(\frac{\sigma}{\sigma_{ys}}\right)^2 \tag{3.45}$$

The shape factor has also been reported to have the form [36]

$$Q = 1 + 1.464\left(\frac{a}{c}\right)^{1.65} \quad \text{for } a/c \le 1 \tag{3.45a}$$

Manipulating Eq. (3.45) and rearranging it yields a very important equation in the fracture mechanics for characterizing cracked pressure vessels. Hence,

$$\frac{a}{2c} = \left[\frac{2}{\pi}\sqrt{Q + \frac{7}{33}\left(\frac{\sigma}{\sigma_{ys}}\right)^2} - \frac{3}{4}\right]^{1/2} \tag{3.46}$$

Fig. 3.6 Shape factor Q for internal and surface flaws

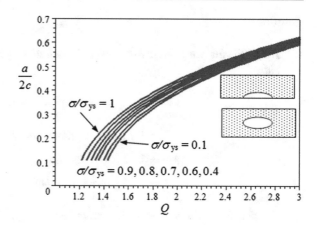

Subsequently, Eq. (3.46) can be used to plot a series of curves as depicted in Fig. 3.6. Furthermore, if a back free-surface correction factor of 1.12 and plastic deformation are considered, then Eq. (3.41) is further corrected as

$$K_I = 1.12 M M_k \sigma \sqrt{\pi(a+r)/Q} \quad \text{For } \lambda = \pi/2 \tag{3.47}$$

In general, a correction factors are used to improve the accuracy of a particular model and they must contain physical meaning. As indicated in Eq. (3.47), the correction factors are combined to account for the increase or decrease in the stress intensity factor.

So far only cracks in the interior of the cylindrical part of a pressure vessel have been presented. A pressure vessel has welded nozzles which may be troublesome in certain occasions. For instance, corner cracks are dangerous and may be found at intersections of the pressure vessel and the nozzle. Obviously, the fracture analysis of nozzle corner cracks in pressure vessels is very imperative and critical for assessing structural integrity of pressure vessels. Consequently, a reliable pressure vessel design must include all possible flaw-generation sites to ensure the structural integrity of the pressure vessels.

In general, pressure vessels can theoretically have any shape, but a cylindrical shade pressure vessel is common, but the presence of internal cracks leads to the determination of further corrected (K_I) using the shape factor Q. It is common to correct the (K_I) expression, which indirectly corrects (G_I) using Eq. (2.39).

Example 3.4. *A steel thin-wall pressure vessel (Fig. 3.5) is subjected to a stress of 420 MPa ($\sigma_h + P$) perpendicular to the crack depth. The vessel has an internal semielliptical surface crack of dimensions $a = 3\,mm$, $2c = 10\,mm$. (a) Use Eq. (3.42) to calculate K_I. (b) Will the pressure vessel leak? Explain. (c) determined the pressure P and the hoop stress σ_h. Use the following data set needed for the required computations: $\sigma_{ys} = 700\,MPa$ and $K_{IC} = 60\,MPa\sqrt{m}$, $B = 6\,mm$, $d_i = 500\,mm$.*

Solution.

(a) *The following parameters are needed for calculating the stress intensity factor. Thus,*

$$\frac{a}{2c} = 0.30, \quad \frac{a}{B} = 0.50, \quad \frac{\sigma}{\sigma_{ys}} = 0.60$$

$$Q = 1.7 \text{ [from Eq. (3.45) and Fig. 3.6]}$$

$$M_k = 1.12 \text{ (from Fig. 3.5)}$$

$$M = 1.00 \text{ [from Eq. (3.44)]}$$

The plastic zone size can be determined using Eq. (3.1) when $\sigma_{yy} = \sigma_{ys}$. This implies that a plastic zone develops as long as the material yields ahead of the crack tip. Collecting all correction factor gives

$$\alpha = 1.12 M M_k / \sqrt{Q} = 0.9621$$

$$K_I = \alpha \sigma \sqrt{\pi a} = 39.23 \text{ MPa}\sqrt{m}$$

(b) *The pressure vessel will not leak because $K_I < K_{IC}$, but extreme caution should be taken because $K_I = K_{IC}/S_F = 0.6538 K_{IC}$. Thus, the safety factor is $S_F = 60/39.23 = 1.53$. If $K_I = K_{IC}$, then leakage would occur.*

(c) *Furthermore, the pressure can be determined using Eqs. (3.43c) and (3.43d). Thus,*

$$\sigma = \sigma_h + P_i = \frac{P_i d_i}{2B} + P_i$$

$$\sigma = \left(\frac{d_i}{2B} + 1\right) P_i$$

Then,

$$P_i = \sigma \left(\frac{d_i}{2B} + 1\right)^{-1} = (420 \text{ MPa}) \left(\frac{500 \text{ mm}}{2 \times 6 \text{ mm}} + 1\right)^{-1}$$

$$P_i = 9.84 \text{ MPa}$$

$$\sigma_h = 420 \text{ MPa} - 9.84 \text{ MPa} = 410.16 \text{ MPa}$$

Example 3.5.

(a) *Derive the stresses associated with internal and external pressures for a cylindrical pressure vessel, which is schematically shown in the figure below.*

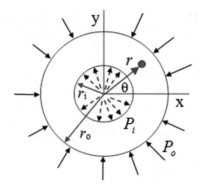

Use the modified Airy stress function published by Dally [12]

$$\phi = a_o + b_o \ln r + c_o r^2$$

The boundary conditions for this problem are

$$\sigma_r = -P_i \ \& \ \tau_{r\theta} = 0 @ \ r = r_i$$

$$\sigma_r = -P_o \ \& \ \tau_{r\theta} = 0 \ r = r_o$$

(b) *Derive Eq. (3.43d).*

Solution.

(a) *Let*

$$\phi = a_o + b_o \ln r + c_o r^2 \tag{a}$$

Combining Eqs. (1.58) and (a) yields the radial stress (σ_r) and the tangential or hoop stress ($\sigma_\theta = \sigma_h$)

$$\sigma_r = \frac{1}{r}\frac{\partial\phi}{\partial r} + \frac{1}{r^2}\frac{\partial^2\phi}{\partial\theta^2} = \frac{b_o}{r^2} + 2c_o$$

$$\sigma_\theta = \frac{\partial^2\phi}{\partial r^2} = -\frac{b_o}{r^2} + 2c_o \tag{b}$$

$$\tau_{r\theta} = \frac{1}{r^2}\frac{\partial\phi}{\partial\theta} - \frac{1}{r}\frac{\partial^2\phi}{\partial r\partial\theta} = 0$$

The constants a_o, b_o, and c_o are determined using the boundary conditions on the inside and outside wall surfaces of the circular model. These are

$$\sigma_r = -P_i \qquad \tau_{r\theta} = 0 \qquad @ \ r = r_i \tag{c}$$

$$\sigma_r = -P_o \qquad \tau_{r\theta} = 0 \qquad @ \ r = r_o$$

Combining Eqs. (b) and (c) yields the coefficients a_o, b_o, and c_o, and the resultant expressions substituted back into (b) give the radial stress (σ_r) and the tangential or hoop stress ($\sigma_\theta = \sigma_h$) as

$$\sigma_t = \sigma_r = \frac{r_i^2 r_o^2 (P_o - P_i)}{(r_o^2 - r_i^2) r^2} + \frac{r_i^2 P_i - r_o^2 P_o}{r_o^2 - r_i^2}$$

$$\sigma_h = \sigma_\theta = -\frac{r_i^2 r_o^2 (P_o - P_i)}{(r_o^2 - r_i^2) r^2} + \frac{r_i^2 P_i - r_o^2 P_o}{r_o^2 - r_i^2} \qquad (d)$$

$$\tau_{r\theta} = 0$$

Denote that shear stress $\tau_{r\theta}$ does not have any contribution to the stress state. For closed-end pressure vessel, the hoop stress is actually the uniform axial stress in the wall of the pressure vessel.

If $P_o = 0$ or $P_i \gg P_o$ and $r = r_i$, then Eq. (d) gives the stress distribution over the wall thickness for pipes and cylinders. Hence, the compressive and tension elastic stress acting on the crack are uniform across the thickness and take the form

$$\sigma_t = \left[\frac{1 - (r_o/r_i)^2}{(r_o/r_i)^2 - 1} \right] P_i < 0 \ \ (Compressive) \qquad (e)$$

$$\sigma_h = \left[\frac{(r_o/r_i)^2 + 1}{(r_o/r_i)^2 - 1} \right] P_i > 0 \ \ (Tension) \qquad (f)$$

(b) *Equation (3.43d) has been already derived in terms of radii. If the diameter of a perfect circle is $d = 2r$, then, Eq. (f) becomes the hoop stress as the crack opening stress for stable crack growth*

$$\sigma_h = \left[\frac{(d_o/d_i)^2 + 1}{(d_o/d_i)^2 - 1} \right] P_i \ \ For \ thick \ walls \qquad (3.43d)$$

Rearrange Eq. (f) along with $B = r_o - r_i$ so that

$$\sigma_h = \left[\frac{r_o^2 + r_i^2}{r_o^2 - r_i^2} \right] P_i$$

$$\sigma_h = \left[\frac{r_o^2 + r_i^2}{(r_o + r_i) B} \right] P_i \qquad (h)$$

Mathematically, letting $r_o = r_i$ in Eq. (h) leads to (3.43c) along with $r_i = d_i/2$. Thus, the hoop stress expression takes the common mathematical definition encountered in many textbooks. Hence,

$$\sigma_h = \frac{r_i P_i}{B} = \frac{d_i P_i}{2B} \quad For\ thin\ walls \tag{3.43c}$$

The current pressure vessel analysis in this example is devoted to large-scale industrial applications to store liquids and gases in containers of diverse geometries at a pressure P.

Example 3.6. *A long steel pipe having a diameter ratio of 1.25 and an internal diameter of 80 mm is subjected to a pressure of 60 MPa. Assume that a semielliptical surface crack, a = 2 mm and 2c = 4 mm, develops during service and that it is perpendicular to the hoop stress as shown in Fig. 3.5. Assume that the design lifetime of the pressurized steel pipe is 20 years and that an undesired pressure surge will burst the pipe. Calculate (a) the possible fracture pressure if the steel has K_{IC} = 30 MPa $\sqrt{m}$ and σ_{ys} = 550 MPa and (b) the time it takes for a pressure surge if the crack velocity is define as $v = c_1 K_I^n$, where v is in m/s, K_I is in MPa $\sqrt{m}$, $c_1 = 2x10^{-13.6}$ (m/sec)(MPa $\sqrt{m}$)$^{-n}$, and n = 1.5.*

Solution.

(a) *From d_o/d_i = 1.25 and d_i = 80 mm, the outside diameter and the vessel thickness are d_o = 100 mm and B = $(d_o - d_i)/2$ = 10 mm, respectively. The pipe is a thick-wall pressure vessel because the diameter ratio is $d_o/d_i = 1.25 >$ 1.1 and B > $d_i/20$ = 4 mm, respectively.*
   The applied stress σ and stress intensity factor K_I equations are

$$\sigma_h = \left[\frac{(d_o/d_i)^2 + 1}{(d_o/d_i)^2 - 1} \right] P = \left[\frac{1.25^2 + 1}{1.25^2 - 1} \right] P = 4.56P \tag{a1}$$

$$\sigma = \sigma_h + P = 4.56P + P = 5.56P \tag{a2}$$

$$K_I = \sigma \sqrt{\frac{\pi a}{Q}} = 5.56P \sqrt{\frac{\pi a}{Q}} \tag{a3}$$

The stress ratio needed to calculate the shape factor Q is

$$\frac{\sigma}{\sigma_{ys}} = \frac{5.56(60\ MPa)}{550\ MPa} = 0.61 \tag{a4}$$

Using Fig. 3.6 or Eq. (3.44) with a/2c = 2/4 = 0.5 yields

$$Q = \left(\frac{\pi}{2} \right)^2 \left[\frac{3}{4} + \left(\frac{a}{2c} \right)^2 \right]^2 - \frac{7}{33} \left(\frac{\sigma}{\sigma_{ys}} \right)^2 \tag{a5}$$

$$Q = \left(\frac{\pi}{2} \right)^2 \left[\frac{3}{4} + (0.5)^2 \right]^2 - \frac{7}{33} (0.61)^2 = 2.39 \tag{a6}$$

From Eqs. (a2) and (a3), the initial applied stress and the initial stress intensity factor are

$$\sigma = (5.56)\,(60\ \text{MPa}) = 333.60\ \text{MPa} \tag{a7}$$

$$K_I = \sigma\sqrt{\frac{\pi a}{Q}} = (333.60\ \text{MPa})\sqrt{\frac{\pi\,(2 \times 10^{-3}\ \text{m})}{2.41}} = 17.03\ \text{MPa}\sqrt{\text{m}} \tag{a8}$$

Therefore, fracture will not occur because $K_I < K_{IC}$. Now, the fracture pressure P_f can be calculated by letting $K_I = K_{IC}$ in Eq. (b) and solve for $P = P_f$. Thus,

$$P_f = \frac{K_{IC}}{5.56}\sqrt{\frac{Q}{\pi a}} = \frac{30\ \text{MPa}\sqrt{\text{m}}}{5.56}\sqrt{\frac{2.39}{\pi\,(2 \times 10^{-3}\ \text{m})}} = 105.23\ \text{MPa} \tag{a9}$$

(b) *The crack velocity can be redefined using the chain rule as follows:*

$$v = \frac{da}{dt} = \frac{da}{dK_I}\frac{dK_I}{dt} \tag{b1}$$

$$v = c_1 K_I^n \tag{b2}$$

$$K_I = \sigma\sqrt{\frac{\pi a}{Q}} \tag{b3}$$

$$K_{IC} = \sigma_f\sqrt{\frac{\pi a}{Q}} \tag{b4}$$

$$a = \frac{Q K_I^2}{\pi\sigma^2} \tag{b5}$$

$$\frac{da}{dK_I} = \frac{2Q K_I}{\pi\sigma^2} \tag{b6}$$

Combining Eqs. (b1) and (b6) yields

$$v = \frac{2Q K_I}{\pi\sigma^2}\frac{dK_I}{dt} \tag{b7}$$

from which

$$dt = \frac{2Q K_I}{\pi v\sigma^2}dK_I \tag{b8}$$

Substituting Eq. (b2) into (b8) gives

$$dt = \frac{2Q K_I}{\pi c_1 K_I^n \sigma^2}dK_I = \frac{2Q K_I^{1-n}}{\pi c_1\sigma^2}dK_I \tag{b9}$$

Integrating yields

$$\int_o^t dt = \frac{2Q}{\pi c_1 \sigma^2} \int_{K_I}^{K_{IC}} K_I^{1-n} dK_I \tag{b10}$$

$$t = -\frac{2Q}{\pi (n-2) c_1 \sigma^2} \left[K_{IC}^{2-n} - K_I^{2-n} \right] \tag{b11}$$

Thus, the time for a pressure surge to occur is calculated using Eq. (b11) as

$$t = -\frac{2(2.39)}{\pi (1.5 - 2)(2 \times 10^{-13.6})(333.60)^2} \left[30^{2-1.5} - 17.03^{2-1.5} \right]$$

$$t = 7.3505 \times 10^8 \text{ s} = 23.31 \text{ years}$$

The stable crack velocity is

$$v = c_1 K_I^n = \left[2 \times 10^{-13.6} (\text{m/s})(\text{MPa}\sqrt{\text{m}})^{-1.5} \right] (17.03 \text{ MPa}\sqrt{\text{m}})^{1.5}$$

$$v = 3.53 \times 10^{-12} \text{ m/s}$$

Example 3.7. *A cylindrical vessel made of AISI 4147 is pressurized at 80 MPa, and it has an internal semielliptical surface crack 5-mm deep and 10-mm long. The diameter and thickness are 50 and 25 cm, respectively. In order to account for the effects of both the applied pressure and the hoop stress, calculate* **(a)** *the stress intensity factor. Will the pressure vessel fracture? If not, determine* **(b)** *the maximum pressure that would cause fracture. Data: $K_{IC} = 120 \text{ MPa}\sqrt{\text{m}}$, $\sigma_{ys} = 945 \text{ MPa}$, and $\sigma_{ts} = 1062 \text{ MPa}$.*

Solution. *Given data: $a = 5 \text{ mm}$, $2c = 10 \text{ mm}$, $d_i = 0.5 \text{ m}$ and $B = 25 \text{ cm}$*

$$P_i = 80 \text{ MPa } and \; a/2c = 0.50$$

Which cylindrical pressure vessel theory should be used?

$$B = \frac{(d_o - d_i)}{2}$$

$$d_o = d_i + 2B = 50 \text{ cm} + 2 \times 25 \text{ cm} = 100 \text{ cm}$$

$$\frac{d_o}{d_i} = \frac{100}{50} = 2$$

Therefore, the thick-wall theory must be used because $d_o/d_i > 1.1$.

(a) *Using Eq. (3.43d) yields the hoop stress*

$$\sigma_h = \left[\frac{(d_o/d_i)^2 + 1}{(d_o/d_i)^2 - 1} \right] P_i = \left[\frac{4+1}{4-1} \right] P_i$$

$$\sigma_h = \frac{5}{3} P_i = \frac{5}{3} (80 \text{ MPa}) = 133.33 \text{ MPa}$$

$$\sigma = \sigma_h + P_i = \frac{5}{3} P_i + P_i = \frac{8}{3} P_i = \frac{8}{3} (80 \text{ MPa})$$

$$\sigma = 213.33 \text{ MPa}$$

Then, the stress ratio is

$$\frac{\sigma}{\sigma_{ys}} = \frac{213.33 \text{ MPa}}{945 \text{ MPa}} = 0.23$$

Using Eq. (3.45) yields the shape factor

$$Q = \left(\frac{\pi}{2} \right)^2 \left[\frac{3}{4} + \left(\frac{a}{2c} \right)^2 \right]^2 - \frac{7}{33} \left(\frac{\sigma}{\sigma_{ys}} \right)^2$$

$$Q = \left(\frac{\pi}{2} \right)^2 \left[\frac{3}{4} + (0.5)^2 \right]^2 - \frac{7}{33} (0.23)^2$$

$$Q = 2.46$$

Using Eq. (3.42) gives the total stress intensity factor

$$K_I = \sigma \sqrt{\frac{\pi a}{Q}} = \frac{8}{3} P_i \sqrt{\frac{\pi a}{Q}}$$

$$K_I = \frac{8}{3} (80 \text{ MPa}) \sqrt{\frac{\pi (5 \times 10^{-3} \text{ m})}{2.46}}$$

$$K_I = 17.05 \text{ MPa} \sqrt{m}$$

Therefore, fracture will not occur because $K_I < K_{IC} = 120 \text{ MPa} \sqrt{m}$.
(b) *The maximum pressure is*

$$P_{max} = \frac{3K_{IC}}{8} \sqrt{\frac{Q}{\pi a}}$$

$$P_{max} = \frac{3}{8} (120 \text{ MPa} \sqrt{m}) \sqrt{\frac{2.46}{\pi (5 \times 10^{-3} \text{ m})}}$$

$$P_{max} = 563.14 \text{ MPa}$$

This implies that the critical stress becomes

$$\sigma_c = \sigma_h = \frac{5}{3}P_{max} \approx 939\,\text{MPa} \approx \sigma_{ys}$$

This would cause catastrophic failure if a pressure surge occurs. However, if the applied pressure is constant and crack growth is time dependent, then leakage would occur due to the lack of sufficient thickness for the crack to run through. This can be determined by calculating the critical crack length as

$$a_c = \frac{Q}{\pi}\left(\frac{3K_{IC}}{8P_i}\right)^2$$

$$a_c = \frac{2.46}{\pi}\left(\frac{3 \times 120}{8 \times 80}\right)^2$$

$$a_c = 24.78 \text{ cm} \approx B = 25 \text{ cm}$$

This results indicate that the remaining ligament, b, of the cylindrical vessel wall is very small. Thus,

$$b = B - a_c = 25 \text{ cm} - 24.78 \text{ cm}$$

$$b = 0.22 \text{ cm} = 2.2 \text{ mm}$$

Therefore, these results, $a_c = 24.78\,mm$ and $b = 2.2\,mm$, suggest that the cylindrical pressure vessel is near catastrophe. It either can burst or leak because $a_c \rightarrow B$ and $b \rightarrow 0$. This means that the crack nearly grew through the entire wall thickness. The technical terminology for this situation is that there exists a through-wall crack, which can grow to unstable size.

3.4.4 Leak-Before-Break Criterion

In general, internal surface cracks in pressure vessels may grow by fatigue or stress corrosion at faster rate than external cracks [7]. In any event, fracture mechanics methods can be used to provide the foundation for assessing the structural integrity of pressure vessels, which can be designed under thin-wall ($B \leq d_i/20$) or thick-wall ($B > d_i/20$) technique. Thus, the leak-before-break (LBB) criterion requires a stable vessel and implies that the failure mode of a cracked pressure vessel is a leaking through-thickness crack, having length $2c$ and depth a. This simply means that a semielliptical or semicircular crack penetrates the vessel wall thickness before a catastrophic break occurs. Technically, the LBB concept is widely used in the nuclear industry, and useful information on this topic can be found elsewhere [2].

The applied internal stress that causes crack growth is based on the superposition principle; $\sigma = \sigma_h + P$, where σ_h is the hoop (circumferential) stress and P is the internal pressure acting the crack surfaces. This stress is a fraction of the yield strength of the material used to make pressure vessels. Thus, $\sigma = \beta \sigma_{ys}$, where $0 < \beta < 1$ and it can be defined as the inverse of a safety factor ($\beta = 1/S_F$). Once the crack reaches the thickness of the pressure vessel, $a = B$, it becomes a leaking through-thickness crack having a specific leakage rate. For a semielliptical crack, the length ($2c$) of a leak can be expected to be greater than the vessel thickness; $2c > B$.

Designing thin-wall pressure vessels to store fluids is a common practice in engineering. If curved plates are welded to make pressure vessels, the welded joints may become the weakest areas of the structure since weld defects can be the source of cracks during service. Accordingly, the internal pressure P acts in the radial direction (Fig. 3.5), and the total force for rupturing the vessel is $P(d_iL)$, where dL is the projected area. Assuming that the stress across the thickness is the hoop stress σ_h and that the cross-sectional area is $A = 2BL$, then the force balance is $P(dL) - \sigma_h(2BL) = 0$ which gives Eq. (3.43c). The hoop stress σ_h is the longitudinal stress and it is twice the transverse or radial stress; as $\sigma_h = 2\sigma_t$. Hence, σ_h and σ_t are principle stresses in designing against yielding. Normally, the design stress is $\sigma_d = \sigma_h/S_F$, where S_F is the safety factor in the range of $1 < S_F < 5$.

For welded joints in pressure vessels, a welding efficiency, $0.80 \leq \varepsilon \leq 1$, can be included in the hoop stress expression to account for weak welded joints [15]. Thus, Eq. (3.43c) becomes

$$\sigma_h = \frac{Pd}{2\varepsilon B} \tag{3.48}$$

If the limiting stress σ_h is the yield strength σ_{ys}, then the thickness of the pressure vessel should be

$$B \geq \frac{Pd}{2\varepsilon \sigma_{ys}} \tag{3.48a}$$

and letting $a = B$ yields the hoop stress and the internal pressure requirements to be defined as

$$\sigma_h \leq \frac{K_{IC}}{\alpha \sqrt{\pi B}} \tag{3.48b}$$

$$P \leq \frac{2}{\pi d} \left(\frac{K_{IC}}{\alpha \sigma_{ys}} \right)^2 \tag{3.48c}$$

Thus, Eq. (3.48c) sets the criterion for LBB using the parameter $(K_{IC}/\sigma_{ys})^2$. This implies the maximum pressure to assure safety when selecting a material for a pressure vessel with the highest $(K_{IC}/\sigma_{ys})^2$.

Let us assume that internal surface cracks develop at welded joints or at any other area in a pressure vessel. In such a scenario, the leak-before-break criterion proposed by Irwin et al. [22] can be used to predict the fracture toughness of pressure vessels. This criterion allows an internal surface crack to grow through the thickness of the vessel so that $a = B$ for leakage to occur.

The fracture toughness relationship between plane stress and plane-strain conditions to establish the leak-before-break criterion may be estimated using an empirical relationship developed by Irwin et al. [22]. Thus, the plane stress fracture toughness ($K_I = K_C$) equation is

$$K_C = K_{IC}\sqrt{1 + \frac{7}{5B^2}\left(\frac{K_{IC}}{\sigma_{ys}}\right)^4} \qquad (3.49a)$$

Irwin's expression [22] for plane stress fracture toughness is derived in Chap. 5, and it is given here for convenience. Thus, the plane stress intensity factor is of the form

$$K_C^2 = \frac{\pi a \sigma^2}{1 - 0.5\left(\sigma/\sigma_{ys}\right)^2} \qquad (3.49b)$$

Combining Eqs. (3.49a) and (3.49b) along with $\sigma = \sigma_{ys}$ as the critical condition and $a = B$ (through-wall crack) yields the leak-before-break criterion for plane-strain condition

$$K_{IC}^6 + 0.71B^2\sigma_{ys}^4 K_{IC}^2 - 4.49B^3\sigma_{ys}^6 = 0 \qquad (3.50)$$

The use of this criterion requires that the vessel thickness meets the ASTM E399 thickness requirement for a cracked pressure vessel in order to withstand the internal pressure.

In summary, the applicability of the leak-before-break (LBB) concept to pressure vessel (closed container designed to hold gases or liquids at a pressure $P > P_{atm}$) is based on the concept that a crack would grow through the thickness of the vessel wall, resulting in a leak when crack length equals the wall thickness ($a = B$).

Typically, pressurized water reactors for the main coolant line in nuclear reactor are commonly designed using LBB, provided that severe degradation due to corrosion, erosion, creep, and fatigue is not a concern.

Example 3.8. *A thin-wall pressure vessel made of Ti-6Al-4V alloy using a welding fabrication technique is used in rocket motors as per Faires [15]. Helium (He) is used to provide pressure on the fuel and lox (liquid oxygen). The vessel internal diameter and length are 0.5 m and 0.6 m, respectively, and the internal pressure is 28 MPa. Assume that a semicircular crack develops; use a welding efficiency of 100 % and a safety factor of 1.5 to calculate the uniform thickness of the vessel. Use the resultant thickness to calculate the fracture toughness according to Eq. (3.50). Select a yield strength of $\sigma_{ys} = 900$ MPa.*

Solution. *The design stress against general yielding is*

$$\sigma = \frac{\sigma_{ys}}{S_F} = \frac{900 \text{ MPa}}{1.5} = 600 \text{ MPa}$$

From Eq. (3.48),

$$B = \frac{Pd}{2\varepsilon\sigma} = \frac{(28 \text{ MPa}) (0.5 \text{ m})}{(2) (1) (600 \text{ MPa})} \tag{3.48}$$

$$B = 1.17 \times 10^{-2} \text{ m} = 1.17 \text{ cm} = 0.46 \text{ in.}$$

From Eq.(3.50),

$$0 = K_{IC}^6 + 0.71 \left(1.17 \times 10^{-2} \text{ m}\right)^2 (900 \text{ MPa})^4 K_{IC}^2 \tag{3.50}$$
$$- 4.49 \left(1.17 \times 10^{-2} \text{ m}\right)^3 (900 \text{ MPa})^6$$
$$K_{IC} = 119.49 \text{ MPa}\sqrt{\text{m}}$$

From Eq. (3.49a),

$$K_C = K_{IC} \sqrt{1 + \frac{7}{5B^2} \left(\frac{K_{IC}}{\sigma_{ys}}\right)^4} \tag{3.49a}$$

$$K_C = \left(119.49 \text{ MPa}\sqrt{\text{m}}\right) \sqrt{1 + \frac{7 (39.16)^4}{5 (1.17 \times 10^{-2})^2 (900)^4}}$$

$$K_C = 121.66 \text{ MPa}\sqrt{\text{m}}$$

In summary, select the proper heat treatment for Ti-6Al-4V alloy so that $\sigma_{ys} = 900\,MPa$ and $K_{IC} = 119.49\,MPa\sqrt{m}$. According to these results, AISI 4147 (Table 3.2) meets the design requirements. However, the ASTM E399 thickness requirement is not met since

$$B_{ASTM} = (2.5) \left(\frac{119.49 \text{ MPa}\sqrt{\text{m}}}{900 \text{ MPa}}\right)^2 = 4.41 \times 10^{-2} \text{ m}$$

$$B_{ASTM} = 4.41 \text{ cm} > B$$

Therefore, the pressure vessel plates are not under plane-strain conditions.

3.4.5   Radial Cracks Around Cylinders

Another commonly encountered surface crack configuration under a remote applied tension, torsion, or a combined loading system is shown in Fig. 3.7. The mixed-mode interaction is of great interest in this section.

Fig. 3.7 Radial crack around a cylinder

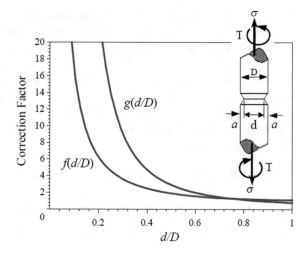

These stress intensity factors are geometrically corrected using $f(d/D)$ and $g(d/D)$ depicted in Fig. 3.7.

The stress intensity factors for the mixed loading system (mode I and mode III) depicted in Fig. 3.7 are

$$K_I = f(d/D)\, \sigma \sqrt{\pi a} \tag{3.51}$$

$$K_{III} = g(d/D)\, \tau \sqrt{\pi a} \tag{3.52}$$

For a gradually applied torque T, the torsional shear stress becomes

$$\tau = \frac{16T}{\pi D^3} \tag{3.53}$$

The correction factors, $f(d/D)$ and $g(d/D)$, were derived by Koiter and Benthem [25] as

$$f(d/D) = \frac{1}{2}\sqrt{\frac{D}{d}}\left[\frac{D}{d} + \frac{1}{2} + \frac{3}{8}\left(\frac{d}{D}\right) - \frac{5}{14}\left(\frac{d}{D}\right)^2 + \frac{11}{15}\left(\frac{d}{D}\right)^3\right] \tag{3.54}$$

$$g(d/D) = \frac{3}{8}\sqrt{\frac{D}{d}}\left[\begin{array}{c}\left(\frac{D}{d}\right)^2 + \frac{1}{2}\left(\frac{D}{d}\right) + \frac{3}{8} + \frac{5}{16}\left(\frac{d}{D}\right) \\ -\frac{35}{128}\left(\frac{d}{D}\right)^2 + \frac{13}{62}\left(\frac{d}{D}\right)^3\end{array}\right] \tag{3.55}$$

The crack length (size) is estimated as

$$a = (D - d)/2 \tag{3.56}$$

Example 3.9. *This example includes the state of stress ahead of a crack based on mixed-mode I and III interactions. Problems of this type are encountered in multi-phase materials such as welded structures, reinforced concrete structures and the like. In order to simulate this type of mixed-mode interactions, consider two identical high-strength steel rods which are prepared for a tension test at 106 MPa and torsion at 69 MPa. Calculate K_I and K_{III}. The rod dimensions are $d = 4$ mm and $D = 8$ mm. If $K_{IIIC} = \sqrt{3/4} K_{IC}$, (a) Will the rods fracture? Explain. (b) Calculate the theoretical fracture tensile and torsion stresses if fracture does not occur in part (a). Use $K_{IC} = 25$ MPa$\sqrt{m}$.*

Solution. *Given data:* $\sigma = 106$ MPa $\tau = 69$ MPa $K_{IC} = 25$ MPa$\sqrt{m}$

(a) *The crack length and the diameter ratios are, respectively,*

$$a = \frac{D-d}{2} = \frac{8 \text{ mm} - 4 \text{ mm}}{2} = 2 \text{ mm}$$

$$\frac{d}{D} = \frac{4 \text{ mm}}{8 \text{ mm}} = 0.5$$

$$\frac{D}{d} = \frac{8 \text{ mm}}{4 \text{ mm}} = 2$$

From Eqs. (3.54) and (3.55) or Fig. 3.7, $f(d/D) = 1.90$ and $g(d/D) = 2.91$. Hence, the applied stress intensity factors are calculated using Eqs. (3.51) and (3.52)

$$K_I = f(d/D) \sigma \sqrt{\pi a} = 16 \text{ MPa}\sqrt{m}$$

$$K_{III} = g(d/D) \tau \sqrt{\pi a} = 16 \text{ MPa}\sqrt{m}$$

$$K_{IIIC} = \sqrt{3/4} \cdot K_{IC} = 21.65 \text{ MPa}\sqrt{m}$$

Therefore, neither rod will fracture since both stress intensity factors are below their critical values; that is, $K_I < K_{IC}$ and $K_{III} < K_{IIIC}$.

(b) *Correspondingly, the state of stress ahead of the crack is given by the fracture stresses. Thus,*

$$\sigma_f = \frac{K_{IC}}{f(d/D) \sqrt{\pi a}} = 166 \text{ MPa}$$

$$\tau_f = \frac{K_{IIIC}}{g(d/D) \sqrt{\pi a}} = 94 \text{ MPa}$$

It has been shown that both tensile and shear fracture mechanisms can occur at the crack tip.

3.5   Fracture Control

Structures usually have inherent flaws or cracks introduced during 1) welding process due to embedded slag, holes, porosity, and lack of fusion and 2) service due to fatigue, stress corrosion cracking (*SCC*), impact damage, and shrinkage.

A fracture-control practice is vital for design engineers in order to assure the integrity of particular structure. This assurance can be accomplished by a close control of:

1) Design constraints   4) Maintenance
2) Fabrication   5) Nondestructive evaluation (NDE)
3) General yielding   6) Environmental effects

The pertinent details for the above elements depend on codes and procedures that are required by a particular organization. However, the suitability of a structure to brittle fracture can be evaluated using the concept of fracture mechanics, which is the main subject in this section. For instance, the elapsed time for crack initiation and crack propagation determines the useful life of a structure, for which the combination of an existing crack size, applied stress, and loading rate may cause the stress intensity factor reach a critical value.

In order to describe the technical aspects of a fracture-control plan, consider a large plate (infinite plate) with a certain plane-strain fracture toughness K_{IC} so that $K_I < K_{IC}$ for a stable crack. Thus, the typical design philosophy [18] uses Eq. (3.28) or (3.29) as the general mathematical model in which $a = a_{max}$ is the maximum allowable crack size in a component, σ is the design stress, and K_I is the applied stress intensity factor. However, the minimum detectable crack size depends on the available equipment for conducting nondestructive tests, but the critical crack size (a_c) can be predicted when the stress intensity factor reaches a critical value, which is commonly known as the plane-strain fracture toughness ($K_I = K_{IC}$) for thick plates. In fact, $K_I < K_{IC}$ can be taken as the material fracture constraint; otherwise, the crack becomes unstable when it reaches a critical length, $a = a_c$, which is strongly controlled by K_{IC}. Thus, solving Eq. (3.29) for the critical crack length yields

$$a_c = \frac{1}{\pi} \left(\frac{K_{IC}}{\alpha \sigma} \right)^2 \tag{3.57}$$

This expression implies that the maximum allowable crack length depends on the magnitude of K_{IC} and the applied stress $\sigma < \sigma_{ys}$. Conclusively, crack propagation occurs when the applied stress intensity factor is equal or greater than fracture toughness, $K_I \geq K_{IC}$ for plane strain or $K_I \geq K_C$ for plane stress condition.

The literature has a vast amount of fracture toughness data for many materials used in engineering construction. For convenience, Table 3.2 is included in this chapter to provide the reader with typical fracture toughness data for some common materials.

Table 3.2 Mechanical properties at room temperature

Material	σ_{ys} (MPa)	σ_{ts} (MPa)	K_{IC} MPa$\sqrt{m}$	E (GPa)	%EL	%RA	Ref.
AerMet 100	1724	1965	126	207	14	65	[11]
AerMet 310	1896	2172	71	207	14	60	[11]
Marage 300	2000	2034	77	207	11	57	[11]
Ti-6Al-4V	869	958	87	117	11	16	[43]
Ti-6Al-4V	1007	1034	40	130	14	29	[43]
AISI 4340	1089	1097	110	207	14	49	[6]
AISI 4147	945	1062	120	207	15	49	[6]
AISI 4340	1476	1896	81	207	10		[32]
Inconel	1172	1404	96		15	18	[39]
18Ni(250)	1290	1345	176		15	66	[43]
2014-T651	455		24				[18]
2024-T3	345		44				[18]
7075-T6	572	641	24	72	12		[38]
$Ni_{49}Fe_{29}$ $P_{14}B_6 Si_2$	800	800	12				[26]
$Ni_{69}B_{14}Si_8$ Cr_6Fe_3	1268	1268	42				[26]
SiC	460	460	3.7	72			[4]
Al_2O_3			4.5	380			[10]

A typical fracture-control plan includes the following:

- The applied stress intensity factor must be $K_I < K_{IC}$ so that it can be used as a constraint to assure structural integrity since crack propagation is restricted.
- The inequality $B \geq 2.5(K_{IC}/\sigma_{ys})^2$ for brittle materials assures that the thickness of designed parts do not fall below a minimum thickness.
- If use of welding is necessary, then it must be used very cautiously since it can degrade the toughness of the welded material, especially in the heat-affected zone (HAZ) which may become brittle as a consequence of rapid cooling leading to smaller grains.
- Control environment to avoid degradation of the structure due to stress corrosion cracks (SCC).
- Limitations of the allowable crack size can be predicted by Eq. (3.57).
- Use of nondestructive test (NDT) techniques must be employed in order to detect flaws or cracks.

3.6 Significance of Thickness

It is clearly shown in Fig. 3.8 how fracture toughness is strongly dependent on the material thickness up to a limiting value. For a thin plate, plane stress condition ($\sigma_z = 0$) governs the fracture process because the plate is too thin to sustain through-thickness stress. For a thick plate, plane-strain condition ($\sigma_z \neq 0$) prevails in which K_{IC} becomes a material's property. It is this property, K_{IC}, that the designer must use to assure structural integrity. The characteristics of a fracture surface, as schematically indicated in Fig. 3.8, vary between both plane stress and plane-strain modes of fracture. The former fracture mode shows a slant fracture (shear lips at approximately 45°) as an indication of partial ductile fracture, and the latter exhibits a flat fracture surface as a representation of brittle fracture. Any combination of these modes of fracture leads to a mixed-mode fracture surface.

In addition, plane stress fracture toughness (K_C) is related to metallurgical features and specimen geometry, and plane-strain fracture toughness (K_{IC}) depends only on metallurgical features, strain rate, and temperature.

The effect of plastic zone size (r) to plate thickness and macroscopic fracture surface appearance is also taken into account. For instance, plane stress state is associated with a maximum toughness and slant fracture, and plane-strain state is related to a minimum toughness and flat fracture. Therefore, plane stress or plane-strain condition depends on σ_{ys}.

Since the plane-strain fracture toughness K_{IC} is a property for a given material, the applied stress level exhibits a dependency on the crack size. This is schematically shown in Fig. 3.9 for two hypothetical materials. Notice that both K_{IC} and a_c influence the stress and the curve is shifted upward at higher K_{IC} level. This means that ductility also has a major influence on the stress and K_{IC}. Let curves **A** and **B** represent failure trends at two different conditions. The interpretation of Fig. 3.9 indicates that there exists an initial crack size (a_o) for material **A** or **B**.

This fracture phenomenon proceeds in a stable manner, provided that both crack size and applied stress are within a controllable range such as $a_o < a < a_c$ and

Fig. 3.8 Effect of specimen thickness on fracture toughness [17]

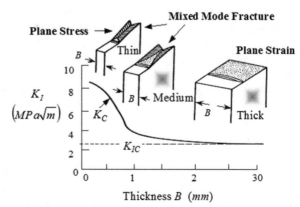

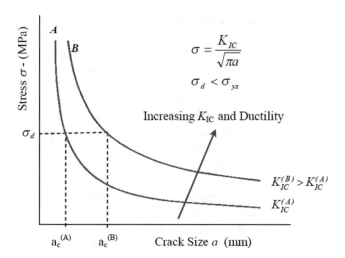

Fig. 3.9 Influence of crack length on fracture stress

$\sigma_d < \sigma < \sigma_{ys}$. Furthermore, crack instability occurs when $a \rightarrow a_c$ and $\sigma \rightarrow \sigma_d$. This schematic representation of crack growth applies to both **A** and **B** curves, which represent the fracture behavior of two hypothetical materials.

In fact, the higher K_{IC}, the greater a_c since the resistance to fracture is controlled by the level of the plane-strain fracture toughness. Hence, for the hypothetical materials included in Fig. 3.9, $a_c(B) > a_c(A)$ and $K_{IC}(B) > K_{IC}(A)$ at $\sigma = \sigma_d$. This implies that material **A** allows a smaller crack extension than material **B**. Although material **B** is the most attractive for engineering applications, its mechanical behavior can be significantly affected by changes in temperature and a corrosive environment. In addition, the solid curves represent ideal elastic behavior, but most materials exhibit plasticity due to the yielding phenomenon.

In general, fracture toughness is one of the most important properties of a solid material for design applications. Despite that plane-strain fracture toughness of a material in the form of a plate depends on the plate thickness, it is also influenced by microstructural features, strain rate, environment, and temperature. Of specific interest is the microstructure of a polycrystalline material since it can be manipulated by using a heat treatment procedure in which the heat treatment temperature, heat treatment time, and cooling medium play a very important role in obtaining a specific microstructure and its corresponding fracture toughness.

3.7   Fracture Tests

The reader should consult the American Society for Testing and Materials (ASTM) test methods for precise details on procedures and guidelines to prepare metallic specimens with well-defined cracks. The goal is to conduct experiments that provide

Fig. 3.10 Load-line displacement for plane-strain K_{IC} test

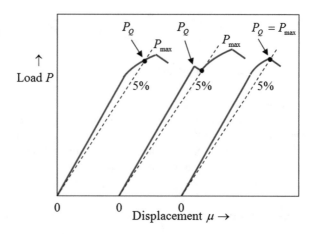

repeatable fracture toughness data having a small data scatter. Briefly, fracture toughness is a material property that is determined when initial crack growth occurs at a specific quasi-static or dynamic loading rate in a particular environment.

Several standard testing methods are available nowadays for determining fracture toughness of cracked metallic materials. Among them, ASTM E399 and ASTM E1820 require that all specimens containing notches be sharpened with fatigue cracks.

3.7.1 ASTM E399 Plane-Strain K_{IC} Test

The reader should consult the ASTM E399 Standard Test Method for precise details. Figure 3.10 schematically displays three typical load-displacement curves used to determine a valid K_{IC} value. The tests are conducted at slow loading rates on specimens having rather strict crack length to width ratios in the range of $0.45 \leq a/w \leq 0.55$. The specimens must exhibit brittle behavior for which the plastic zone, if formed, must be very small compared with the crack length; that is, $r \ll a$.

The procedure calls for the following steps:

- Machine a C(T), SE(B), or disk-shaped compact DC(T) specimen with a suitable notch
- Sharpen the specimen notch with a fatigue crack. Also, $0.45 \leq a/w \leq 0.55$
- Load the specimen at convenient load rate to obtain a load-displacement curve (Fig. 3.10)
- Draw a 5 % secant line (dash line) and determine the value for P_Q and P_{max}.
- If $P_{max}/P_Q < 1.1$, then calculate $K_Q = \alpha \sigma \sqrt{\pi a}$ where $\sigma = P_Q/A_o$. Otherwise, the test is not valid.

- If $a, b, B \geq 2.5 \left(K_Q / \sigma_{ys} \right)^2$, then $K_Q = K_{IC}$; otherwise the test is not valid. Here, $b = w - a$ is the specimen ligament size.

Fracture mechanics testing techniques are based on the ASTM standard test procedures for evaluating the effects of specimen geometry, crack configuration, microstructural features, environment, and metallurgical variables, such as heat treatment time at temperature T or temperature at time t. Among several specimen geometries, the compact tension C(T) specimen prevails. In fracture testing in the field of corrosion, the material specimen geometry plays an important factor in order to evaluate the stress corrosion cracking (SCC) mechanism using suitable specimen geometries, but the most important factors are the loading mode and environment.

Example 3.10. *The load-displacement curve to determine K_{IC} for an annealed, quenched, and tempered compact tension C(T) specimen made out of a hypothetical steel having $\sigma_{ys} = 900\,MPa$, $\nu = 1/3$, and $E = 207\,GPa$. Assume that during fatigue cracking the machine notch produced a very sharp crack prior to testing the specimen. The total original crack length, including the notch depth and the fatigue crack, is 29 mm. The specimen dimensions are $B = 30\,mm$, $w = 60\,mm$, and $h/w = 0.6$. Find K_{IC}.*

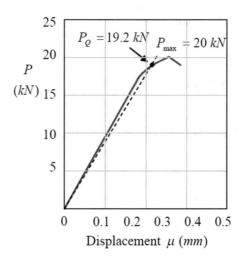

Solution. *Specimen dimensions:* $0.45 \leq a/w \leq 0.55$, $b = w - a = 31\,mm$, $B/w = 0.52$ and $h = 0.6w = 36\,mm$

From the given figure,

$$\frac{P_{max}}{P_Q} = \frac{20}{19.2} = 1.04$$

$$\frac{P_{max}}{P_Q} < 1.1$$

Proceed the computation since $P_{max}/P_Q < 1.1$. The applied stress is

$$\sigma = \frac{P_Q}{A_o} = \frac{19.2 \times 10^{-3} \text{ MN}}{(30 \times 10^{-3} \text{ m}) (60 \times 10^{-3} \text{ m})}$$

$$\sigma = 10.67 \text{ MPa}$$

From Table 3.1 and $x = a/w = 29/60 = 0.48333$, the finite specimen geometry correction factor is

$$\alpha = \frac{(2+x)}{(1-x)^{3/2}} \left(0.886 + 4.64x - 13.32x^2 + 14.72x^3 - 5.6x^4\right)$$

$$\alpha = 9.1837$$

Then, the possible fracture toughness becomes

$$K_Q = \alpha\sigma\sqrt{\pi a} = (9.1837)(10.67 \text{ MPa})\sqrt{\pi (29 \times 10^{-3} \text{ m})}$$

$$K_Q = 29.58 \text{ MPa}\sqrt{\text{m}}$$

The size requirements, including $b = w - a$, are

$$(a, b, B)_{ASTM} \geq 2.5 \left(K_Q/\sigma_{ys}\right)^2 = 2.5 \left(\frac{29.58 \text{ MPa}\sqrt{\text{m}}}{900 \text{ MPa}}\right)^2$$

$$(a, b, B)_{ASTM} \geq 2.70 \text{ mm } (minimum \ requirement)$$

$$Original: \ a = 29 \text{ mm}, \ b = 31 \text{ mm } and \ B = 30 \text{ mm}$$

Finally, the plane-strain fracture toughness is

$$K_{IC} = K_Q = 29.58 \text{ MPa}\sqrt{\text{m}}$$

since both ASTM E399 requirements, $P_{max}/P_Q < 1.1$ and $a, b, B \geq 2.70 \text{ mm}$, are met; therefore, the test is valid. Details on the procedure for obtaining a valid fracture mechanics test using different specimen geometries can be found in a corresponding ASTM standard book. Hence, fracture toughness testing on solid specimens can be evaluated using linear-elastic fracture mechanics (LEFM) as well as the elastic-plastic fracture mechanics (EPFM). This implies that fracture toughness is a generic property that measures the resistance to stable crack growth.

3.7.2 ASTM E1820 K_I, J_I, and $CTOD$ Tests

This is a standard that includes tension (mode I) test procedures for determining linear fracture toughness (K_{IC}), the elastic-plastic fracture toughness (J_{IC}) in terms

Fig. 3.11 Schematic crack
length and load-displacement
curves

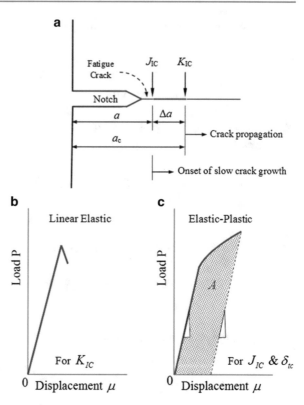

of the J-integral , and the crack tip opening displacement ($\delta_{tc} = CTOD$) of metallic
materials. This test method is less strict than ASTM E399 standard test method since
it allows a wider a/w range, that is, $0.45 \leq a/w \leq 0.70$, and provides means to test
a particular specimen geometry without knowing the type of test needed. Details on
J_{IC} and δ_c fracture mechanics theories will be dealt with in later chapters; however,
it is convenient at this moment to schematically show the crack lengths used for
determining J_{IC} and K_{IC}. This is shown in Fig. 3.11a.

Figures 3.11b and c schematically show two possible load-displacement (P-μ)
curves [5]. Denote that Fig. 3.11b is for a K_{IC} test because the load-line displacement
curve (P-μ) or plot shows a linear behavior. On the other hand, Fig. 3.11c is for J_{IC}
and δ_c tests due to the partial P-μ nonlinearity, which is the graphical representation
of plasticity ahead of the crack tip at a macroscale.

In addition, measuring fracture toughness of elastic-plastic materials requires a
J resistance curve; $J = f(\Delta a)$. This can be done by measuring the crack length
a_i and, subsequently, computing the J-integral J_i and the crack extension $\Delta a_i =
a_i - a_o$, where $a = a_o$ is the original crack length. Figure 3.12 shows a schematic J_I
resistance curve as per ASTM E1820 method for mode I. The ultimate goal of this
procedure is to determine the critical value of the J-integral (J_{IC}) as a measure of
fracture toughness when the ASTM E399 method gives unacceptable results.

Fig. 3.12 Schematic
J-resistance curve for
measurement of fracture
toughness as per ASTM
E1820 Standard Test Method
for mode I loading

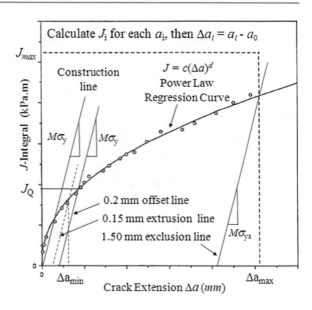

For a single specimen technique under mode I loading, the construction line in
Fig. 3.12 is given by

$$J_I = M\sigma_y \Delta a \tag{3.58}$$

$$\sigma_y = 0.5 \left(\sigma_{ys} + \sigma_{ts}\right) \tag{3.59}$$

where σ_y = Flow stress or effective yield strength
σ_{ys} = Yield strength
σ_{ts} = Ultimate tensile strength

Then, construct the other straight lines with the same slope ($M\sigma_y$) as shown in
Fig. 3.12. Here, σ_y is the flow stress and usually $M = 2$. The term J_{max} is the largest
of the values given by

$$J_{max} = b\sigma_y/10 \tag{3.60}$$

$$J_{max} = B\sigma_y/10 \tag{3.61}$$

and calculate the maximum crack extension as

$$\Delta a_{max} = 0.25b \tag{3.62}$$

where $b = w - a$ = Ligament, B = Thickness and w = Width.

Once J_{max} and Δa_{max} are determined, curve fitting is done on the data for
$J_I = f(\Delta a)$ between Δa_{min} and Δa_{max} using a power-law equation such as

$$J_I = C_1 (\Delta a)^{C_2} \tag{3.63}$$

where C_1 and C_2 are nonlinear regression constants. Then, graphically locate the value of the J_Q-integral on Fig. 3.12 and compute the size requirements defined by

$$B, b \geq \frac{25 J_Q}{\sigma_y} \tag{3.64}$$

If the size requirements are satisfied, assign the value of J_Q to J_{IC}; that is, $J_Q = J_{IC}$. Explicitly, J_{IC} represents the total J-integral for the onset of crack growth, and it is the sum of the elastic (J_{Ie}) and plastic (J_{Ip}) J-integral components. According to the ASTM E1820 "Calculations for the Basic Test Method," the generalized total J-integral and its components for mode I quasi-static loading configuration take the form

$$J_I = J_{Ie} + J_{Ip} \tag{3.65}$$

$$J_{Ie} = \frac{K_I^2}{E'} \tag{3.66}$$

$$J_{Ip} = \frac{\eta A}{bB} \tag{3.67}$$

Here, B is the specimen thickness of a rectangular plate, and A is the plastic area under the load-displacement curve shown in Fig. 3.11b. If a load-line displacement record is used, then $\eta = 1.9$, and for a crack mouth opening displacement record, η takes the form

$$\eta = 3.667 - 2.199\,(a/w) + 0.437\,(a/w)^2 \quad [\text{SE(B)}] \tag{3.68}$$

$$\eta = 2 + 0.522\,(b/w) \quad \text{C(T) and DC(T)} \tag{3.69}$$

where [SE(B)] = Single-edge bend specimen [14]

    C(T) = Compact Tension specimen

    DC(T) = Disk-shaped compact specimen

The J-integral is an energy term for assessing fracture toughness, especially, of brittle materials, but it can also be applied to evaluate cracked ductile solids. The latter topic is dealt with in a later chapter. Nonetheless, the J-integral is determined in accordance with the J-controlled or J-dominance approach for considering theoretical constraints ahead of the crack tip.

The general resultant mathematical treatment for determining the total J-integral, Eq. (3.65), includes the correction factor due to plasticity in front of the crack tip as indicated in Eq. (3.67) through the n factor.

Fig. 3.13 Crack plane
identification for plastically
deformed plates as per ASTM
E399 standard [3]

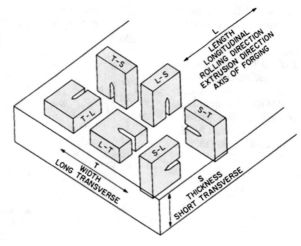

3.7.3 Crack Plane Identification

Figure 3.13 schematically shows specimens having mechanically induced preferred orientation of cracks in plastically deformed plates [3]. A two-letter code is used for crack plane identification. The first letter designates the loading direction and the second letter represents the crack plane. The reader should consult the ASTM Designation E399 for crack plane identification in solid cylinders.

For isotropic polycrystalline materials (solids that are composed of many grains of varying size and orientation) being evaluated at a macroscale, K_{IC} or J_{IC} becomes independent of crack plane orientation. Thus,

$$K_{IC}(T\text{-}L) \simeq K_{IC}(L\text{-}T) \simeq K_{IC}(S\text{-}L) \tag{3.70}$$

For anisotropic materials, the identification of the crack plane in plates is very important due to variations in mechanical properties. For instance, the inhomogeneity of plastic deformation is due to grain deformation and microstructural defects (dislocations, voids, and the like). For instance, cold rolled plates exhibit anisotropy in mechanical properties, such as the plane-strain fracture toughness (K_{IC}), the critical J-integral (J_{IC}), yield strength (σ_{ys}), and other mechanical properties when determined per crack plane orientation. Thus,

$$K_{IC}(T\text{-}L) \neq K_{IC}(L\text{-}T) \neq K_{IC}(S\text{-}L) \tag{3.71}$$

Therefore, material properties depend on the loading direction normal to the crack plane. Moreover, most single crystals exhibit anisotropy in mechanical properties according to the direction along which they are measured. In general, if anisotropy is a problem in selecting a material for a particular application, then heat treatment becomes a suitable option to make an anisotropic plate into an

Fig. 3.14 Experimental
$J - \Delta a$ curves for ASTM
A285 Grade B low-carbon
steel tested at 60°C in air.
L-T and T-L compact tension
C(T) specimens with B =
0.475 in. were used as per
ASTM E1820 Standard Test
Method. Author's specimen
identification codes: AD-5,
QS and AD-9, QS [35]

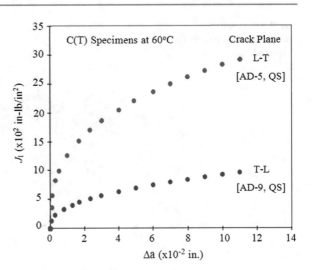

isotropic material due to atomic diffusion at relatively high temperature. The final microstructural condition in a plate depends on the cooling method being used. Unfortunately, this topic is out of the scope of this section.

In addition, the characterization of a solid material is usually done by developing resistance curves or R-curves, and if the material is anisotropic, then properties become dependent on the crack plane orientation. For instance, Fig. 3.14 shows experimental data for two crack plane orientations, L-T and T-L, in C(T) specimens made out of ASTM A285 Grade B [35]. Denote that this particular J-Δa experimental data sets exhibit, to an extent, a power-law behavior. The significance of the J-Δa curves in Fig. 3.14 is that L-T specimens show higher J-integral values than the T-L configuration; therefore, the crack plane orientation has a strong effect on fracture toughness of anisotropic solid materials.

Another R-curve is shown in Fig. 3.15 for an ASTM A588 Grade A steel plate being calcium treated (CaT) containing 0.003 % sulfur level, normalized at 899 °C, and air cooled [40]. This solid material clearly exhibits crack orientation dependency. In addition, Fig. 3.16 shows fracture toughness experimental data for the ALCOA 5xxx series wrought aluminum alloy, C557 ALCOA Al-Mn-Mg-Sc alloy. This Al-alloy exhibits slight variations on plane-strain fracture toughness K_{JIC} at the anticipated service temperatures using the T-L and S-L crack plane orientations [13]. Denote that K_{JIC} in Fig. 3.16 was determined using the J-integral approach [13] on T-L and S-L specimens.

It is clear from Figs. 3.14 through 3.16 that fracture toughness (J_I or K_I) is sensitive to crack plane orientation, specifically in cold worked polycrystalline plates. This can be attributed to the microstructure with preferred orientation of crystallographic planes. Therefore, it is significant to document the crack plane orientation designations in fracture toughness measurements relative the specimen configuration (geometry) [1].

Fig. 3.15 Variation in K_{JIC} of C557 sheet with crack plane and test temperature. The J-integral was used to determine the plane strain fracture toughness K_{JIC} [13]

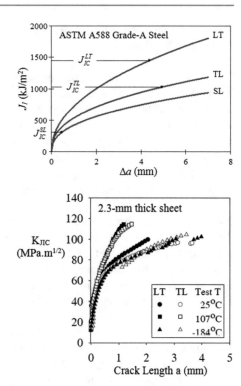

Fig. 3.16 Variation in K_{JIC} of C557 sheet with crack plane and test temperature [13]

Anyway, the crack plane orientation should be identified because of the anisotropy effects on material's properties. Conversely, isotropic materials are considered to have microstructural homogeneity when tested at a macroscale for determining the mechanical properties in a particular environment.

Furthermore, the ASTM E399 standard has sketches of crack plane orientation code for rectangular and for bar and hollow cylinder sections. Also available is the ASTM method B645 for fracture toughness testing of aluminum alloys.

3.8 Problems

3.1. A steel strap 1-mm thick and 20-mm wide with a through-thickness center crack 4-mm long is loaded to failure. **(a)** Determine the critical load if $K_{IC} = 80\,\text{MPa}\sqrt{\text{m}}$ for the strap material. **(b)** Use an available correction factor, $\alpha = f(a/w)$, for this crack configuration and calculate the critical stress as $\sigma_c = (Fraction)\sigma_\infty$.

3.2. A steel tension bar 8-mm thick and 50-mm wide with an initial single-edge crack of 10-mm long is subjected to an uniaxial stress $\sigma = 140\,\text{MPa}$. Determine **(a)** the stress intensity factor K_I. Is the crack stable? Calculate **(b)** the critical crack size and **(c)** the critical load. Data: $K_{IC} = 60\,\text{MPa}\sqrt{\text{m}}$. [Solution: (a) 34 MPa, (b) 31.1 mm, and (c) 98.84 kN].

3.3. A very sharp penny-shaped crack with a diameter of 22-mm is completely embedded in a highly brittle solid. Assume that catastrophic fracture occurs when a stress of 600 MPa is applied. **(a)** What is the fracture toughness for this solid? Assume that this fracture toughness is for plane-strain conditions. **(b)** If a sheet 5-mm-thick plate is prepared for fracture toughness testing, would the fracture toughness value calculated in part **(a)** be an acceptable number according to the ASTM E399 standard? Use $\sigma_{ys} = 1342\,\text{MPa}$. **(c)** What thickness would be required for the fracture toughness test to be valid?

3.4. A guitar steel string has a miniature circumferential crack of 0.009 mm deep. This implies that the radius ratio is almost unity, $d/D \simeq 1$. Another string has a localized miniature surface crack (single-edge crack like) of 0.009 mm deep. Assume that both strings are identical with an outer diameter of 0.28 mm. If a load of 49 N is applied to the string when being tuned, will it break? Given data: $K_{IC} = 15\,\text{MPa}\sqrt{\text{m}}$ and $\sigma_{ys} = 795\,\text{MPa}$.

3.5. A 7075-*T*6 aluminum alloy is loaded in tension. Initially the 10-mm-thick, 100-mm-wide, and 500-mm-long plate has a 4-mm single-edge through-thickness crack. **(a)** Is this a valid test? **(b)** Calculate the maximum allowable tension stress this plate can support. **(c)** Is it necessary to correct K_I due to crack tip plasticity? Why or why not? **(d)** Calculate the design stress and then stress intensity factor if the safety factor is 1.5. Use the following properties: $\sigma_{ys} = 586\,\text{MPa}$ and $K_{IC} = 33\,\text{MPa}\sqrt{\text{m}}$. [Solution: (a) It is a valid test because $B_{ASTM} < B_{actual}$, (b) $\sigma = 261\,\text{MPa}$, (c) It is not necessary because $K_I \simeq K_{IC}$, and (d) $\sigma_d = 174\,\text{MPa}$ and $K_{I,d} = 22\,\text{MPa}\sqrt{\text{m}}$].

3.6. A steel plate (30-mm thick, 1.2-m wide, and 2.5-m long) is under tension. It is operated below its ductile-to-brittle transition temperature (with $K_{IC} = 28.3\,\text{MPa}\sqrt{\text{m}}$). If a 65-mm long through the thickness central crack is present, calculate **(a)** the tensile stress for catastrophic failure. Compare this stress with the yield strength of 240 MPa. **(b)** Determine the safety factor.

3.7. Show that the following inequality $dK_I/da > 0$ is valid for crack instability in a large plate under a remote external tension stress.

3.8. The plate below has an internal crack subjected to a pressure P on the crack surface. The stress intensity factors at points A and B are

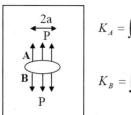

$$K_A = \int \frac{P}{\sqrt{\pi a}} \sqrt{\frac{a+x}{a-x}} \cdot dx$$

$$K_B = \int \frac{P}{\sqrt{\pi a}} \sqrt{\frac{a-x}{a+x}} \cdot dx$$

Use the principle of superposition to show that the total stress intensity factor is defined by $K_I = P\sqrt{\pi a}$.

3.9. A pressure vessel is to be designed using the **leak-before-break criterion** based on the circumferential wall stress and plane-strain fracture toughness. The design stress is restricted by the yield strength σ_{ys} and a safety factor (S_F). Derive expressions for **(a)** the critical crack size and **(b)** the maximum allowable pressure when the crack size is equal to the vessel thickness.

3.10. A stock of steel plates with $G_{IC} = 130\,\text{kJ/m}^2$, $\sigma_{ys} = 2200\,\text{MPa}$, $E = 207\,\text{GPa}$, $v = 1/3$ are used to fabricate a cylindrical pressure vessel ($d_i = 5\,\text{m}$ and $B = 25.4\,\text{mm}$). The vessel fractured at a pressure of 20 MPa. Subsequent failure analysis revealed an internal semielliptical surface crack of $a = 2.5\,\text{mm}$ and $2c = 10\,\text{mm}$. **(a)** Use a fracture mechanics approach to predict the critical crack length this steel would tolerate. **(b)** Based on this catastrophic failure, another vessel was constructed with $d_o = 5.5\,\text{m}$ and $d_i = 5\,\text{m}$. Will this new vessel fracture at a pressure of 20 MPa if there is an internal semielliptical surface crack having the same dimensions as in part **(a)**?

3.11. A cylindrical pressure vessel with $B = 25.4\,\text{mm}$ and $d_i = 800\,\text{mm}$ is subjected to an internal pressure P_i. The material has $K_{IC} = 31\,\text{MPa}\sqrt{\text{m}}$ and $\sigma_{ys} = 600\,\text{MPa}$. **(a)** Use a safety factor to determined the actual pressure P_i. **(b)** Assume that there exists an internal semielliptical surface crack with $a = 5\,\text{mm}$ and $2c = 25\,\text{mm}$ and that a pressure surge occurs causing fracture of the vessel. Calculate the fracture internal pressure P_f and **(c)** the critical crack length.

3.12. This is a problem that involves strength of materials and fracture mechanics. An AISI 4340 steel is used to design a cylindrical pressure having an inside diameter and an outside diameter of 6.35 cm and 12.07 cm, respectively. The hoop stress is not to exceed 80 % of the yield strength of the material. **(a)** Is the structure a thin-wall vessel or a thick-wall pipe? **(b)** What is the internal pressure? **(c)** Assume that an internal semielliptical surface crack exist with $a = 2\,\text{mm}$ and $2c = 6\,\text{mm}$. Will the vessel fail? **(d)** Will you recommend another steel for the pressure vessel? Why? Or Why not? **(e)** What is the maximum crack length AISI 4340 steel can tolerate? Explain.

3.13. A simple supported beam made of soda glass ($E = 71$ MPa and $G_{IC} = 12$ J/m^2) is subjected to a bending load as shown in the figure below. Assume an initial crack length of 0.1 mm due to either stress corrosion cracking (SCC) or a mechanical defect introduced during fabrication or handling. The design stress (working stress) and the crack velocity equation are 0.10 MPa and $v = \beta K_I^m = (6.24$ m/s$) K^{15.83}$, respectively. Use this information to calculate the lifetime of the beam if the maximum bending stress is given by $\sigma = (MB/2)/I$ and $M = \sigma_d L^2/8$, where the second moment of area is $I = wB^3/12$.

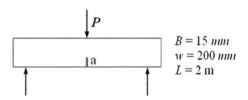

$$P$$

$$B = 15 \; mm$$
$$w = 200 \; mm$$
$$L = 2 \; m$$

3.14. Plot the given data for a hypothetical solid SE(T) specimen and determine the plane-strain fracture toughness K_{IC}. Here, $\alpha = f(a/w)\sqrt{\pi}$ is the modified geometry correction factor.

$a^{-1/2}$ (m$^{-1/2}$)	6	9.8	15	18	22	25	35	38
$\alpha\sigma$ (MPa)	30	50	70	90	116	122	175	190

3.15. A pressure vessel ($L \gg B = 15$ mm, $d_i = 2$ m) is to be made out of a weldable steel alloy having $\sigma_{ys} = 1200$ MPa and $K_{IC} = 85$ MPa. If an embedded elliptical crack ($2a = 5$ mm and $2c = 16$ mm) as shown below is perpendicular to the hoop stress, due to welding defects, the given data correspond to the operating room temperature, and the operating pressure is 8 MPa; then calculate the applied stress intensity factor. Will the pressure vessel explode?

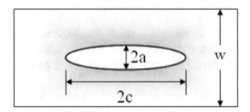

$$2a$$

$$w$$

$$2c$$

3.16. A steel alloy pressure vessel subjected to a constant internal pressure of 20 MPa contains an internal semielliptical surface crack with dimensions shown below. Calculate K_I using **(a)** Q and σ/σ_{ys} as per Eq. (3.45) and **(b)** $Q = 1 + 1.464\,(a/c)^{1.65}$ as per reference [30]. **(c)** Compare results and explain. **(d)** If fracture does not occur, calculate the safe factors $S_F^{(a)}$ and $S_F^{(b)}$. Explain the results. Data: $K_{IC} = 92$ MPa$\sqrt{m}$, $\sigma_{ys} = 900$ MPa, $v = 0.30$, and $B = 30$ mm.

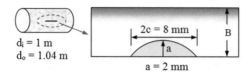

3.17. An aluminum alloy plate has a plane-strain fracture toughness of $30\,\text{MPa}\sqrt{\text{m}}$. Two identical single-edge cracked specimens are subjected to tension loading. **(a)** One specimen having a 2-mm crack fractures at a stress level of 330 MPa. Calculate the geometry correction factor $\alpha = f(a/w)$. **(b)** Will the second specimen having a 1-mm crack fracture when loaded at 430 MPa? [Solution: (a) $\alpha = 1.1469$, (b) $K_I = 27.64\,\text{MPa}\sqrt{\text{m}}$].

References

1. T.L. Anderson, *Fracture Mechanics: Fundamentals and Applications* (Taylor & Francis Group, LLC/ CRC Press, Boston, 1991), pp. 301–303
2. Applicability of the Leak Before Break Concept, Report of the IAEA Extrabudgetary Programme on the Safety of WWER-440 Model 230 Nuclear Power Plants, International Atomic Energy Agency (IAEA), IAEA-TECDOC-710 (June 1993), pp. 1011–4289. http://www-pub. iaea.org/MTCD/Publications/PDF/te_710_web.pdf [ISSN 1011–4289]
3. ASTM Designation: E 399–09, Standard test method for linear-elastic plane-strain fracture toughness K_{IC} of metallic materials (2009)
4. M.W. Barsoum, *Fundamentals of Ceramics* (McGraw-Hill, New York, 1997), p. 401
5. J.M. Barsom, S.T. Rolfe, *Fracture and Fatigue Control in Structures: Application of Fracture Mechanics*, 3rd edn. (Butterworth Heinemann, Woburn, 1999)
6. J.M. Barsom, S.T. Rolfe, *Fracture and Fatigue in Structure: Application of fracture Mechanics*, 3rd edn. American Society for Testing and Materials Philadelphia, PA, (1999)
7. G.S. Bhuyan, *Leak-Before-Break and Fatigue Crack Growth Analysis of All-Steel On-Board Natural Gas Cylinders*. ASTM STP 1189-EB (1993)
8. D. Broek, *Elementary Engineering Fracture Mechanics*, 4th edn. (Kluwer Academic Publisher, Boston, 1986)
9. W.F. Brown Jr., J.E. Strawley, *Plane Strain Crack Toughness Testing of High Strength Metallic Materials*. ASTM STP, vol. 410 (ASTM, Philadelphia, 1966), pp. 133–198
10. Y.M. Chiang et al., *Physical Ceramics: Principles for Ceramic Science and Engineering* (Wiley, New York, 1997)
11. J.M. Dahl, P.M. Novotny, Airframe and landing gear alloy. Adv. Mater. Process **155**(3), 23–25 (1999)
12. J.W. Dally, W.F. Riley, *Experimental Stress Analysis*, 3rd edn. (McGraw-Hill, New York, 1991)
13. M.S. Domack, D.L. Dicus, *Evaluation of Sc-Bearing Aluminum Alloy C557 for Aerospace Applications*. NASA/TM-2002-211633, National Aeronautics and Space Administration, Washington, DC 20546-0001, April 2002, pp. 1–14
14. H.A. Ernst, P.C. Paris, J.D. Landes, Estimations of the j-integral and tearing modulus t from a single test record, in *Fracture Mechanics, in Thirteenth Conference*, ed. by R. Roberts. ASTM STP, vol. 743 (ASTM, New York, 1981), pp. 476–502
15. V.M. Faires, *Design of Machine Elements* (The Macmillan Company, New York, 1965), pp. 34–35
16. C.E. Feddersen, Discussion, ASTM STP 410, (1967), pp. 77–79
17. H.O. Fuchs, R.I. Stephens, *Metal Fatigue in Engineering* (Wiley, New York, 1980)

18. R.W. Hertzberg, *Deformation and Fracture Mechanics of Engineering Materials*, 3rd edn. (Wiley, New York, 1989)
19. C.E. Inglis, Stresses in a plate due to the presence of cracks and sharp corners. Trans. Inst. Nav. Arch. **55**, 219–241 (1913)
20. G.R. Irwin, in *Fracture I* ed. by S. Flugge. Handbuch der physik, vol. VI (Springer, New York, 1958), pp. 558–590
21. G.R. Irwin, The crack extension force for a part-through crack in a plate. Trans. ASME J. Appl. Mech. **29**(4), 651–654 (1962)
22. G.R. Irwin, J.M. Krafft, P.C. Paris, A.A. Wells, Basic concepts of crack growth and fracture. NRL Report 6598, Naval Research Laboratory, Washington, DC, Nov 21, 1967
23. A.S. Kobayashi, Fracture mechanics, in *Experimental Techniques in Fracture Mechanics*, ed. by A.S. Kobayashi. Society for experimental Stress Analysis, SESA Monograph, vol. 1 (The Iowa State University Press, Ames, 1973)
24. A.S. Kobayashi, M. Zii, L.R. Hall, Approximate stress intensity factor for an embedded elliptical crack near to parallel free surfaces. Int. J. Fract. Mech. **1**, 81–95 (1965)
25. W.T. Koiter J.P. Benthem, in *Mechanics of Fracture*, ed. by G.C. Sih, vol. 1 (Noordhoff, Leyden, 1973)
26. J.C.M. Li, Mechanical properties of amorphous metals and alloys, in *Treatise on Materials Science and Technology. Ultrarapid Quenching of Liquid Alloys*, vol. 20, ed. by H. Herman (Academic, New York, 1981), pp. 326–360
27. M.A. Meyers, K.K. Chawla, *Mechanical Metallurgy Principles and Applications*, vol. 161 (Prentice-Hall, Englewood Cliffs, 1984), pp. 452–454
28. R.K. Nalla, J.J. Kruzic, J.H. Kinney, R.O. Ritchie, Mechanistic aspects of fracture and R-curve behavior in human cortical bone. Biomaterials **26**, 217–231 (2004)
29. R.K. Nalla, J.J. Kruzic, J.H. Kinney, R.O. Ritchie, Aspects of in vitro in human cortical bone: time and cycle dependent crack growth. Biomaterials **26**, 2183–2195 (2005)
30. J. Newman Jr., I.S> Raju, An empirical stress-intensity factor equation for the surface crack. Eng. Fract. Mech. **15**(1–2), 185–192 (1981)
31. C.P. Paris, G.C. Sih, Stress analysis of cracks, in *Fracture Toughness Testing and its Applications*. ASTM STP, vol. 381 (ASTM, Philadelphia, 1965), pp. 30–83
32. R.C. Shah, ASTM STP 560 (1971), p. 29
33. G.C. Sih, *Handbook of Stress-Intensity Factors*. Institute of Fracture and Solid Mechanics, Lehigh University, George C. Sih, Bethlehem, Pennsylvania (1973)
34. I.N. Sneddon, The distribution of stress in neighborhood of a crack in a elastic solid. Proc. R. Soc. Lond. A **187**, 229–260 (1946)
35. K.H. Subramanian, A.J. Duncan, R.L. Sindelar, Report WSRC-MS-2002-00259, Contract No. DE-AC09-96SR18500 with the U.S. Department of Energy (2002), pp. 1–7
36. W.F. Brown Jr., J.E. Strawley, *Plane Strain Crack Toughness Testing of High Strength Metallic Materials*. ASTM STP, vol. 410 (ASTM, Philadelphia, 1966), pp. 133–198
37. H. Tada, P.C. Paris, G.R. Irwin, *The Stress Analysis of Cracks Handbook*, 3rd edn. (ASME Press, New York, 2000)
38. T.E. Tietz, I.G. Palmer, ASM materials science seminar, in *Advances in Power Metallurgy*, vol. 189 (1981)
39. R.L. Tobler, Low temperature effects on the fracture behavior of a nickel base superalloy. Cryogenics **16**(11), 669–674 (1976)
40. A.D. Wilson, Fracture mechanics, in *Fifteenth Symposium, ASTM STP* ed. by R.J. Sanford, vol. 833, Philadelphia (1984), pp. 412–435
41. S.X. Wu, Fracture toughness determination of bearing steel using chevron-notch three point bend specimen. Eng. Fract. Mech. **19**, 221–232 (1984)
42. S.X. Wu, *Chevron-Notched Specimens: Testing and Analysis*, ed. by J.H. Underwood, S.W. Freiman, F.I. Baratta, vol. 176 (ASTM, Philadelphia, 1984)
43. G.R. Yoder et al., ASTM STP 801 (1983), p. 159

Linear-Elastic Field Equations

4

4.1 Introduction

Linear-elastic fracture mechanics (LEFM) and a quasi-static load action are considered in this chapter in order to derive the stress, strain, and displacement field equations in two dimensions adjacent to a crack tip of an arbitrarily shaped crack. The field equations can be derived in series form using rectangular and polar coordinates. Any dynamic or local unloading is neglected in the foregoing mathematical procedures for modeling the plastic zone being as a circle with different radius r. It is intended henceforward to demonstrate that the trigonometric functions $f(\theta)$ and $g(\theta)$ have increasing terms as the radius r increases. However, the irreversible action that takes place at the crack tip suggests that a few terms in a series expansion may be needed for characterizing the crack tip field equations, which ought to be manipulated using a symbolic software design for this purpose.

4.2 Field Equations: Mode I

The analytical approach used in this chapter assumes a very small plastic zone, where higher-order stress terms are neglected. It is also assumed that the field equations are valid and exact for $r \to 0$. The aim here is to derive the stress field equations as functions of the stress intensity factor (K_I), the plastic zone size (r), and trigonometric functions $f_{ij}(\theta)$. Recall that $K_I = \sigma \sqrt{\pi a}$ is defined by Eqs. (2.34) and (3.28) for a cracked specimen under tension load (mode I) and in general, $K_I = f(\sigma, a)$. For an uncracked specimen in tension, the applied stress is defined by Eq. (1.2) as $\sigma = P/A$, where P is the applied load, A is the cross-sectional area, and $\sigma = f(P)$. Actually, K_I is limited to its critical value known as fracture toughness K_C for plane stress and K_{IC} for plane-strain conditions, and σ is limited to the yield strength of the material. For an elastic perfectly plastic material, a small plastic zone develops ahead of the crack tip; otherwise, the material would be perfectly elastic or pure brittle such as glass.

© Springer International Publishing Switzerland 2017
N. Perez, *Fracture Mechanics*, DOI 10.1007/978-3-319-24999-5_4

Fig. 4.1 Crack tip at the
center of the circular plastic
zone having a contour C in
the domain D

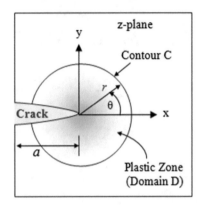

Consider a domain D containing a single-edge crack and a round plastic zone as
depicted in Fig. 4.1. Let the crack tip be located at the center of the plastic zone with
an inclined radius r at an angle θ.

The goal here is to derive the elastic stresses near the crack tip using complex
variable theory. Generalizing Eq. (3.1) in series form along with $A_o = K_I/\sqrt{2\pi}$
yields the Westergaard stress function Z and Z' as [21]

$$Z = \sum_{m=o}^{M} A_m z^{m-1/2} \tag{4.1}$$

$$Z' = \frac{dZ}{dz} = \sum_{m=o}^{M}\left(m - \frac{1}{2}\right) A_m z^{m-3/2} \tag{4.2}$$

For convenience, redefine the complex variable z and Z' using Euler's formula for a
real angle θ and a real coefficient m. Hence,

$$Z = \sum_{m=o}^{M} A_m r^{m-1/2}\left[\cos\left(m - \frac{1}{2}\right)\theta + i\sin\left(m - \frac{1}{2}\right)\theta\right] \tag{4.3}$$

$$Z' = \sum_{m=o}^{M}\left(m - \frac{1}{2}\right) A_m r^{m-3/2}\left[\cos\left(m - \frac{3}{2}\right)\theta + i\sin\left(m - \frac{3}{2}\right)\theta\right] \tag{4.4}$$

The general definitions of the complex variable z and the dependent variable y in
polar coordinates are given below:

$$z = x + iy = re^{i\theta} \tag{4.4}$$

$$e^{\pm im\theta} = \cos m\theta \pm i\sin m\theta \tag{4.5}$$

$$y = r\sin\theta = 2r\sin\frac{\theta}{2}\cos\frac{\theta}{2} \tag{4.6}$$

The real and imaginary parts [consult Eq. (3.9)] for the Airy stress function defined by Eq. (3.6) are

$$\text{Re}\,Z = \sum_{m=o}^{M} A_m r^{m-1/2} \cos\left(m - \frac{1}{2}\right)\theta \qquad (4.7)$$

$$\text{Re}\,Z' = \sum_{m=o}^{M} \left(m - \frac{1}{2}\right) A_m r^{m-3/2} \cos\left(m - \frac{3}{2}\right)\theta \qquad (4.8)$$

$$\text{Im}\,Z = \sum_{m=o}^{M} A_m r^{m-1/2} \sin\left(m - \frac{1}{2}\right)\theta \qquad (4.9)$$

$$\text{Im}\,Z' = \sum_{m=o}^{M} \left(m - \frac{1}{2}\right) A_m r^{m-3/2} \sin\left(m - \frac{3}{2}\right)\theta \qquad (4.10)$$

The stresses in complex form as per Eqs. (1.40), (3.18), or (3.20) are

$$\sigma_x = \text{Re}\,Z - y\,\text{Im}\,Z'$$
$$\sigma_y = \text{Re}\,Z + y\,\text{Im}\,Z' \qquad (4.11)$$
$$\tau_{xy} = -y\,\text{Re}\,Z'$$

Inserting Eq. (4.7) through (4.10) into (4.11) yields the stresses in series form with $0 \le m \le M$

$$\sigma_x = \sum_{m=o}^{M} A_m r^{m-1/2} \cos\left(m - \frac{1}{2}\right)\theta$$

$$\qquad - \sum_{m=o}^{M} 2\left(m - \frac{1}{2}\right) A_m r^{m-1/2} \cos\frac{\theta}{2} \sin\frac{\theta}{2} \sin\left(m - \frac{3}{2}\right)\theta$$

$$\sigma_y = \sum_{m=o}^{M} A_m r^{m-1/2} \cos\left(m - \frac{1}{2}\right)\theta \qquad (4.12)$$

$$\qquad + \sum_{m=o}^{M} 2\left(m - \frac{1}{2}\right) A_m r^{m-1/2} \cos\frac{\theta}{2} \sin\frac{\theta}{2} \sin\left(m - \frac{3}{2}\right)\theta$$

$$\tau_{xy} = - \sum_{m=o}^{M} 2\left(m - \frac{1}{2}\right) A_m r^{m-1/2} \cos\frac{\theta}{2} \sin\frac{\theta}{2} \cos\left(m - \frac{3}{2}\right)\theta$$

(SSY) Assume that a small-scale yielding phenomenon takes place in elastic solids and that the crack can be treated as a semi-infinite defect. Thus, Eq. (4.1) or (4.2) indicates that the number of terms in the series and the amount of experimental data decrease when $M \to 0$ as $r \to 0$.

Let $m = M = 0$ in Eq. (4.12) so that the first-order stress field equations for mode I become

$$\sigma_x = \frac{K_I}{\sqrt{2\pi r}} \cos \frac{\theta}{2} \left(1 - \sin \frac{\theta}{2} \sin \frac{3\theta}{2} \right)$$

$$\sigma_y = \frac{K_I}{\sqrt{2\pi r}} \cos \frac{\theta}{2} \left(1 + \sin \frac{\theta}{2} \sin \frac{3\theta}{2} \right) \tag{4.13}$$

$$\tau_{xy} = \frac{K_I}{\sqrt{2\pi r}} \cos \frac{\theta}{2} \sin \frac{\theta}{2} \cos \frac{3\theta}{2}$$

The stress in the z-direction is of particular interest because it defines plane conditions as express by Eqs. (1.34) and (1.35). Hence, $\sigma_z = 0$ for plane stress and for plane strain

$$\sigma_z = v \left(\sigma_x + \sigma_y \right) \tag{4.14}$$

$$\sigma_z = \frac{2vK_I}{\sqrt{2\pi r}} \cos \frac{\theta}{2} \tag{4.15}$$

Figure 4.2 depicts the trigonometric trends of the functions given by Eqs. (4.13) and (4.15). This is to show the reader how these functions vary with increasing angle θ.

Observe that most curves exhibit relatively nonuniform trends or shape as the angle θ increases. Therefore, these curves show the elastic stress behavior near the crack tip.

Additionally, letting the angle be $\theta = 0$ implies that the crack grows along the x-axis (crack plane) in a self-similar manner and so that only one term is needed for characterizing the crack tip stress field, which in turn, is independent of specimen

Fig. 4.2 Normalized stress distribution for mode I

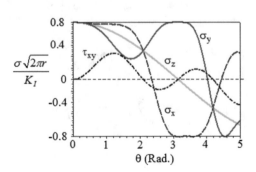

size and geometry. In this case, Eqs. (4.13) and (4.15) for plane-strain conditions in Cartesian coordinates become

$$\sigma_x = \frac{K_I}{\sqrt{2\pi r}} \qquad \text{for } \theta = 0$$

$$\sigma_y = \frac{K_I}{\sqrt{2\pi r}} \qquad \text{for } \theta = 0 \tag{4.16}$$

$$\tau_{xy} = 0 \qquad \text{for } \theta = 0$$

$$\sigma_z = \frac{2\nu K_I}{\sqrt{2\pi r}} \qquad \text{for } \theta = 0$$

For instance, loading a cracked specimen in tension one can deduce from Eq. (4.16) that the stress intensity factor is simply dependent on the stress (σ_y) perpendicular to the crack plane and the plastic zone size (r). Thus,

$$K_I = \sigma_y \sqrt{2\pi r} \qquad \text{for } \theta = 0 \tag{4.16a}$$

If the crack length and the applied stress are $a = 2r$ and $\sigma = \sigma_y$, respectively, then the stress intensity factor takes the same mathematical definition as Eq. (3.28). Hence,

$$K_I = \sigma \sqrt{\pi a} \qquad \text{for } \theta = 0 \tag{4.16b}$$

Substituting Eq. (4.13) into (1.10) and (1.11) and the resultant expressions into (1.4) yields the strain and displacement field equations

$$\begin{bmatrix} \epsilon_x \\ \epsilon_y \\ \gamma_{xy} \end{bmatrix} = \frac{K_I}{\sqrt{2\pi r}} \left\{ \begin{array}{c} \cos \frac{\theta}{2} \left[(1-v) - (1+v) \sin \frac{\theta}{2} \sin \frac{3\theta}{2} \right] \\ \cos \frac{\theta}{2} \left[(1-v) + (1+v) \sin \frac{\theta}{2} \sin \frac{3\theta}{2} \right] \\ \sin \frac{\theta}{2} \cos \frac{3\theta}{2} \end{array} \right\} \tag{4.17}$$

and

$$\begin{bmatrix} \mu_x \\ \mu_y \\ \mu_z \end{bmatrix} = \frac{2K_I}{\sqrt{2\pi} E} \left\{ \begin{array}{c} \sqrt{r} \cos \frac{\theta}{2} \left[(1-v) + (1+vv) \sin^2 \frac{\theta}{2} \right] \\ (1+v)^{-1} \sqrt{r} \sin \frac{\theta}{2} \left[2 - (1+v) \cos^2 \frac{\theta}{2} \right] \\ -\frac{vB}{\sqrt{r}} \cos \frac{\theta}{2} \end{array} \right\} \tag{4.18}$$

Here, Eq. (4.18) indicates that the displacement μ_z is singular since $\mu_z \to \infty$ as $r \to 0$ whereas $\mu_x = \mu_y = 0$. Therefore, μ_z is the only displacement singularity in the order of $r^{-1/2}$. In general, Weertman [50] describes that a crack displacement occurs when the theoretical tensile stress σ_{ts} causes a theoretical separation $y \simeq 2b$, where b is the length of the Burgers vector of a dislocation. If σ_{ts} reaches its theoretical fracture value, then atomic bonds break and the crack faces are no longer subjected to traction forces. The latter condition is referred to as stress-free crack.

Basically, these displacement distributions are just plots of the trigonometric functions, which show the trend of each elastic displacement for an ideal isotropic solid subjected to an external or infinite uniaxial loading . Denote that the displacement μ_x drastically decreases reaching negative values at $\theta \leq 0$, μ_y fluctuates, and μ_z increases very slightly with increasing angle θ.

In fact, fracture toughness is strongly influenced by microstructure, thickness, strain rate, environment, and temperature, but in particular, the fracture toughness transition from plane stress to plane strain is remarkably due to increasing plate thickness. This is depicted in Fig. 3.8 for plates having through-thickness cracks. However, part through-thickness cracks encountered in pressure vessel and pipe walls behave differently. This type of defect is in the form of internal, embedded, or external cracks, which may eventually grow through the thickness.

In addition, the thickness of a plate is defined below using the real variable x_3 along the z-axis since z is utilized as a complex function in Chaps. 3 and 4. Thus, the thickness of a plate can be defined by the integral

$$\int \partial x_3 = B \tag{4.19}$$

Further, the above equations for Mode I loading have been derived assuming that solid materials are isotropic and brittle and that crack growth occurs along its own plane, namely the x-axis. However, the stress field at the crack tip can be broken up into three components, mode I, mode II, and mode III, as schematically shown in Fig. 3.1. This implies that the main crack in mode I branches out, specifically in crystalline materials due to microstructural features such as secondary phases, voids, grain boundaries, dislocations, and a possible combination of these defects.

The above field equations, Eqs. (4.17) and (4.18), give an approximation to the stress intensity factor K_I and strongly depend on trigonometric functions. For instance, Fig. 4.3 shows the distribution of the normalized displacements as functions of the angle θ in radians for $v = 0.3$.

Theoretically, an applied quasi-static mode I causes the crack to open and grow along the x-axis at an angle $\theta = 0$. This means that crack growth occurs in a self-similar manner even if the applied stress intensity factor K_I is less than the

Fig. 4.3 Distribution of normalized displacements for mode I

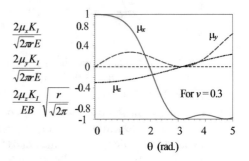

plane-strain fracture toughness K_{IC} of the material. In certain cases, a straight crack under tension may change growth direction due to microstructural features such as particles, grain-boundary resistance, and the like. In the latter case, fracture can occur at an angle relative to the original crack plane.

4.3   Field Equations: Mode II

Deriving the stress and displacement field equations in shear mode II is of particular importance for analyzing mixed-mode systems. Assuming that the crack is loaded with a remote shear stress τ interesting results can be determined. Specifically, two-dimensional problems in cracked structural components can be evaluated using the Westergaard stress functions. For instance, the stress function Z for an infinite sheet containing a central crack of length $2a$ has been introduced in Chap. 3.

Let the Westergaard stress function Z, Z', and $\overline{Z}$ along with $B_o = K_{II}/\sqrt{2\pi}$ be defined by

$$Z = \sum_{m=o}^{M} B_m z^{m-1/2} \tag{4.20}$$

$$Z' = \frac{dZ}{dz} = \sum_{m=o}^{M} \left(m - \frac{1}{2} \right) B_m z^{m-3/2} \tag{4.21}$$

$$\overline{Z} = \int Z dz = \sum_{m=o}^{M} \frac{B_m z^{m+1/2}}{m + 1/2} \tag{4.22}$$

These functions are essential in determining the elastic stresses at the crack tip being introduced in Chap. 3. Thus, the development of the stress equations is presented first due to its simplistic approach.

Combining Eqs. (4.4) and (4.5) into (4.20) through (4.22) gives

$$Z = \sum_{m=o}^{M} B_m r^{m-1/2} \left[\cos \left(m - \frac{1}{2} \right) \theta + i \sin \left(m - \frac{1}{2} \right) \theta \right] \tag{4.23}$$

$$Z' = \sum_{m=o}^{M} \left(m - \frac{1}{2} \right) B_m r^{m-\frac{3}{2}} \left[\cos \left(m - \frac{3}{2} \right) \theta + i \sin \left(m - \frac{3}{2} \right) \theta \right] \tag{4.24}$$

$$\overline{Z} = \sum_{m=o}^{M} \frac{B_m r^{m+1/2}}{m + 1/2} \left[\cos \left(m + \frac{1}{2} \right) \theta + i \sin \left(m + \frac{1}{2} \right) \theta \right] \tag{4.25}$$

From which the real (Re) and imaginary (Im) parts are, respectively,

$$\operatorname{Re} Z = \sum_{m=o}^{M} B_m r^{m-1/2} \cos\left(m - \frac{1}{2}\right)\theta \tag{4.26}$$

$$\operatorname{Re} Z' = \sum_{m=o}^{M}\left(m - \frac{1}{2}\right) B_m r^{m-3/2} \cos\left(m - \frac{3}{2}\right)\theta \tag{4.27}$$

$$\operatorname{Re} \overline{Z} = \sum_{m=o}^{M} \frac{B_m r^{m+1/2}}{m + 1/2} \cos\left(m + \frac{1}{2}\right)\theta \tag{4.28}$$

and

$$\operatorname{Im} Z = \sum_{m=o}^{M} B_m r^{m-1/2} \sin\left(m - \frac{1}{2}\right)\theta \tag{4.29}$$

$$\operatorname{Im} Z' = \sum_{m=o}^{M}\left(m - \frac{1}{2}\right) B_m r^{m-3/2} \sin\left(m - \frac{3}{2}\right)\theta \tag{4.30}$$

$$\operatorname{Im} \overline{Z} = \sum_{m=o}^{M} \frac{B_m r^{m+1/2}}{m + 1/2} \sin\left(m + \frac{1}{2}\right)\theta \tag{4.31}$$

The Airy stress function for the sliding shear mode can be defined by [21]

$$\phi = -y \operatorname{Re} \overline{Z} \tag{4.32}$$

According to Eqs. (3.16) through (3.18) without body forces ($\Omega = 0$) and Eq. (4.32), the stress equations become

$$\sigma_x = \frac{\partial^2 \phi}{\partial y^2} = y \operatorname{Re} Z' + 2 \operatorname{Im} Z$$

$$\sigma_y = \frac{\partial^2 \phi}{\partial x^2} = -y \operatorname{Re} Z' \tag{4.33}$$

$$\tau_{xy} = -\frac{\partial^2 \phi}{\partial x \partial y} = \operatorname{Re} Z - y \operatorname{Im} Z'$$

or

$$\sigma_x = 2\sum_{m=o}^{M}\left(m - \frac{1}{2}\right) B_m r^{m-1/2} \sin\frac{\theta}{2} \cos\frac{\theta}{2} \cos\left(m - \frac{3}{2}\right)\theta$$

$$+ 2\sum_{m=o}^{M} B_m r^{m-1/2} \sin\left(m - \frac{1}{2}\right)\theta$$

$$\sigma_y = -2 \sum_{m=o}^{M} \left(m - \frac{1}{2}\right) B_m r^{m-1/2} \sin\frac{\theta}{2} \cos\frac{\theta}{2} \cos\left(m - \frac{3}{2}\right)\theta \qquad (4.34)$$

$$\tau_{xy} = \sum_{m=o}^{M} B_m r^{m-1/2} \cos\left(m - \frac{1}{2}\right)\theta$$

$$-2 \sum_{m=o}^{M} \left(m - \frac{1}{2}\right) B_m r^{m-1/2} \sin\frac{\theta}{2} \cos\frac{\theta}{2} \sin\left(m - \frac{3}{2}\right)\theta$$

The first terms in the series are sufficient to obtain accurate results. Thus, let $m = 0$ in Eq. (4.34) along with $B_o = K_{II}/\sqrt{2\pi}$ to get

$$\sigma_x = -\frac{K_{II}}{\sqrt{2\pi r}} \sin\frac{\theta}{2}\left[2 + \cos\frac{\theta}{2}\cos\frac{3\theta}{2}\right]$$

$$\sigma_y = \frac{K_{II}}{\sqrt{2\pi r}} \sin\frac{\theta}{2}\cos\frac{\theta}{2}\cos\frac{3\theta}{2} \qquad (4.35)$$

$$\tau_{xy} = \frac{K_{II}}{\sqrt{2\pi r}} \cos\frac{\theta}{2}\left[1 - \sin\frac{\theta}{2}\sin\frac{3\theta}{2}\right]$$

Substituting Eq. (4.35) into (1.10) and (1.11) and the resultant expressions into (1.4) yields the strains and displacements functions. Thus, the strains for a two-dimensional analysis are

$$\epsilon_x = -\frac{K_{II}}{\sqrt{2\pi r E}} \sin\frac{\theta}{2}\left[2 + (1 + v)\cos\frac{\theta}{2}\cos\frac{3\theta}{2}\right]$$

$$\epsilon_y = \frac{K_{II}}{\sqrt{2\pi r E}} \sin\frac{\theta}{2}\left[2v + (1 + v)\cos\frac{\theta}{2}\cos\frac{3\theta}{2}\right] \qquad (4.36)$$

$$\gamma_{xy} = \frac{2(1 + v)K_{II}}{\sqrt{2\pi r E}} \cos\frac{\theta}{2}\left[1 - \sin\frac{\theta}{2}\sin\frac{3\theta}{2}\right]$$

and the displacements are

$$\mu_x = \frac{2(1 + v)K_{II}}{E}\sqrt{\frac{r}{2\pi}} \sin\frac{\theta}{2}\left[\frac{2}{1 + v} + \cos^2\frac{\theta}{2}\right]$$

$$\mu_y = \frac{2K_{II}}{E}\sqrt{\frac{r}{2\pi}} \cos\frac{\theta}{2}\left[(v - 1) + (1 + v)\sin^2\frac{\theta}{2}\right] \qquad (4.37)$$

$$\mu_z = \frac{2vBK_{II}}{E}\sqrt{\frac{r}{2\pi}} \sin\frac{\theta}{2}$$

Obviously, these displacement expressions are strongly dependent on the trigono-metric functions. The reader should plot these functions to observe how divergent they are for a constant Poisson's ratio v.

Evaluating Eq. (4.37) at $\theta = 0$ yields one nonzero displacement as

$$\mu_x = 0$$

$$\mu_y = \frac{2\,(v-1)\,K_{II}}{E}\sqrt{\frac{r}{2\pi}} \tag{4.37a}$$

$$\mu_z = 0$$

Apparently, mode II seems to be a very simple case since only one displacement is needed Cartesian coordinates at $\theta = 0$. The Mode II characterization, as well as other modes of loadings, must include the determination of the stress field, and related strains and displacements, and the driving force G_{II} that defines the fracture toughness of any elastic or elastic perfectly plastic material. In fact, mode II tests can be done under quasi-static and fatigue loading conditions.

4.4   Polar Coordinate Formulation

4.4.1   Mode I and Mode II Cases

Consider an elastic plate containing a single-edge crack subjected to a quasi-static tension loading shown in Fig. 4.4 [5, 33].

In the absence of body forces, equilibrium is satisfied through the Airy stress function in polar coordinates. In a two-dimensional analysis, the general mathemat-

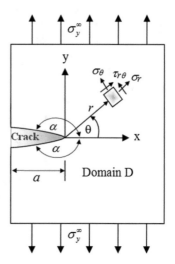

Fig. 4.4 Single-edge crack and stresses in polar coordinates at a point in the domain D

ical definition of the elastic stress field in terms of the Airy stress function ϕ in polar coordinates is

$$\sigma_r = \frac{1}{r}\frac{\partial \phi}{\partial r} + \frac{1}{r^2}\frac{\partial^2 \phi}{\partial \theta^2}$$

$$\sigma_\theta = \frac{\partial^2 \phi}{\partial r^2} \tag{4.38}$$

$$\tau_{r\theta} = \frac{1}{r^2}\frac{\partial \phi}{\partial \theta} - \frac{1}{r}\frac{\partial^2 \phi}{\partial r \partial \theta}$$

These equations were initially derived in Chap. 1, Sect. 1.5, but they are included here for convenience. According to the crack configuration illustrated in Fig. 4.4, the boundary conditions for a single-edge crack are defined as [33]

$$\sigma_\theta = \tau_{r\theta} = 0 \quad \text{for } \theta = \pm\alpha \tag{4.39}$$

Now, it is convenient to use the Airy stress function as follows [52]:

$$\phi = r^{\lambda+1} f(\theta) \tag{4.40}$$

Substituting Eq. (4.40) into (1.62) yields an expression in terms of the cigenvalue λ. The resultant expression is a fourth-order partial differential homogeneous equation with $f = f(\theta)$ [33]. Thus,

$$\frac{\partial^4 f}{\partial \theta^4} + 2\left(\lambda^2 + 1\right)\frac{\partial^2 f}{\partial \theta^2} + \left(\lambda^2 - 1\right)^2 f = 0 \tag{4.41}$$

The general solution of this high-order polynomial is

$$f = C_1 \cos(\lambda - 1)\theta + C_2 \sin(\lambda - 1)\theta + C_3 \cos(\lambda + 1)\theta + C_4 \sin(\lambda + 1)\theta \tag{4.42}$$

with the following boundary conditions: $f = 0$ and $df/d\theta = 0$ for $\theta = \pm\alpha$. Using a homogeneous equation with constant coefficient technique, the following general function is used to determine the eigenvalue λ

$$f = C_i e^{r_i \theta} \tag{4.43}$$

where C_i = Constants
   r_i = Roots
   Hence, Eq. (4.43) gives

$$\frac{\partial f}{\partial \theta} = C_i r_i e^{r_i \theta}$$

$$\frac{\partial^2 f}{\partial \theta^2} = C_i r_i^2 e^{r_i \theta} \tag{4.44}$$

$$\frac{\partial^3 f}{\partial \theta^3} = C_i r_i^3 e^{r\theta}$$

$$\frac{\partial^4 f}{\partial \theta^4} = C_i r_i^4 e^{r\theta}$$

Substituting these derivatives into Eq. (4.41) yields a fourth-order polynomial

$$r^4 + 2\left(\lambda^2 + 1\right) r^2 + \left(\lambda^2 - 1\right)^2 = 0 \qquad (4.45)$$

The solution of Eq. (4.45) gives four roots defined by

$$r_1 = i\left(\lambda - 1\right)$$
$$r_2 = -i\left(\lambda - 1\right) \qquad (4.46)$$
$$r_3 = i\left(\lambda + 1\right)$$
$$r_4 = -i\left(\lambda + 1\right)$$

For convenience, the symmetric and antisymmetric parts of Eq. (4.42) for mode I and II, respectively, are given in matrix form [33]

$$\begin{bmatrix} \cos{(\lambda - 1)\alpha} & \cos{(\lambda + 1)\alpha} \\ (\lambda - 1)\sin{(\lambda - 1)\alpha} & (\lambda + 1)\sin{(\lambda + 1)\alpha} \end{bmatrix} \begin{bmatrix} C_1 \\ C_3 \end{bmatrix} = \begin{bmatrix} 0 \\ 0 \end{bmatrix} \qquad (4.47)$$

$$\begin{bmatrix} \sin{(\lambda - 1)\alpha} & \sin{(\lambda + 1)\alpha} \\ (\lambda - 1)\cos{(\lambda - 1)\alpha} & (\lambda + 1)\cos{(\lambda + 1)\alpha} \end{bmatrix} \begin{bmatrix} C_2 \\ C_4 \end{bmatrix} = \begin{bmatrix} 0 \\ 0 \end{bmatrix} \qquad (4.48)$$

Setting the determinants of Eqs. (4.47) and (4.48) to zero along with the trigonometric function $2 \sin A \cos B = \sin{(A + B)} \sin{(A - B)}$ yields

$$\sin{(2\lambda\alpha)} \pm \lambda \sin{(2\alpha)} = 0 \qquad (4.49)$$

If $\alpha = \pi$, then $\sin{(2\pi)} = \sin{(360°)} = 0$ and the solution of Eq. (4.49) is a characteristic equation of the form

$$\sin{(2\lambda\alpha)} = 0 \qquad (4.50)$$

which has only real roots for $\lambda = n/2$ where $n = 1, 2, 3, 4, \ldots$ The constants in Eqs. (4.47) and (4.48) take the form [33]

$$\begin{array}{lll} C_{3n} = -\frac{n-2}{n+2}C_{1n} & C_{4n} = -C_{2n} & \text{For } n = 1, 3, 5, 7\ldots \\ C_{3n} = -C_{1n} & C_{4n} = -\frac{n-2}{n+2}C_{2n} & \text{For } n = 2, 4, 6, 8\ldots \end{array} \qquad (4.51)$$

Substitute Eq. (4.51) into (4.42), and if the resultant expression along with $f = f(\theta)$ is substituted back into (4.40), the Airy stress function in power series form along with $\lambda = n/2$ and $\alpha = \theta$ becomes

$$
\phi = \sum_{n=1,3\ldots}^{N} r^{1+n/2} \left\{ \begin{array}{l} C_{1n}\left[\cos\left(\frac{n}{2}-1\right)\theta - \frac{n-2}{n+2}\cos\left(\frac{n}{2}+1\right)\theta\right] \\ +C_{2n}\left[\sin\left(\frac{n}{2}-1\right)\theta - \sin\left(\frac{n}{2}+1\right)\theta\right] \end{array} \right\}
$$

$$
+ \sum_{n=2,4\ldots}^{N} r^{1+n/2} \left\{ \begin{array}{l} C_{1n}\left[\cos\left(\frac{n}{2}-1\right)\theta - \cos\left(\frac{n}{2}+1\right)\theta\right] \\ +C_{2n}\left[\sin\left(\frac{n}{2}-1\right)\theta - \frac{n-2}{n+2}\sin\left(\frac{n}{2}+1\right)\theta\right] \end{array} \right\}
$$

(4.52)

Substituting Eq. (4.52) into (4.38) yields the elastic stresses for mode I

$$
\sigma_r = \sum_{n=1,3,\ldots}^{N} \frac{1}{4}r^{n/2-1}C_{1n}\left[\begin{array}{l} n(6-n)\cos\left(\frac{n}{2}-1\right)\theta \\ -n(2-n)\cos\left(\frac{n}{2}+1\right)\theta \end{array} \right]
$$

$$
+ \sum_{n=1,3,\ldots}^{N} \frac{1}{4}r^{n/2-1}C_{2n}\left[\begin{array}{l} n(6-n)\sin\left(\frac{n}{2}-1\right)\theta \\ -n(2+n)\sin\left(\frac{n}{2}+1\right)\theta \end{array} \right]
$$

$$
+ \sum_{n=1,3,\ldots}^{N} \frac{1}{4}r^{n/2-1}C_{1n}\left[\begin{array}{l} n(6-n)\cos\left(\frac{n}{2}-1\right)\theta \\ -n(2-n)\cos\left(\frac{n}{2}+1\right)\theta \end{array} \right]
$$

$$
+ \sum_{n=2,4,\ldots}^{N} \frac{1}{4}r^{n/2-1}C_{2n}\left[\begin{array}{l} n(6-n)\sin\left(\frac{n}{2}-1\right)\theta \\ -n(2-n)\sin\left(\frac{n}{2}+1\right)\theta \end{array} \right]
$$

(4.53)

$$
\sigma_\theta = \sum_{n=1,3,\ldots}^{N} \frac{1}{4}r^{n/2-1}C_{1n}\left[\begin{array}{l} n(2+n)\cos\left(\frac{n}{2}-1\right)\theta \\ -n(2-n)\cos\left(\frac{n}{2}+1\right)\theta \end{array} \right]
$$

$$
+ \sum_{n=1,3,\ldots}^{N} \frac{1}{4}r^{n/2-1}C_{2n}\left[\begin{array}{l} n(2+n)\sin\left(\frac{n}{2}-1\right)\theta \\ -n(2+n)\sin\left(\frac{n}{2}+1\right)\theta \end{array} \right]
$$

$$
+ \sum_{n=2,4,\ldots}^{N} \frac{1}{4}r^{n/2-1}C_{1n}\left[\begin{array}{l} n(2+n)\cos\left(\frac{n}{2}-1\right)\theta \\ -n(2+n)\cos\left(\frac{n}{2}+1\right)\theta \end{array} \right]
$$

$$
+ \sum_{n=2,4,\ldots}^{N} \frac{1}{4}r^{n/2-1}C_{2n}\left[\begin{array}{l} n(2+n)\sin\left(\frac{n}{2}-1\right)\theta \\ -n(2-n)\sin\left(\frac{n}{2}+1\right)\theta \end{array} \right]
$$

(4.54)

$$\tau_{r\theta} = \sum_{n=1,3,...}^{N} \frac{1}{4} r^{n/2-1} C_{1n} \left[\begin{array}{c} n(2-n)\sin\left(\frac{n}{2}-1\right)\theta \\ +n(2-n)\sin\left(\frac{n}{2}+1\right)\theta \end{array} \right]$$

$$+ \sum_{n=1,3,...}^{N} \frac{1}{4} r^{n/2-1} C_{2n} \left[\begin{array}{c} n(2-n)\cos\left(\frac{n}{2}-1\right)\theta \\ -n(2+n)\cos\left(\frac{n}{2}+1\right)\theta \end{array} \right]$$

$$+ \sum_{n=2,4,...}^{N} \frac{1}{4} r^{n/2-1} C_{1n} \left[\begin{array}{c} n(n-2)\sin\left(\frac{n}{2}-1\right)\theta \\ -n(n+2)\sin\left(\frac{n}{2}+1\right)\theta \end{array} \right]$$

$$+ \sum_{n=2,4,...}^{N} \frac{1}{4} r^{n/2-1} C_{2n} \left[\begin{array}{c} n(2-n)\cos\left(\frac{n}{2}-1\right)\theta \\ +n(n-2)\cos\left(\frac{n}{2}+1\right)\theta \end{array} \right] \tag{4.55}$$

with

$$C_{11} = \frac{K_I}{\sqrt{2\pi}} \tag{4.56}$$

$$C_{21} = \frac{K_{II}}{\sqrt{2\pi}} \tag{4.57}$$

Simplifying Eqs. (4.53) through (4.57) with $n = 1$ gives the dominant stresses adjacent to the crack tip for $r \to 0$. This means that a single term in the series is sufficient to determine the stress field equations. Thus, the stresses in polar coordinates for mode I are [11]

$$\sigma_r = \frac{K_I}{\sqrt{2\pi r}} \left(\frac{5}{4}\cos\frac{\theta}{2} - \frac{1}{4}\cos\frac{3\theta}{2} \right) = \frac{K_I}{\sqrt{2\pi r}} \cos\frac{\theta}{2} \left(1 + \sin^2\frac{\theta}{2} \right)$$
$$\tag{4.58a}$$

$$\sigma_\theta = \frac{K_I}{\sqrt{2\pi r}} \left(\frac{3}{4}\cos\frac{\theta}{2} + \frac{1}{4}\cos\frac{3\theta}{2} \right) = \frac{K_I}{\sqrt{2\pi r}} \cos\frac{\theta}{2} \left(1 - \sin^2\frac{\theta}{2} \right)$$
$$\tag{4.58b}$$

$$\tau_{r\theta} = \frac{K_I}{\sqrt{2\pi r}} \left(\frac{1}{4}\sin\frac{\theta}{2} + \frac{1}{4}\sin\frac{3\theta}{2} \right) = \frac{K_I}{\sqrt{2\pi r}} \sin\frac{\theta}{2}\cos^2\frac{\theta}{2} \tag{4.58c}$$

and those for mode II take the form

$$\sigma_r = \frac{K_{II}}{\sqrt{2\pi r}} \left(-\frac{5}{4}\sin\frac{\theta}{2} + \frac{3}{4}\sin\frac{3\theta}{2} \right) = \frac{K_{II}}{\sqrt{2\pi r}} f_r(\theta)_{II} \tag{4.59a}$$

$$\sigma_\theta = \frac{K_{II}}{\sqrt{2\pi r}} \left(-\frac{3}{4}\sin\frac{\theta}{2} - \frac{3}{4}\sin\frac{3\theta}{2} \right) = \frac{K_{II}}{\sqrt{2\pi r}} f_\theta(\theta)_{II} \tag{4.59b}$$

$$\tau_{r\theta} = \frac{K_{II}}{\sqrt{2\pi r}} \left(\frac{1}{4}\cos\frac{\theta}{2} + \frac{3}{4}\cos\frac{3\theta}{2} \right) = \frac{K_{II}}{\sqrt{2\pi r}} f_{r\theta}(\theta)_{II} \tag{4.59c}$$

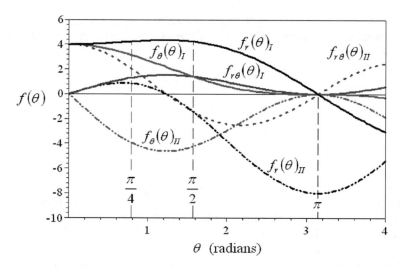

Fig. 4.5 Plots of $f(\theta)$ functions representing the distribution of normalized stresses for modes I and II as per Eqs. (4.58) and (4.59)

Fig. 4.6 Displacements at a point near the single-edge crack in the domain D

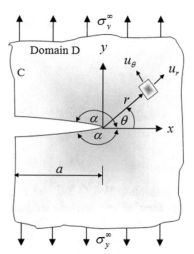

The distribution of normalized stresses, such as $4\sigma_r\sqrt{2\pi r}/K_I = f_r(\theta)_I$, given in Eqs. (4.58) and (4.59) is governed by the trigonometric functions, which are plotted in Fig. 4.5 for convenience.

Consider a single-edge crack model shown in Fig. 4.6 and related radial and circumferential displacements μ_r and μ_θ, respectively. Accordingly, the strains are defined by

$$\epsilon_r = \frac{\partial \mu_r}{\partial r}$$

$$\epsilon_\theta = \frac{\mu_r}{r} + \frac{1}{r}\frac{\partial \mu_\theta}{\partial \theta} \tag{4.60}$$

$$\gamma_{r\theta} = \frac{1}{r}\frac{\partial \mu_r}{\partial r} + \frac{\partial \mu_\theta}{\partial r} - \frac{\mu_\theta}{r}$$

Furthermore, Hooke's law for plane conditions states that the strain (amount by which a solid body is deformed) is linearly related to the stress (force causing the deformation). Thus, those materials that exhibit linear-elastic deformation are referred to as Hookean materials, and the amount of deformation for plane stress and plane-strain conditions are approximated by the following strain equations:

$$E\epsilon_r = \sigma_r - \nu\sigma_\theta$$

$$E\epsilon_\theta = \sigma_\theta - \nu\sigma_r \quad \text{for plane stress} \tag{4.61}$$

$$G\gamma_{r\theta} = \tau_{r\theta}$$

and

$$2G\epsilon_r = (1-\nu)\sigma_r - \nu\sigma_\theta$$

$$2G\epsilon_\theta = (1-\nu)\sigma_\theta - \nu\sigma_r \quad \text{for plane strain} \tag{4.62}$$

$$G\gamma_{r\theta} = \tau_{r\theta}$$

Substitute Eqs. (4.58) and (4.59) into (4.61) and (4.62), and solve for ϵ_r and ϵ_θ. Then, substitute these strains into Eq. (4.60) and integrate the strains to get the displacement expressions. Thus,

$$\mu_r = \frac{K_I}{4G}\sqrt{\frac{r}{2\pi}}\left[(2\kappa-1)\cos\frac{\theta}{2} - \cos\frac{3\theta}{2}\right] \quad \text{for mode I} \tag{4.63}$$

$$\mu_\theta = \frac{K_I}{4G}\sqrt{\frac{r}{2\pi}}\left[-(2\kappa-1)\sin\frac{\theta}{2} + \sin\frac{3\theta}{2}\right] \quad \text{for mode I}$$

and

$$\mu_r = \frac{K_{II}}{4G}\sqrt{\frac{r}{2\pi}}\left[-(2\kappa-1)\sin\frac{\theta}{2} + 3\sin\frac{3\theta}{2}\right] \quad \text{for mode II} \tag{4.64}$$

$$\mu_\theta = \frac{K_{II}}{4G}\sqrt{\frac{r}{2\pi}}\left[-(2\kappa-1)\cos\frac{\theta}{2} + 3\cos\frac{3\theta}{2}\right] \quad \text{for mode II}$$

where

$$\kappa = \frac{3 - v}{1 + v} \quad \text{for plane stress} \tag{4.65}$$

$$\kappa = 3 - 4v \quad \text{for plane strain} \tag{4.66}$$

If $\theta = 0$ and $v = 1/3$, Eqs. (4.63) and (4.64) yield

$$\mu_r = \frac{K_I}{4G} \sqrt{\frac{r}{2\pi}} [(2\kappa - 1) - 1] > 0 \qquad \text{for mode I}$$

$$\mu_\theta = 0 \qquad \text{for mode I} \tag{a}$$

$$\mu_r = v0 \qquad \text{for mode II}$$

$$\mu_\theta = \frac{K_{II}}{4G} \sqrt{\frac{r}{2\pi}} [-(2\kappa - 1) + 3] > 0 \quad \text{for mode II}$$

Thus,

$$\mu_r > 0 \ \text{ and } \ \mu_\theta = 0 \ \text{ for mode I}$$

$$\mu_r = 0 \ \text{ and } \ \mu_\theta > 0 \ \text{ for mode II \& plane stress} \tag{b}$$

$$\mu_r = 0 \ \text{ and } \ \mu_\theta > 0 \ \text{ for mode II \& plane strain}$$

Therefore, the displacements in polar coordinates vary according to the type of loading case.

4.4.2 Mode III Loading Case

This section includes the mathematical procedure for determining the elastic displacements and elastic shear stresses for mode III loading. This particular stress mode involves the tearing fracture process for plates (Fig. 3.1) and the torsional fracture process for shafts. The mathematical treatment that follows is for isotropic and homogeneous materials subjected to quasi-static mode III loading [33].

Consider the crack model given in Fig. 4.7 which represents a single-edge crack in a plate or a radial face on a shaft as illustrated in Fig. 4.8.

Thus, a two-dimensional stress analysis indicates that the displacements and stresses are independent of the coordinate $z = x_3$ [33]. The resulting mechanical deformation generates anti-plane shear or mode III entities such that $\mu_x = \mu_y = 0$ and $\sigma_x = \sigma_y = \sigma_z = \tau_{xy} = 0$ along the line of interaction. Consequently, the nonzero shear stresses and displacement in Cartesian coordinates are $\tau_{xz} \neq 0$, $\tau_{yz} \neq 0$, and $\mu_z \neq 0$. In polar coordinates, $\tau_{rz} \neq 0$ and $\tau_{\theta z} \neq 0$.

Hooke's law in polar coordinates (Fig. 4.7) gives the shear stresses as

$$\tau_{rz} = G\gamma_{rz} = G\frac{\partial \mu_z}{\partial r} \tag{4.67}$$

$$\tau_{\theta z} = G\gamma_{\theta z} = \frac{G}{r}\frac{\partial \mu_z}{\partial \theta}$$

Fig. 4.7 Anti-plane stress
components in the domain D
surrounded by the contour C

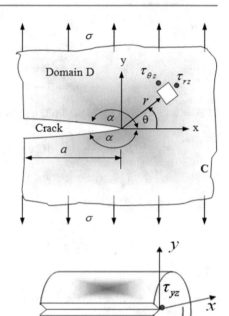

Fig. 4.8 Surface cracks on
cylinders subjected to mode
III by torsion

The equilibrium equation, the bipotential equation and the boundary condition
for μ_z are, respectively

$$\frac{\partial (r\tau_{rz})}{\partial r} + \frac{\partial \tau_{\theta z}}{\partial \theta} = 0 \tag{4.68}$$

$$\nabla^2 \mu_z = \frac{\partial^2 \mu_z}{\partial r^2} + \frac{1}{r}\frac{\partial \mu_z}{\partial r} + \frac{1}{r^2}\frac{\partial^2 \mu_z}{\partial \theta^2} \tag{4.69a}$$

$$\nabla^2 \mu_z = 0 \tag{4.69b}$$

$$\tau_{\theta z} = \frac{\partial \mu_z}{\partial \theta} = 0 \quad \text{For } \theta = \pm\alpha \tag{4.70}$$

Now assume that the out-of-plane displacement is defined by [33]

$$\mu_z = r^\lambda f(\theta) \tag{4.71}$$

Letting $f = f(\theta)$ and taking the derivatives needed in Eq. (4.69a), and simplifying
the resultant expression, yields the governing second-order differential equation:

$$\frac{\partial^2 f}{\partial \theta^2} + \lambda^2 f = 0 \tag{4.72}$$

The solution of Eq. (4.72) can be determined by letting the function f be defined by

$$f = De^{r\theta} \tag{4.73}$$

Furthermore, Eq. (4.72) yields an equation that has no real solution

$$r^2 + \lambda^2 = 0 \tag{4.74}$$

$$r = -i\lambda \tag{4.75}$$

where i is loosely defined as $i = \sqrt{-1}$ is the imaginary unit in complex variable theory and λ is referred to as the eigenvalue term.

For the sake of clarity, the prefix *eigen* in eigenvalue is the German word for innate (inherent). Nonetheless, inserting Eq. (4.75) into (4.73) gives the characteristic equation as the solution of Eq. (4.72) needed for determining the antisymmetric displacement μ_z when $\theta = 0$. Hence,

$$f = \mathrm{Im}\left(De^{i\lambda\theta}\right) = \mathrm{Im}\,D\left(\cos\lambda\theta + i\sin\lambda\theta\right) \tag{a}$$

$$f = D\sin\lambda\theta \tag{4.76}$$

Differentiating Eq. (4.76) generates the boundary condition expression, which in turn is the characteristic equation

$$\frac{df}{d\theta} = D\lambda\cos\lambda\theta = 0 \quad \text{for } \theta = \pm\alpha \tag{4.77}$$

so that

$$\cos\lambda\theta = 0 \tag{4.78}$$

Let $\alpha = \pi$ in Eq. (4.78) in order to simulate a crack in a solid body (Fig. 4.7) under mode III loading. Consequently, the eigenvalues λ take the form

$$\lambda = \frac{n}{2} \quad \text{for } n = 1, 3, 5, 7, \ldots. \tag{4.79}$$

Clearly, Eq. (4.78) becomes

$$\cos\lambda\pi = 0 \tag{4.80}$$

$$\cos\frac{\pi}{2} = \cos\frac{3\pi}{2} = \cos\frac{5\pi}{2} = \cos\frac{7\pi}{2} = 0 \tag{4.81}$$

Consequently, the antisymmetric displacement along the crack line and the respective shear stresses defined by Eqs. (4.71) and (4.67), respectively, become

$$\mu_z = D_n r^{n/2}\sin\frac{n\theta}{2} \tag{4.82}$$

$$\tau_{rz} = G\frac{\partial\mu_z}{\partial r} = \frac{nG}{2}D_n r^{n/2-1}\sin\frac{n\theta}{2} \tag{4.83}$$

Fig. 4.9 Distribution of
normalized shear stresses for
mode III and $n = 1$

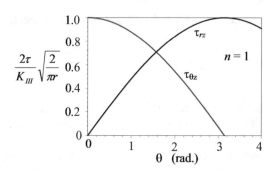

$$\tau_{\theta z} = \frac{G}{r}\frac{\partial \mu_z}{\partial \theta} = \frac{nG}{2}D_n r^{n/2-1}\cos\frac{n\theta}{2} \tag{4.84}$$

The normalized shear stress distribution is depicted in Fig. 4.9. Notice that these stresses, Eqs. (4.83) and (4.84), have an opposite distribution because of the trigonometric terms in the equations.

Furthermore, the constant D_n for $n = 1$ becomes [33]

$$D_1 = \frac{K_{III}}{G}\sqrt{\frac{2}{\pi}} \tag{4.85}$$

The parameter D_1 in Eqs. (4.82) through (4.84) corresponds to the dominant term in the series. So far the mechanics framework for including higher-order terms in the stress series has been described for a purely linear-elastic stress field at the crack tip. This analytical treatment allows a small plastic zone (r) and predicts the shear stress field. In fact, K_{III} is the anti-plane stress intensity factor or the stress intensity factor for mode III loading condition which is applicable to torsion tests. Recall that G is the shear modulus of rigidity defined by Eq. (1.9). Thus, mode III is found by the solution of an anti-plane shear problem which the simplest mode of loading to solve in the field of fracture mechanics.

Evaluating Eqs. (4.82) through (4.84) along with Eq. (4.85) at $\theta = 0$ yields the following results

$$\tau_{\theta z} = \tau_{yz} = \frac{K_{III}}{\sqrt{2\pi r}} \quad \text{for } \theta = 0 \tag{4.86}$$

$$\tau_{rz} = \tau_{xz} = 0 \quad \text{for } \theta = 0 \tag{4.87}$$

$$\mu_z = 0 \quad \text{for } \theta = 0 \tag{4.88}$$

Denote that Eq. (4.86) estimates the out-of-plane shear stress $\tau_{\theta z}$ being proportional to the inverse square root of r ($\tau_{\theta z} \propto r^{-1/2}$) and predicts the shear stress field singularity, which means that the shear stress becomes infinitely large as the plastic zone size approaches zero ($\tau_{\theta z} \to \infty$ as $r \to 0$).

Despite the singularity stress case, it is apparent that the theory of linear elasticity is applicable at a very small distance from the crack tip, where $\tau_{\theta z} \to \infty$ as $r \to 0$.

4.5 Higher-Order Stress Field

Assume that the damage ahead of the crack tip is characterized by second-order local stresses, which are dependent of specimen size and geometry as opposed to the first-order terms. Consequently, the stress field is no longer singular as $r \to 0$, and the second-order term in the series of expansion is known in the literature as the T-stress (T_x) for elastic behavior which accounts for effects of stress biaxiality. Several T-stress solutions are available in the literature [9, 17, 25–29, 42].

For elastic-plastic and fully plastic materials, the second-order term is also known as the J-Q approach [9]. In particular, O'Dowd and Shih [40, 41] can be consulted for obtaining details of the J-Q theory which describes the fundamentals that provide quantitative measures of the crack tip deformation. Nevertheless, the term Q accounts for plasticity in the triaxiality state crack tip stress field.

Considering a mixed-mode fracture process and the effects of the T-stress in cracked bodies, the asymptotic stress state at the crack tip can be determined by adding Eqs. (4.13) and (4.35) and the T-stress. Thus, for mode I and II interaction, the stresses are

$$
\begin{bmatrix} \sigma_x & \tau_{xy} \\ \tau_{xy} & \sigma_y \end{bmatrix} = \frac{K_I}{\sqrt{2\pi r}} \begin{bmatrix} f_x(\theta) & f_{xy}(\theta) \\ f_{xy}(\theta) & f_y(\theta) \end{bmatrix} \tag{4.89}
$$

$$
+ \frac{K_{II}}{\sqrt{2\pi r}} \begin{bmatrix} g_x(\theta) & g_{xy}(\theta) \\ g_{xy}(\theta) & g_{xy}(\theta) \end{bmatrix} + \begin{bmatrix} T_x & 0 \\ 0 & 0 \end{bmatrix}
$$

With regard to the T-stress theory, Larsson and Carlsson [36] and Sherry et al. [47] defined T_x as the non-singular stress that acts in the direction parallel to the crack plane, and it is given as [46]

$$
T_x = \sigma_x - \frac{K_I}{\sqrt{2\pi r}} f_x(\theta) \tag{4.90}
$$

For $\theta = 0$, the stress state Eq. (4.89) becomes

$$
\begin{bmatrix} \sigma_x & \tau_{xy} \\ \tau_{xy} & \sigma_y \end{bmatrix} = \frac{K_I}{\sqrt{2\pi r}} \begin{bmatrix} 1 & 0 \\ 0 & 1 \end{bmatrix} + \frac{K_{II}}{\sqrt{2\pi r}} \begin{bmatrix} 0 & 1 \\ 1 & 0 \end{bmatrix} \tag{4.91}
$$

$$
+ \begin{bmatrix} T_x & 0 \\ 0 & 0 \end{bmatrix}
$$

and for pure mode I loading, $K_{II} = 0$, the T-stress becomes

$$T_x = \sigma_x - \frac{K_I}{\sqrt{2\pi r}} \tag{4.92}$$

$$T_x = \sigma_x - \sigma_y$$

In fact, this equation is the modified σ_x stress function in Eq. (4.13). Moreover, Leevers and Radon [37] defined T_x to be independent of the stress intensity factor as

$$T_x = \beta\sigma \tag{4.93}$$

Here, σ is the remote applied stress (MPa). The geometric correction factor $\alpha = f(a/w)$ introduced in Chap. 3 can be included in Eq. (4.93); that is, $T_x = \alpha\beta\sigma$. Similarly, for pure mode II, the T-stress is [17]

$$T_x = \sigma_x + \frac{K_{II}}{\sqrt{2\pi r}} \tag{4.94}$$

$$T_x = \sigma_x + \tau_{xy} \tag{4.95}$$

In addition, the parameter β in Eq. (4.93) is the dimensionless stress biaxiality ratio defined by [37]

$$\beta = \frac{T_x\sqrt{\pi a}}{K_I} \tag{4.96}$$

For convenience, the geometry correction and the biaxiality factors for some common specimen configurations are

- Single-edge cracked plate (SET) in tension (Table 3.1) with $L/w \geq 1.5$ and $x = a/w$ [25]:

$$\frac{T_x}{\sigma} = \frac{1}{(1-x)^2}\left[\begin{array}{c} -0.526 + 0.641x + 0.2049x^2 \\ +0.755x^3 - 0.7974x^4 + 0.1966x^5 \end{array}\right] \tag{4.97}$$

$$\beta = \frac{1}{\sqrt{1-x}}\left[\begin{array}{c} -0.469 + 0.1414x + 1.433x^2 + 0.0777x^3 \\ -1.6195x^4 + 0.859x^5 \end{array}\right] \tag{4.98}$$

- Double-edge cracked plate DE(T) in tension (Table 3.1) with $L/w \geq 1.5$ and $x = a/w$ [25]:

$$\frac{T_x}{\sigma} = -0.526 - 0.0438x + 0.0444x^2 + 0.12194x^3 \tag{4.99}$$

$$\beta = -0.469 - 0.071x + 0.1196x^2 + 0.2801x^3 \tag{4.100}$$

- Double-edge cracked plate DE(T) in tension (Table 3.1) with $L/w = 1$ and $x = a/w$ [25]:

$$\frac{T_x}{\sigma} = -0.526 + 0.1804x - 2.7241x^2 + 9.5966x^3 - 6.3883x^4 \quad (4.101)$$

$$\beta = -0.469 + 0.1229x - 1.2256x^2 + 6.0628x^3 - 4.4983x^4 \quad (4.102)$$

- Double cantilever beam DCB(T) in tension (Fig. 6.5) [25]:

$$T_x = \frac{\beta K_I}{\sqrt{\pi a}} = P\sqrt{\frac{12}{\pi a h}}\left[\frac{a/h + 0.68}{0.681\,(h/a) + 0.0685}\right] \quad (4.103)$$

$$\beta = \frac{1}{0.681\,(h/a) + 0.685} \quad \text{For } h/a < 1.5 \quad (4.104)$$

$$K_I = P\sqrt{\frac{12}{h}}\,(a/h + 0.68) \quad (4.105)$$

where $P =$ Load per unit length (MN/m)

- Compact tension C(T) specimen in tension (Table 3.1) with $L/w = 0.6$ and $a/w < 1$ [47]:

$$\frac{T_x}{\sigma} = 6.063 - 78.987x + 380.46x^2 - 661.70x^3 + 428.45x^4 \quad (4.106)$$

The significance of the T-stress (T_x) can be explained by plotting, say, Eq. (4.101) along with (4.102) as shown in Fig. 4.10 for a double-edge cracked plate DE(T) (see Table 3.1) with $L/w = 1$ and $x = a/w$ [25].

$$\frac{T_x}{\sigma} = -0.526 + 0.1804x - 2.7241x^2 + 9.5966x^3 - 6.3883x^4 \quad (4.101)$$

$$\beta = -0.469 + 0.1229x - 1.2256x^2 + 6.0628x^3 - 4.4983x^4 \quad (4.102)$$

The T_x/σ profile shown in Fig. 4.10 exhibits a transition for the double-edge cracked plate DE(T) at approximately $a/w = 0.70$. Similarly, the dimensionless stress biaxiality ratio β also exhibits nearly equal trend as the T_x/σ curve. For $T_x/\sigma < 0$, the local crack tip stresses are below the limits predicted by the small-scale yielding since the T_x is in the compressive state stabilizing the crack path [9]. The opposite occurs for $T_x/\sigma > 0$, leading to a high degree of triaxiality in the crack tip stresses since T_x is in the tensile state [35]. These two cases may be interpreted as if a mode I crack in homogeneous solids tends to grow along its plane when the T_x in front of the crack is negative $(T_x < 0)$, whereas it deviates from its original plane or becomes directionally unstable if the T_x is positive $(T_x > 0)$ [19].

These observations, $T_x < 0$ and $T_x > 0$, have also been reported [3, 4, 31] to occur in an adhesive bond under mode I loading. In addition, the T-stress has also

Fig. 4.10 Normalized
T-stress as a function of
normalized crack length for a
double-edge specimen loaded
in tension

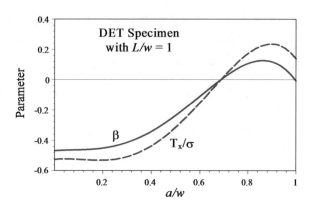

been reported [34] that it has a significant effect on crack initiation angles in brittle
fracture of functionally graded materials [34].

Initially, the T-stress is below the yield limit of the material; $T_x < \sigma_{ys}$, and it is
negative because is in compression. After some crack growth has occurred, $T_x < \sigma_{ys}$
become positive since it changes to a tension entity. Figure 4.10 depicts the $T_x < \sigma_{ys}$
transition from compression to tension. Moreover, these observations means that the
relations $T_x < \sigma_{ys}$ and $T_x < \sigma_{ys} = 0$ imply stable crack growth as the external load
is increasing. As a result, the plastic zone increases and more plastic deformation is
accumulated.

According to Williams [52], the T-stress (T_x) can be derived using an Airy stress
function in polar coordinates for a cracked body having the coordinates origin at the
crack tip. This equation can also be found in a compendium of the T-stress solutions
reported by Fett [25]. Thus, the symmetric Airy stress function for mode I loading is

$$\phi_I = \sigma w^2 \sum_{m=0}^{\infty} \left(\frac{r}{w}\right)^{m+3/2} A_m \left[\cos\left(m + \frac{1}{2}\right)\theta - \frac{m + \frac{3}{2}}{m - \frac{1}{2}} \cos\left(m - \frac{1}{2}\right)\theta\right]$$

$$+ \sigma w^2 \sum_{m=0}^{\infty} \left(\frac{r}{w}\right)^{m+2} B_m \left[\cos\left(m + 2\right)\theta - \cos m\theta\right] \qquad (4.107)$$

and the antisymmetric part for mode II becomes

$$\phi_{II} = \sigma w^2 \sum_{m=0}^{\infty} \left(\frac{r}{w}\right)^{m+3/2} C_m \left[\sin\left(m + \frac{1}{2}\right)\theta - \sin\left(m - \frac{1}{2}\right)\theta\right] \qquad (4.108)$$

$$+ \sigma w^2 \sum_{m=0}^{\infty} \left(\frac{r}{w}\right)^{m+2} D_m \left[\sin\left(m + 3\right)\theta - \frac{m + 3}{m + 1} \sin\left(m + 1\right)\theta\right]$$

where w is the characteristic dimension.

According to Fett [25], Eq. (4.107) can be used to determine the T-stress as

$$T_x = -4\sigma B_o - \sigma_x^{(o)} \tag{4.109}$$

Here, $\sigma_x^{(o)}$ is the stress contribution in the crack-free body. The effect of the T-stress on crack path in crystalline solids and adhesively bonded joints is an interesting subject in fracture mechanics. This implies that the directionality of cracks is significantly influenced by the magnitude of the T-stress, which can describe the stress field near the crack tip for mode I or II under certain assumptions [29].

A mathematical treatment for determining the T-stress can be based on the Westergaard stress function, Williams stress function based on Airy's approach, Green's function method, Boundary collocation method, and principle of superposition [25]. Further details on this topic can be found elsewhere [25, 35, 42, 46].

4.6 Method of Conformal Mapping

This section is an extension of conformal mapping being introduced in Chap. 1. Basically, conformal mapping is a powerful method to transform a region having an arbitrary shape in the z-plane into a region in the ζ-plane using complex variables. This means that the interior or exterior of a particular domain D can be mapped into a domain R by using two complex potentials or complex stress functions, $\gamma(z)$ and $\psi(z)$, which have certain mathematical forms to be determined henceforward. These complex potentials are known as *Kolosov–Muskhelishvili potentials*, which have found wide application in several boundary problems.

Despite the large number of conformal mapping being developed for different applications [1, 2, 10, 12, 14, 15, 30, 38, 44, 48], the general picture of conformal mapping or transformation of interest in this section is schematically illustrated in Fig. 4.11. This illustration implies that an arbitrary region D can be mapped or transformed into region R through the complex function $z = f(\zeta)$, commonly for a unit circle with $|\zeta| = 1$. Conversely, $\zeta = f(z)$ maps points in the ζ-plane onto the z-plane. The general complex functions that enable interior and exterior mapping onto a circle may be defined as complex power polynomials [44]. Thus,

Fig. 4.11 Conformal mapping of region D in the z-plane into the ζ-plane

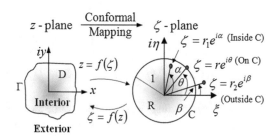

Fig. 4.12 Internal and external conformal mapping of an ellipse into a unit circle

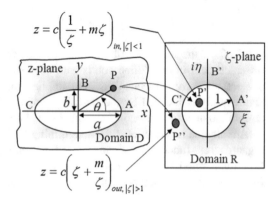

$$z = \sum_{n=0}^{N} c_n \zeta^n \qquad \text{(inside } C\text{)}$$

$$z = re^{i\theta} = r\left(\cos\theta + i\sin\theta\right) \quad \text{(on } C\text{)} \qquad (4.110)$$

$$z = \frac{A}{\zeta} + \sum_{n=0}^{N} c_n \zeta^n \qquad \text{(outside } C\text{)}$$

Here, constants A and c_n depend on the shape of the region R. Details on complex formulation of plane elasticity problems, specifically circular domains, can be found elsewhere [38, 44, 48]. For example, Fig. 4.12 schematically shows the conformal mapping of a point P outside the ellipse into a point P' inside ($|\zeta| < 1$) and P'' outside ($|\zeta| > 1$) the circle. Recall that the absolute value of ζ is defined as $|\zeta| = \pm\sqrt{\xi^2 + \eta^2}$ or $|\zeta| = \rho = 1$ for a unit circle.

The complex functions for mapping point P shown in Fig. 4.12 are defined by the following complex equations [38]

$$z = c\left(\frac{1}{\zeta} + m\zeta\right) \quad \text{(exterior)} \qquad (4.111\text{a})$$

$$z = c\left(\zeta + \frac{m}{\zeta}\right) \quad \text{(interior)} \qquad (4.111\text{b})$$

where $c = (a+b)/2$ and $m = (a-b)/(a+b)$ with $a =$ semimajor axis and $b =$ semiminor axis.

According to Muskhelishvili [38], once a region in the infinite z-plane is mapped onto a unit circle with $|\zeta| < 1$, it can always be transformed back into the infinite plane with a circular hole with $|\zeta_1| > 1$. Thus, it is sufficient to make the substitution $\zeta = 1/\zeta_1$ for accomplishing the transformation. In addition, finite simply connected regions can be mapped onto a unit circle with $|\zeta| < 1$ and infinite simply connected regions onto a region where $|\zeta| > 1$.

Fig. 4.13 Center crack in an infinite plane

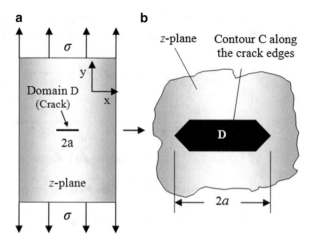

The complex functions for exterior and interior mapping of region D in the z-plane onto the region R in the ζ-plane and conversely [38, 44, 48] can be deduced from the complex function for an elliptical hole.

From Chap. 2, if $b \to 0$, then the ellipse becomes a crack and Eqs. (4.111a) and (4.111b) along with $c = a/2$ and $m = 1$ become the mapping functions for a straight crack [44] as shown in Fig. 4.13a.

Thus, the mapping function for a straight central crack in an infinite plate is of the form

$$z = f(\zeta) = \frac{a}{2}\left(\frac{1}{\zeta} + \zeta\right) \tag{4.112}$$

If $dz/d\zeta = 0$, then $\zeta = \pm 1$ or $|\zeta| = 1$. Solving for ζ in Eq. (4.112) yields a quadratic equation, $\zeta^2 - (2z/a)\zeta + 1 = 0$, and its solutions for exterior conformal mapping of the circular domain R depicted in Fig. 4.12 is

$$\zeta = f(z) = \frac{1}{a}\left(z + \sqrt{z^2 - a^2}\right) \quad \text{(exterior)} \tag{4.113}$$

The application of conformal mapping to fracture mechanics is specifically illustrated in Fig. 4.13 for a stress-free (traction-free crack faces) straight central crack of length 2a in an infinite plate having a domain D in the z-plane.

If the plate is uniformly and remotely loaded in tension along the y-axis, then the loading condition in this particular case is $\sigma_y^\infty = \sigma$, $\sigma_x^\infty = \tau_{xy}^\infty = 0$. It is assumed that the crack is stress-free and bounded by its edges making a contour C, which is expanded for clarity in Fig. 4.13b.

Exterior conformal mapping enables the development of the analytical procedure for deriving the elastic stress field equations near the crack tip. In exterior mapping, the interior of the stress-free crack is not included. This can be accomplished using the complex potentials $\gamma(z)$ and $\psi(z)$ defined by Eqs. (1.83) and (1.84) for mapping an arbitrary shaped region into a circular domain. Accordingly,

$$\gamma(z) = -\frac{\sum_{j=1}^{m} F_j}{2\pi(1+\kappa)} \log(z) + \left(\frac{\sigma_x^\infty + \sigma_y^\infty}{4}\right) z + \gamma^*(z)$$

$$\psi(z) = \frac{\kappa \sum_{j=1}^{m} \overline{F}_j}{2\pi(1+\kappa)} \log(z) + \left(\frac{\sigma_y^\infty - \sigma_x^\infty + 2i\tau_{xy}^\infty}{2}\right) z + \psi^*(z)$$

(4.113)

where $\gamma^*(z)$ and $\psi^*(z)$ are analytic and single-valued functions in the solid body [24].

The plane conditions factor κ is based on the Poisson's ratio ν. Thus,

$$\kappa = 3 - 4\nu \quad \text{for plane strain} \tag{a}$$

$$\kappa = \frac{3 - \nu}{1 + \nu} \quad \text{for plane stress} \tag{b}$$

According to Muskhelishvili [38] and Sadd [44], the complex mapping function $z = f(\zeta)$ defined by Eq. (4.112) for a circular domain has the general form

$$f(\zeta) = A\zeta^{-1} + \varphi \tag{c}$$

$$\log(z) = \log[f(\zeta)] = -\log(\zeta) + \varphi_1 \tag{d}$$

where A is a constant and φ and φ_1 are analytic functions.

This suggests that $f(\zeta)$ approaches ζ^{-1} and consequently,

$$\log(z) = \log f(\zeta) \rightarrow \log\left(\zeta^{-1}\right) = -\log(\zeta)$$

Accordingly, the complex potentials defined by Eq. (4.113) can be expressed in terms of the complex variable ζ. Hence, [38, 44]

$$\gamma(z) = \frac{F}{2\pi(1+\kappa)} \log(\zeta) + \left(\frac{\sigma_x^\infty + \sigma_y^\infty}{4}\right) \zeta + \gamma^*(\zeta)$$

$$\psi(z) = -\frac{\kappa \overline{F}}{2\pi(1+\kappa)} \log(\zeta) + \left(\frac{\sigma_y^\infty - \sigma_x^\infty + 2i\tau_{xy}^\infty}{2}\right) \zeta + \psi^*(\zeta)$$

(4.114)

Assume that the resultant forces in the interior boundary C vanish; $F = \overline{F} = 0$ and that the complex potentials, $\gamma^*(z)$ and $\psi^*(z)$, do not contribute to the stresses and displacements for exterior mapping of the crack onto the exterior of a unit circle. Making use of these assumptions and applying the loading condition, $\sigma_y^\infty = \sigma$ and $\sigma_x^\infty = \tau_{xy}^\infty = 0$, to Eq. (4.113) yields the *first set* of complex potentials for exterior mapping of a straight crack. In this case, $\gamma(z) \rightarrow \gamma_o(z)$ and $\psi(z) \rightarrow \psi_o(z)$ such that

$$\gamma_o(z) = \frac{\sigma}{4}z$$

$$\psi_o(z) = \frac{\sigma}{2}z \tag{4.115}$$

In fact, the force components, $F = F_x + iF_y$ and $\overline{F} = F_x - iF_y$, and the complex potential $\gamma^*(z)$ and $\psi^*(z)$ in Eq. (4.113) vanish since the crack is internally stress-free and the conformal mapping of interest is external or exterior.

Further, a *second set* of complex potentials which are referred to as the image complex potentials, say $\gamma_1(z) = \gamma_{im}(z)$ and $\psi_1(z) = \psi_{im}(z)$, must be determined assuming that the boundary of the crack or hole becomes traction-free under the uniaxial tension mode [30]. Make use of the superposition principle so that the net response caused by the two stimuli (external loading condition and traction at the crack or hole) leads to the complete set of complex potentials $\gamma(z) = \gamma_o(z) + \gamma_1(z)$ and $\psi(z) = \psi_o(z) + \psi_1(z)$. Hence, $\gamma_1(z)$ and $\psi_1(z)$ are assumed to be analytic outside the boundary of a unit circle and can be found using Cauchy integral formula.

4.6.1   Cauchy Integral Formula

A complete development of this technique [38] is out of the scope in this book, but a brief description of this topic is included in Chap. 1, and for convenience, an additional material is covered herein for determining the stress field equations ahead of a crack tip.

It is apparent that the Cauchy integral formula, which is also referred to as Cauchy's differentiation formula, provides an analytic approach to find appropriate complex potentials, $\gamma(z)$ and $\psi(z)$, for deriving the elastic displacements and stresses near a crack tip in two dimensions.

The Cauchy integral formula describes contour integrals by letting a contour C be a closed curve traversed counterclockwise. It is required that the complex function $f(z)$ be analytic everywhere inside or outside C so that its derivative is nonzero, $f'(z) \neq 0$.

For any point $z = x + iy$ inside C, the following integral holds [38]

$$f(z) = \frac{1}{2\pi i} \int_C \frac{f(\zeta)}{\zeta - z} d\zeta \tag{4.116}$$

which is not analytically continued when z passes from region D^+ to region D^- or vice versa as shown in Fig. 4.14 [38]. Further, $f(z) \to 0$ as $z \to \infty$ or simply put $f(\infty) = 0$.

Recall that the Cauchy integral formula introduced in Chap. 1, Eq. (1.109), is used in this section in order to find closed forms of the complex potentials $\gamma(z)$ and $\psi(z)$. The Cauchy integral formula are

Fig. 4.14 Central crack in an infinite plane subjected to a tension loading

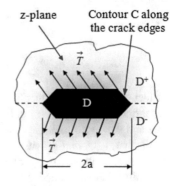

z-plane   Contour C along
                  the crack edges

$$\left[\frac{1}{2\pi i}\int_C \frac{\gamma(\zeta)}{\zeta-z}d\zeta\right]_1 + \left[\frac{1}{2\pi i}\int_C \frac{z\overline{\gamma'(\zeta)}}{\zeta-z}d\zeta\right]_2 + \left[\frac{1}{2\pi i}\int_C \frac{\overline{\psi(\zeta)}}{\zeta-z}d\zeta\right]_3$$

$$= \left[\frac{1}{2\pi i}\int_C \frac{h(\zeta)}{\zeta-z}d\zeta\right]_4 \qquad (4.117)$$

The solutions of these integrals, as defined by Eq. (1.124), are conveniently summarized as [13, 44]

$$\gamma(z) = \frac{1}{2\pi i}\int_C \frac{h(\zeta)}{\zeta-z}d\zeta - \overline{a_1}z$$

$$\psi(z) = \int_C \frac{\overline{h(\zeta)}}{\zeta-z}d\zeta - \frac{\gamma'(z)}{z} + \frac{a_1}{z} \qquad (4.118)$$

$$a_1 + \overline{a_1} = \frac{1}{2\pi i}\int_C \frac{h(\zeta)}{\zeta^2}d\zeta$$

$$a_2 = \frac{1}{2\pi i}\int_C \frac{h(\zeta)}{\zeta^3}d\zeta$$

The application of the Cauchy integral formula to fracture mechanics requires further work. For instance, Fig. 4.14, which is an extension of Fig. 4.13, schematically illustrates two regions to be used for determining the crack tip field equations at a distance r. The z-plane is divided into upper D^+ and lower D^- regions. In general, this is referred to as the Riemann-Hilbert problem which requires two continuous and analytic boundary functions $\chi^+(z)$ and $\chi^-(z)$ for regions D^+ and D^-, respectively. These boundary functions, $\chi^+(z)$ and $\chi^-(z)$, are referred to as Plemelj functions which form part of the Cauchy integral formula.

Now, the combination of stress expressions determined in Chap. 1, Eq. (1.95), for conformal mapping a domain D onto a unit circle may be modified by considering equal traction vectors $\overrightarrow{T^+}$ and $\overrightarrow{T^-}$, but opposite in direction as shown by the arrows in Fig. 4.14. This is the case when a solid is half-space in the region D^+ where $y \geq 0$ with the system boundary at $y = 0$. The resultant stress combination expressions are [38, 44]:

- For unit circle, the stresses and the displacement function, $U = u + iv$, are

$$\sigma_x + \sigma_y = 2\left[\gamma'(z) + \overline{\gamma'(z)}\right] = 4\,\mathrm{Re}\left[\gamma'(z)\right]$$

$$\sigma_y - \sigma_x + 2i\tau_{xy} = 2\left[\bar{z}\gamma''(z) + \psi'(z)\right] \qquad (4.119)$$

$$2GU = \kappa\gamma(z) - z\overline{\gamma'(z)} - \overline{\psi(z)}$$

- For a half-space in the region D^+,

$$\sigma_x + \sigma_y = 4\,\mathrm{Re}\left[\gamma'(z)\right]$$

$$\sigma_y - i\tau_{xy} = \gamma'(z) + \overline{\gamma'(z)} + (z - \bar{z})\,\overline{\gamma''(z)} \qquad (4.120)$$

$$2GU = \kappa\gamma(z) - z\overline{\gamma'(z)} - \overline{\psi(z)}$$

Due to displacement continuity and traction conditions outside the crack, the last expression of Eq. (4.120) may be defined as

$$\gamma(z) - z\overline{\gamma'(z)} - \overline{\psi(z)} = g(z) \qquad (4.120a)$$

$$\kappa\gamma(z) + z\overline{\gamma'(z)} + \overline{\psi(z)} = h(z) \qquad (4.120b)$$

where the boundary functions $g(z)$ and $h(z)$ are assumed to be continuous and analytic. Adding Eqs. (4.120a) and (4.120b) gives

$$(\kappa + 1)\,\gamma(z) = g(z) + h(z) \qquad (4.121)$$

which is an equation that relates the unknown complex potential $\gamma(z)$ with two unknown complex functions, $g(z)$ and $h(z)$. Further, solving this two-dimensional crack problem requires use of superposition principle of complex variable formulation. Thus, the boundary conditions for the upper D^+ and the lower D^- crack surfaces lead to a relationship between elastic stresses and traction forces as

$$\sigma_y - i\tau_{xy} = i\left(T_x^+ + iT_y^+\right) \qquad (4.122a)$$

$$\sigma_y - i\tau_{xy} = -i\left(T_x^- + iT_y^-\right) \qquad (4.122b)$$

such that

$$\gamma'_+(z) + \gamma'_-(z) - g'_-(z) = i\left(T_x^+ + iT_y^+\right) \qquad (4.123a)$$

$$\gamma'_-(z) + \gamma'_+(z) - g'_-(z) = -i\left(T_x^- + iT_y^-\right) \qquad (4.123b)$$

For the special case where tractions acting on the crack faces are equal and opposite $(T_x^+ = -T_x^-$ and $T_y^+ = -T_y^-)$, the Cauchy integral formula can be used to find a complex potential, say $\gamma_1(z)$, which is referred to as the image term of the complex potential $\gamma_o(z)$ due to crack-free or hole-free plane [30]. In fact, the Cauchy integral formula for the crack configuration shown in Fig. 4.14 is related to boundary values of sectionally analytic functions on opposite sides of the crack line and has the general form [20, 38?]

$$\gamma_1'(z) - g'(z) = \frac{\chi(z)}{2\pi i} \int_C \frac{h(\zeta)}{\chi^+(\zeta)(\zeta - z)} d\zeta + P(z)\chi(z) \tag{4.124}$$

where the Plemelj functions, $\chi(z)$ and $\chi^+(\zeta)$, and the arbitrary polynomial $P(z)$ with unknown constants must be determined from conditions at infinity [20, 38]. Hence,

$$\chi(z) = \frac{1}{\sqrt{z^2 - a^2}}$$

$$\chi^+(\zeta) = \frac{1}{i\sqrt{a^2 - \zeta^2}} \tag{4.124a}$$

$$P(z) = \sum_{n=0}^{\infty} c_n z^n = c_o + c_1 z + \ldots$$

For uniform traction $T_x^+ = -T_x^- = \tau_{xy}^\infty = \tau$, $T_y^+ = -T_y^- = \sigma_y^\infty = \sigma$, the boundary functions and the polynomial in Eq. (4.124) are [20]

$$h(\zeta) = -\left(\sigma_y^\infty + i\tau_{xy}^\infty\right) \tag{4.124b}$$

$$g_-'(z) = P(z) = 0$$

and consequently, Eq. (4.124) reduces to

$$\gamma_1'(z) = \frac{\chi(z)}{2\pi i} \int_C \frac{h(\zeta)}{\chi^+(\zeta)(\zeta - z)} d\zeta \tag{4.125}$$

Further, $h(\zeta)$ in Eq. (4.125) is a function prescribed on the crack contour C defined by $-a < x < a$. Also, $\chi^+(\zeta)$ and $\chi^-(\zeta)$ are boundary values of $\chi(z)$ when z approaches a point ζ along any path on the upper $(+)$ or lower $(-)$ region of the contour C. In this case, $\chi(z)$ is assumed to be continuous at ζ from the upper $(+)$ or lower $(-)$ region [38].

In fact, this integral, Eq. (4.125), is simply a Riemann-Hilbert problem described by Muskhelishvili [38] as a problem of linear relationship for providing analytic (holomorphic) solutions when cracks (cuts) in isotropic solids are considered as discontinuities, which are subjected to force tractions on their faces. The crack line has upper and lower faces (Fig. 4.14), and the boundary conditions for the crack can be placed on these faces independently. Further, the general solution of the homogeneous Riemann-Hilbert problem, as defined by Eq. (4.125), has been used [16, 20] to describe the behavior of incremental displacement and stress fields of interface cracks in prestressed composite materials under anti-plane shear (mode III) using conformal mapping.

Substituting Eqs. (4.124a) and (4.124b) into (4.124) yields

$$f'(z) = -\frac{\left(\sigma_y^\infty + i\tau_{xy}^\infty\right)}{2\pi\sqrt{z^2 - a^2}} \int_C \frac{\sqrt{a^2 - \zeta^2}}{\zeta - a} d\zeta \qquad (a)$$

$$f'(z) = -\frac{\left(\sigma_y^\infty + i\tau_{xy}^\infty\right)}{2\pi\sqrt{z^2 - a^2}} \int_{-a}^{+a} \frac{\sqrt{a^2 - \zeta^2}}{\zeta - a} d\zeta \qquad (b)$$

The radical above may be evaluated using power series as [38]

$$\sqrt{a^2 - \zeta^2} = -i\zeta\sqrt{1 - \frac{a^2}{\zeta^2}} = -i\zeta\left(1 - \frac{a^2}{2\zeta^2} + \ldots\ldots\right) \qquad (c)$$

Instead, evaluate the integral as described by Muskhelishvili [38] on page 501 or take the solution from Azhdarı et al., Eq. (B2b), page 6472 [8] The resultant solution is

$$\int_C \frac{\sqrt{a^2 - \zeta^2}}{\zeta - a} d\zeta = \pi i\left(\sqrt{a^2 - \zeta^2} + i\zeta\right) \qquad (d)$$

which can be written along with $\zeta = z$ as

$$\int_C \frac{\sqrt{a^2 - \zeta^2}}{\zeta - a} d\zeta = \pi\left(\sqrt{z^2 - a^2} - z\right) \qquad (e)$$

This solution can also be found elsewhere [20]. Substituting Eq. (e) into (a), one gets

$$f'(z) = -\frac{\left(\sigma_y^\infty + i\tau_{xy}^\infty\right)}{2}\left(1 - \frac{z}{\sqrt{z^2 - a^2}}\right) \qquad (f)$$

Hence, the sought complex function is determined by integrating Eq. (f)

$$f(z) = \int f'(z)\, dz = -\frac{\left(\sigma_y^\infty - i\tau_{xy}^\infty\right)}{2}\left(z - \sqrt{z^2 - a^2}\right) \qquad (g)$$

Collecting the real and imaginary parts gives

$$f(z) = -\frac{\sigma_y^\infty}{2}\left(z - \sqrt{z^2 - a^2}\right) + i\frac{\tau_{xy}^\infty}{2}\left(z - \sqrt{z^2 - a^2}\right) \qquad (4.126)$$

Then, the sought second set or image term $\gamma_1(z) = \gamma_{im}(z)$ of $\gamma_o(z)$ for the infinite z-plane containing the central crack becomes

$$\gamma_1(z) = \text{Re}[f(z)] = -\frac{\sigma_y^{\infty}}{2}\left(z - \sqrt{z^2 - a^2}\right) \qquad (4.127)$$

Thus, for a unit circular domain with $|\zeta| = 1$, add Eqs. (4.115) and (4.127) to find the complete set of the complex potential $\gamma(z)$ by superposition principle as [13,44]

$$\gamma(z) = \gamma_o(z) + \gamma_1(z)$$

$$= \frac{\sigma_y^{\infty}}{4}z - \frac{\sigma_y^{\infty}}{2}\left(z - \sqrt{z^2 - a^2}\right) \qquad (4.128)$$

$$= \frac{\sigma_y^{\infty}}{4}\left(2\sqrt{z^2 - a^2} - z\right)$$

The second complex potential $\psi(z)$ needed for determining the displacement and stress field equations ahead of the crack tip is found elsewhere [13,44,45], and it is given below along with Eq. (4.128) evaluated at $\sigma_y^{\infty} = \sigma$. Hence,

$$\gamma(z) = \frac{\sigma}{4}\left(2\sqrt{z^2 - a^2} - z\right) \qquad (4.129)$$

$$\psi(z) = \frac{\sigma}{2}\left(z - \frac{a^2}{\sqrt{z^2 - a^2}}\right) \qquad (4.130)$$

In addition, if $\psi(z) = \psi_o(z) + \psi_1(z)$ and $\psi(z)$ is given by Eq. (4.130), then the image term of $\psi_o(z)$, Eq. (4.115), is deduced as

$$\psi_1(z) = \psi_{im}(z) = -\frac{\sigma}{2}\left(\frac{a^2}{\sqrt{z^2 - a^2}}\right) \qquad (4.130a)$$

These multiple-valued complex potentials are made single-valued because of the existing crack and are considered to be discontinuous functions [13]. Applying Eqs. (4.129) and (4.130) to (4.119) yields the sought combination of stress expressions in terms of the complex variable z [44]

$$\sigma_x + \sigma_y = 4\,\text{Re}\left[\gamma'(z)\right] = \sigma\,\text{Re}\left(\frac{2z}{\sqrt{z^2 - a^2}} - 1\right)$$

$$= \sigma\,\text{Re}\left(\frac{2z}{\sqrt{(z + a)(z - a)}} - 1\right) \qquad (4.131)$$

$$\sigma_y - \sigma_x + i2\tau_{xy} = \sigma\left[\frac{\bar{z}}{\sqrt{z - a^2}} - \frac{\bar{z}z^2}{(z^2 - a^2)^{3/2}}\right]$$

$$+ \sigma\left[+\frac{za^2}{(z^2 - a^2)^{3/2}} + 1\right]$$

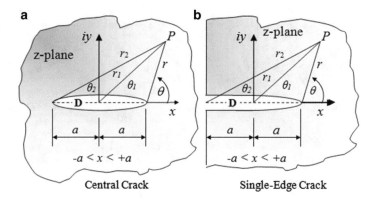

Fig. 4.15 (a) Central elliptical crack and (b) single-edge crack in an infinite solid subjected to a uniform state of stress σ_y^∞ and τ_{xy}^∞ at infinity

The application of Eq. (4.131) to a tension loaded plate containing either a central elliptical crack or a single-edge crack is demonstrated using the crack coordinates shown in Fig. 4.15 [22, 24, 42, 44, 50].

According to the crack coordinates for a point P in Fig. 4.15a, it is convenient to use polar coordinates along with the auxiliary angles, θ_1 and θ_2, because of the following important definitions:

$$\sqrt{z^2 - a^2} = \sqrt{(z + a)(z - a)} = \sqrt{r_2} e^{i\theta_2/2} \sqrt{r} e^{i\theta/2}$$

$$z - a = re^{i\theta}; \quad z = r_1 e^{i\theta_1}; \quad z + a = r_2 e^{i\theta_2} \tag{a}$$

$$r \sin \theta = r_1 \sin \theta_1 = r_2 \sin \theta_2$$

$$-\pi \le \theta \le \pi; \quad 0 \le \theta \le \pi \ \& \ 0 \le \theta_2 \le 2\pi$$

where $\theta = \pi$, $\theta_2 = 0$ at $y = 0^+$, and $-a \le x \le a$ for the upper crack surface. And thus $\theta = \pi$, $\theta_2 = 2\pi$ at $y = 0^-$, and $-a \le x \le a$ for the lower crack surface. These conditions lead to $z^2 - a^2 = -rr_2 = x^2 - a^2$.

Now, assume a traction-free crack surface as the boundary condition and apply the proper definitions given above to Eq. (4.131a) to get

$$\sigma_x + \sigma_y = \sigma \operatorname{Re} \left(\frac{2r_1 e^{i\theta_1}}{\sqrt{r_2 e^{i\theta_2} re^{i\theta}}} - 1 \right) = \sigma \operatorname{Re} \left(\frac{r_1}{\sqrt{r_2 r}} \frac{e^{i\theta_1}}{e^{i\theta_2/2} e^{i\theta/2}} - 1 \right)$$

$$= \sigma \operatorname{Re} \left[\frac{r_1}{\sqrt{r_2 r}} e^{i(\theta_1 - \theta_2/2 - \theta/2)} - 1 \right]$$

$$= \sigma \, \mathrm{Re} \left\{ \frac{r_1}{\sqrt{r_2 r}} \left[\cos \left(\frac{\theta_2 + \theta}{2} - \theta_1 \right) + i \sin \left(\theta_1 - \frac{\theta_2 - \theta}{2} \right) \right] - 1 \right\}$$

$$= \sigma \left[\frac{r_1}{\sqrt{r_2 r}} \cos \left(\theta_1 - \frac{\theta_2 - \theta}{2} \right) - 1 \right] \tag{4.132}$$

For $r_1 = a$ and $r_2 = 2a$ at $\theta_1 = \theta_2 = 0$ and $r > 0$ at $\theta \to \pi$, Eq. (4.132) becomes

$$\sigma_x + \sigma_y = \sigma \left(\frac{a}{\sqrt{2ar}} \cos \frac{\theta}{2} - 1 \right) \tag{4.133}$$

Similarly,

$$\sigma_y - \sigma_x + 2i\tau_{xy} = \sigma + i \frac{2a^2 \sigma r \sin \theta}{(r_2 r)^{3/2}} \left[\cos \frac{3(\theta + \theta_2)}{2} - i \sin \frac{3(\theta + \theta_2)}{2} \right]$$

$$\sigma_y - \sigma_x + 2i\tau_{xy} = \sigma + i \frac{2a^2 \sigma r \sin \theta}{2ar\sqrt{2ar}} \left[\cos \frac{3\theta}{2} - i \sin \frac{3\theta}{2} \right] \tag{4.134}$$

$$\sigma_y - \sigma_x + 2i\tau_{xy} = \sigma + \frac{2a\sigma}{\sqrt{2ar}} \left(\sin \theta \sin \frac{3\theta}{2} + i \sin \theta \cos \frac{3\theta}{2} \right)$$

Extracting the real and imaginary parts of Eq. (4.134) yields along with (4.133) the combination of crack tip stress expressions

$$\sigma_x + \sigma_y = \frac{2\sigma \sqrt{\pi a}}{\sqrt{2\pi r}} \cos \frac{\theta}{2} - \sigma$$

$$\sigma_y - \sigma_x = \frac{2\sigma \sqrt{\pi a}}{\sqrt{2\pi r}} \sin \frac{\theta}{2} \sin \frac{3\theta}{2} + \sigma \tag{4.135}$$

$$i\tau_{xy} = i \frac{\sigma \sqrt{\pi a}}{\sqrt{2\pi r}} \cos \frac{\theta}{2} \sin \frac{\theta}{2} \cos \frac{3\theta}{2}$$

Manipulating and simplifying the stress expressions defined by Eq. (4.135) gives the elastic stresses near the crack tip in a convenient closed form. Hence, the analytical expressions for predicting the individual elastic crack tip stress field in Cartesian coordinates become exactly the same as derived in a previous section. The individual elastic stresses are

$$\sigma_x = \frac{\sigma \sqrt{\pi a}}{\sqrt{2\pi r}} \cos \frac{\theta}{2} \left(1 - \sin \frac{\theta}{2} \sin \frac{3\theta}{2} \right)$$

$$\sigma_y = \frac{\sigma \sqrt{\pi a}}{\sqrt{2\pi r}} \cos \frac{\theta}{2} \left(1 + \sin \frac{\theta}{2} \sin \frac{3\theta}{2} \right) \tag{4.136}$$

$$\tau_{xy} = \frac{\sigma \sqrt{\pi a}}{\sqrt{2\pi r}} \cos \frac{\theta}{2} \sin \frac{\theta}{2} \cos \frac{3\theta}{2}$$

Using the definition of the stress intensity factor for mode I, $K_I = \sigma \sqrt{\pi a}$, provides the elastic stress field near the crack tip as defined by Eq. (4.13). Hence,

$$\sigma_x = \frac{K_I}{\sqrt{2\pi r}} \cos \frac{\theta}{2} \left(1 - \sin \frac{\theta}{2} \sin \frac{3\theta}{2} \right) = \frac{K_I}{\sqrt{2\pi r}} f_x(\theta)_I$$

$$\sigma_y = \frac{K_I}{\sqrt{2\pi r}} \cos \frac{\theta}{2} \left(1 + \sin \frac{\theta}{2} \sin \frac{3\theta}{2} \right) = \frac{K_I}{\sqrt{2\pi r}} f_y(\theta)_I \qquad (4.13)$$

$$\tau_{xy} = \frac{K_I}{\sqrt{2\pi r}} \cos \frac{\theta}{2} \sin \frac{\theta}{2} \cos \frac{3\theta}{2} = \frac{K_I}{\sqrt{2\pi r}} f_{xy}(\theta)_I$$

In polar coordinates,

$$\sigma_r = \frac{K_I}{4\sqrt{2\pi r}} \left(5 \cos \frac{\theta}{2} - \cos \frac{3\theta}{2} \right) = \frac{K_I}{4\sqrt{2\pi r}} f_r(\theta)_I$$

$$\sigma_\theta = \frac{K_I}{4\sqrt{2\pi r}} \left(3 \cos \frac{\theta}{2} + \cos \frac{3\theta}{2} \right) = \frac{K_I}{4\sqrt{2\pi r}} f_\theta(\theta)_I \qquad (4.58)$$

$$\tau_{r\theta} = \frac{K_I}{4\sqrt{2\pi r}} \left(\sin \frac{\theta}{2} + \sin \frac{3\theta}{2} \right) = \frac{K_I}{4\sqrt{2\pi r}} f_{r\theta}(\theta)_I$$

Assume that a solid deforms under plane-strain condition and that it is subjected to bounded stress conditions under mixed-modes I and II at infinity. In such a case, the stress field near the crack tip can be derived using the Airy stress function [13]

$$\phi = \frac{K_I}{3\sqrt{2\pi}} r^{3/2} \left(\cos \frac{3\theta}{2} + 3 \cos \frac{\theta}{2} \right) - \frac{K_I}{\sqrt{2\pi}} r^{3/2} \left(\sin \frac{3\theta}{2} + \sin \frac{\theta}{2} \right) \qquad (4.137)$$

In addition, the complex displacement is $U = u + iv$. Excluding details on the algebra required to manipulate Eq. (4.119), the displacement components in rectangular coordinates are

$$u = \frac{K_I}{G} \sqrt{\frac{r}{2\pi}} \cos \frac{\theta}{2} \left[\frac{\kappa - 1}{2} + \sin^2 \frac{\theta}{2} \right] \qquad (4.138a)$$

$$= \frac{K_I}{2G} \sqrt{\frac{r}{2\pi}} \cos \frac{\theta}{2} \left[\kappa - \cos \theta \right]$$

$$v = \frac{K_I}{G} \sqrt{\frac{r}{2\pi}} \sin \frac{\theta}{2} \left[\frac{\kappa + 1}{2} - \cos^2 \frac{\theta}{2} \right] \qquad (4.138b)$$

$$= \frac{K_I}{2G} \sqrt{\frac{r}{2\pi}} \sin \frac{\theta}{2} \left[\kappa - \cos \theta \right]$$

Therefore, Eq. (4.138) takes into account the effect of Poisson's ratio ν through the parameter κ, which defines either plane stress or plane-strain condition under mode

I loading. Denote that $\mu_x = \mu$ and $\mu_y = v$, and if $\theta = 0$ for crack growth along the x-axis, it gives $\mu_x > 0$ and $\mu_y = 0$.

Example 4.1. *Consider an infinite plate containing a single-edge crack subjected to mode III loading as shown in the sketch given below. Find the elastic shear stress and displacement field equations using the given complex potential, which are used for mapping an arbitrary shaped region into a circular domain.*

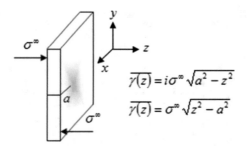

$$\overline{\gamma(z)} = i\sigma^\infty \sqrt{a^2 - z^2}$$

$$\overline{\gamma(z)} = \sigma^\infty \sqrt{z^2 - a^2}$$

Solution. *Use the crack coordinates given in Fig. 4.15b so that*

$$\overline{\gamma(z)} = \sigma^\infty \sqrt{z^2 - a^2} = \sigma^\infty \sqrt{(z+a)(z-a)}$$

$$\overline{\gamma'(z)} = \frac{\sigma^\infty z}{\sqrt{z^2 - a^2}} = \frac{\sigma^\infty z}{\sqrt{(z+a)(z-a)}}$$

Now, let the complex variable z be related to the auxiliary crack angles as

$$z = r_1 e^{i\theta_1}$$

$$z + a = r_2 e^{i\theta_2}$$

$$z - a = r e^{i\theta}$$

then

$$\overline{\gamma'(z)} = \frac{\sigma^\infty r_1 e^{i\theta_1}}{\sqrt{r_2 e^{i\theta_2} r e^{i\theta}}}$$

$$\overline{\gamma'(z)} = \frac{r_1 \sigma^\infty}{\sqrt{r_2 r}} \exp\left[i\left(\theta_1 - \frac{\theta_2 + \theta}{2}\right)\right]$$

Letting $r_1 = a$ and $r_2 = 2a$ at $\theta_1 = 0$ gives the conjugate potential function and the elastic entities. Hence, the potential function becomes

$$\overline{\gamma'(z)} = \frac{a\sigma^\infty}{\sqrt{2ar}} \left[\cos\left(\theta_1 - \frac{\theta_2 + \theta}{2}\right) + i\sin\left(\theta_1 - \frac{\theta_2 + \theta}{2}\right)\right]$$

Then, the real and imaginary parts give the shear stresses as

$$\tau_{yz} = \text{Re}\,\overline{\gamma'(z)} = \frac{a\sigma^\infty}{\sqrt{2ar}}\cos\left(\theta_1 - \frac{\theta_2 + \theta}{2}\right)$$

$$\tau_{xz} = \text{Im}\,\overline{\gamma'(z)} = \frac{a\sigma^\infty}{\sqrt{2ar}}\sin\left(\theta_1 - \frac{\theta_2 + \theta}{2}\right)$$

The out-of-plane displacement is

$$u_z = \frac{1}{G}\,\text{Im}\,\overline{\gamma(z)} = \frac{1}{G}\sigma^\infty\sqrt{r_2 e^{i\theta_2} r e^{i\theta}}$$

$$u_z = \frac{1}{G}\sigma^\infty\sqrt{2ar}\sin\left(\frac{\theta_2 + \theta}{2}\right)$$

For $\theta_1 = \theta_2 = 0$, the shear stresses and the displacement are

$$\tau_{yz} = \frac{a\sigma^\infty}{\sqrt{2ar}}\cos\left(\frac{\theta}{2}\right) = \frac{\sigma^\infty\sqrt{\pi a}}{\sqrt{2\pi r}}\cos\left(\frac{\theta}{2}\right) = \frac{K_{III}}{\sqrt{2\pi r}}\cos\left(\frac{\theta}{2}\right)$$

$$\tau_{rz} = \frac{a\sigma^\infty}{\sqrt{2ar}}\sin\left(\frac{\theta}{2}\right) = \frac{\sigma^\infty\sqrt{\pi a}}{\sqrt{2\pi r}}\sin\left(\frac{\theta}{2}\right) = \frac{K_{III}}{\sqrt{2\pi r}}\sin\left(\frac{\theta}{2}\right)$$

$$u_z = \frac{\sigma^\infty}{G}\sqrt{2ar}\sin\left(\frac{\theta}{2}\right) = \frac{K_{III}}{\pi G}\sqrt{2\pi r}$$

Now, evaluating these entities by letting $\theta = 0$, the shear stresses and the out-of-plane (antisymmetric) displacement become

$$\tau_{yz} = \frac{a\sigma^\infty}{\sqrt{2ar}}$$

$$\tau_{xz} = 0$$

$$u_z = 0$$

This implies that crack growth under mode III is governed by a single elastic shear stress τ_{yz}. Recall that mode III is also referred to as the antisymmetric, tearing or anti-plane shear mode loading. The real and imaginary trigonometric parts, as determined above, can be evaluated at $\theta_1 = \theta_2 = \pi$ Thus,

$$\tau_{yz} = \text{Re}\,\overline{\gamma'(z)} = \frac{a\sigma^\infty}{\sqrt{2ar}}\cos\left(\pi - \frac{\pi + \theta}{2}\right)$$

$$\tau_{xz} = \text{Im}\,\overline{\gamma'(z)} = \frac{a\sigma^\infty}{\sqrt{2ar}}\sin\left(\pi - \frac{\pi + \theta}{2}\right)$$

The normalized shear stress distributions for τ_{yz} and τ_{xz} are depicted in the figure given below.

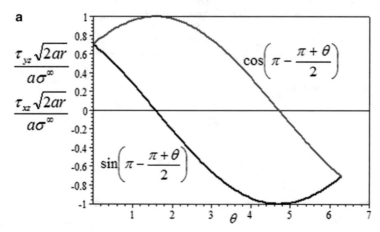

The next figure shows the distribution of the normalized shear τ_{yz} at several values of θ_2 and fixed $\theta_1 = 0$.

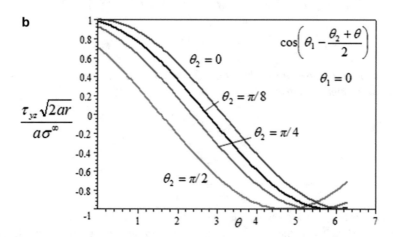

Denote that τ_{yz} uniformly decreases with increasing θ at the given auxiliary angle θ_2.

Example 4.2. *Consider an infinite plate containing an embedded elliptical crack shown below subjected to remote and uniform uniaxial tension (mode I fracture).*

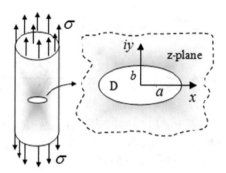

A complete stress analysis must include shear stresses along slip planes due to the applied tension loading condition and shear stresses due to dislocations arbitrarily located along the slip planes. A complete stress analysis is very complicated, and for this reason, transforming or mapping the exterior of the ellipse in the z-plane into the exterior of a unit circle in the ζ-plane is considered. The mapping function for the elliptical crack and its inversion expressions are, respectively,

$$z = c\left(\zeta + \frac{m}{\zeta}\right)$$

$$\zeta = \frac{z}{c} + \sqrt{\left(\frac{z}{2c}\right)^2 - m}$$

where $c > 0$, $0 \leq m \leq 1$, $b = a/2$, $a = 2\,mm$, $b = 1\,mm$ and

$$c = \frac{a+b}{2}$$

$$m = \frac{a-b}{a+b}$$

$$\rho = b^2/a \ (\text{crack tip radius})$$

The required analytical work for determining all necessary stresses is complicated and difficult to obtain. However, conformal mapping of the elliptical crack in the z-plane onto a unit circle in the ζ-plane can alleviate the complicated algebra to an extent using Fischer and Beltz [30] approach, which includes the image terms $\gamma_{im}(\zeta)$ and $\psi_{im}(\zeta)$ of the complex potentials $\gamma_o(z)$ and $\psi_o(z)$, respectively. The complete set of complex potentials are $\gamma(\zeta) = \gamma_o(\zeta) + \gamma_{im}(\zeta)$ and $\psi(\zeta) = \psi_o(\zeta) + \psi_{im}(\zeta)$ where

$$\gamma_o(\zeta) = \frac{c\sigma}{4}\left(\zeta + \frac{m}{\zeta}\right)$$

$$\gamma_{im}(\zeta) = -\frac{c\sigma}{2}\left(\frac{1+m}{\zeta}\right)$$

$$\psi_o(\zeta) = \frac{c\sigma}{2}\left(\zeta + \frac{m}{\zeta}\right)$$

$$\psi_{im}(\zeta) = -\frac{c\sigma}{2}\left(\frac{1+m}{\zeta} + \frac{(1+m)\left(1+m\zeta^2\right)}{\zeta\left(\zeta^2 - m\right)}\right)$$

Determine the elastic stresses ahead of the crack tip using Eq. (1.95) and the applied stress intensity factor K_I. Use an applied stress of $\sigma = \sigma_y^\infty = 100$ MPa and assume that crack growth occurs along the crack plane (x-axis).

Solution. *The constants are*

$$c = \frac{a + a/2}{2} = \frac{3a}{4} = 1.5 \text{ mm} \qquad b = a/2 = 1 \text{ mm}$$

$$m = \frac{c - c/2}{c + c/2} = \frac{1}{3} = 0.33$$

$$\rho = b^2/a = a/4 = 0.5 \text{ mm } (crack \text{ } tip \text{ } radius)$$

Recall that $r = 1$ and $|\zeta| > 1$ for a unit disk and $\theta = 0$ on the x-axis. Therefore,

$$z = re^{i\theta/2} = r(\cos\theta/2 + i\sin\theta/2) = 1$$

$$\bar{z} = re^{-i\theta/2} = r(\cos\theta/2 - i\sin\theta/2) = 1$$

The complex potentials in the ζ-plane are

$$\gamma(\zeta) = \frac{c\sigma}{4}\left(\zeta - \frac{2+m}{\zeta}\right)$$

$$\psi(\zeta) = \frac{c\sigma}{2}\left[\zeta - \frac{1}{\zeta} - \frac{(1+m)\left(1+m\zeta^2\right)}{\zeta\left(\zeta^2 - m\right)}\right]$$

and the complex variable ζ and its derivative $d\zeta/dz$ are

$$\zeta = \frac{z}{c} + \sqrt{[z/(2c)]^2 - m}$$

$$\frac{d\zeta}{dz} = \frac{1}{c} + \frac{z}{2c\sqrt{z^2 - 4c^2m}}$$

Use the chain rule to find the derivatives of $\gamma(\zeta)$ and $\psi(\zeta)$ with respect to z. Due to the complexity of the mathematics, not all details are shown. Hence,

$$\frac{d\gamma\,(\zeta)}{d\zeta} = \frac{c\sigma}{4}\left(1 + \frac{m+2}{\zeta^2}\right) = \frac{c\sigma}{4}\left[1 + \frac{m+2}{\left(z/c + \sqrt{[z/\,(2c)]^2 - m}\right)^2}\right]$$

$$\gamma'\,(z) = \frac{d\gamma\,(\zeta)}{d\zeta}\frac{d\zeta}{dz} = \frac{c\sigma}{4}\left[1 + \frac{m+2}{\left(z/c + \sqrt{[z/\,(2c)]^2 - m}\right)^2}\right]$$

$$\times \left(\frac{1}{c} + \frac{z}{2c\sqrt{z^2 - 4c^2 m}}\right)$$

$$\gamma'\,(z) = (0.25 - 1.0165i)\,\sigma \tag{a}$$

and

$$\gamma''\,(z) = \frac{d\gamma'\,(z)}{d\zeta}\frac{d\zeta}{dz} = -\frac{c\sigma}{4}\left(\frac{2\,(m+2)}{\zeta^3}\right)\left(\frac{1}{c} + \frac{z}{2c\sqrt{z^2 - 4c^2 m}}\right)$$

$$\gamma''\,(z) = -\frac{c\sigma}{4}\left[\frac{2\,(m+2)}{\left(z/c + \sqrt{[z/\,(2c)]^2 - m}\right)^3}\right]\left(\frac{1}{c} + \frac{z}{2c\sqrt{z^2 - 4c^2 m}}\right)$$

$$\gamma''\,(z) = (0.875 + 0.64082i)\,\sigma \tag{b}$$

Then,

$$\psi\,(\zeta) = \frac{c\sigma}{2}\left[\zeta - \frac{1}{\zeta} - \frac{(1+m)\left(1 + m\zeta^2\right)}{\zeta\left(\zeta^2 - m\right)}\right]$$

$$\frac{d\psi\,(\zeta)}{d\zeta} = \frac{c\sigma}{2}\left\{\frac{1}{\zeta^2} + \left(\frac{m+1}{\zeta^2 - m}\right)\left[\frac{\zeta^2 m + 1}{\zeta^2} + \frac{2\left(\zeta^2 m + 1\right)}{\left(\zeta^2 - m\right)} - 2m\right] + 1\right\}$$

$$\psi'\,(z) = \frac{d\psi\,(\zeta)}{d\zeta}\frac{d\zeta}{dz} = (5.2810 + 7.5758i)\,\sigma \tag{c}$$

From Eq. (1.95), the real parts are

$$\sigma_x + \sigma_y = 4\,\mathrm{Re}\left[\gamma'\,(z)\right] = \sigma$$

$$\sigma_y - \sigma_x = 2\,\mathrm{Re}\left[\bar{z}\gamma''\,(z) + \psi'\,(z)\right] = 3.1405\sigma \tag{d}$$

$$\tau_{xy} = \mathrm{Im}\left[\bar{z}\gamma''\,(z) + \psi'\,(z)\right] = 6.281\sigma$$

from which the opening stress and the stress intensity factor become

$$\sigma_y = 2\,\mathrm{Re}\left[\gamma'(z)\right] + \left[\bar{z}\gamma''(z) + \psi'(z)\right]$$

$$K_I = \left\{2\,\mathrm{Re}\left[\gamma'(z)\right] + \left[\bar{z}\gamma''(z) + \psi'(z)\right]\right\}\sqrt{\pi a}$$

Inserting Eqs. (a) through (c) into (d) the above combination of stress equations yields the elastic stresses as

$$\sigma_y = 2.0753\sigma;\ \ \sigma_x = 1.0753\sigma;\ \ \tau_{xy} = 6.281\sigma \tag{e}$$

If $\sigma = 100$ *MPa, then the elastic stresses near the crack tips are*

$$\sigma_y \simeq 207.53\,\mathrm{MPa};\ \ \sigma_x = 107.53\,\mathrm{MPa};\ \ \tau_{xy} = 628.10\,\mathrm{MPa}$$

The stress intensity factor with a correction factor α *can be found in Chap. 3 as*

$$K_I = \alpha\sigma\sqrt{\pi a}$$

Although α *depends on the crack geometry and crack type, Eq. (e) for* σ_y *gives the general correction factor for the* K_I *equation. The conformal mapping used in this example yields* $\alpha = \sigma_y/\sigma = 2.0753$

$$K_I = \sigma_y\sqrt{\pi a} = 2.0753\sigma\sqrt{\pi a}\ \ \&\ \ \alpha = 2.0753\ \ \textit{(Conformal mapping)}$$

$$K_I = (2.0753)\,(100\ \mathrm{MPa})\,\sqrt{\pi\,(1\mathrm{x}10^{-3}\ \mathrm{m})}$$

$$K_I = 11.63\ \mathrm{MPa}\sqrt{\mathrm{m}}$$

Using Eq. (4.13) yields the same value for stress intensity factor along with $\theta = 0$ *such that the crack will grow along its own plane (x-axis). Hence,*

$$K_I = \sigma_y\sqrt{\pi a}\cos\frac{\theta}{2}\left(1 + \sin\frac{\theta}{2}\sin\frac{3\theta}{2}\right)\ \ @\ \theta = 0$$

$$K_I = 2.0753\sigma\sqrt{\pi a}\ \ \&\ \ \alpha = 2.0753$$

$$K_I = (2.0753)\,(100\ \mathrm{MPa})\,\sqrt{\pi\,(1\mathrm{x}10^{-3}\ \mathrm{m})}$$

$$K_I = 11.63\ \mathrm{MPa}\sqrt{\mathrm{m}}$$

For comparison purposes, use an approximation scheme similar to the one intro-duced in Chap. 3 for pressure vessels to get the stress intensity factor K_I *for a infinite plane. Thus, the corrected* K_I *equation is*

$$K_I = \sigma_y\sqrt{\pi a/Q}.f(\lambda)$$

$$f(\lambda) = \left[\sin^2(\lambda) + \left(\frac{a}{c}\right)^2\cos^2(\lambda)\right]^{1/4}$$

$$Q = 1 + 1.464\left(\frac{a}{c}\right)^{1.65} = 1 + 1.464\left(\frac{c/2}{c}\right)^{1.65} = 1.4665$$

where Q is the flaw shape factor, Eq. (3.44) or Fig. 3.6, and the maximum K_I value is obtainable when $\lambda = 90°$ (Fig. 3.5) such that $f(\lambda) = 1$ [6]. Then,

$$K_I = \sigma_y \sqrt{\pi a / Q} f(\lambda) = 2.0753\sigma \sqrt{\pi a / 1.4665}$$

$$K_I = 1.7196\sigma \sqrt{\pi a} \quad \& \quad \alpha = 1.7196 \quad \text{(Crack geometry)}$$

$$K_I = (1.7196)(100 \text{ MPa}) \sqrt{\pi (1 * 10^{-3} \text{ m})}$$

$$K_I = 9.64 \text{ MPa}\sqrt{\text{m}}$$

Despite that conformal mapping and the Airy stress function approaches do not incorporate any correction factor due to the crack geometry, the induced correction factor α for K_I is due to the fundamental combination of stresses. Both complex variable methods provided the same K_I value under the current conditions, and it is approximately 17 % higher than the K_I equation due to the crack shape factor Q (crack geometry factor). Hence,

$$K_I = \left(11.63 \text{ MPa}\sqrt{\text{m}}\right) \cos\frac{\theta}{2}\left(1 + \sin\frac{\theta}{2}\sin\frac{3\theta}{2}\right)$$

$$K_I(\text{max}) = 11.63 \text{ MPa}\sqrt{\text{m}} \quad at \; \theta = 0$$

$$K_I(\text{min}) = 4.45 \text{ MPa}\sqrt{\text{m}} \quad at \; \theta = \pi/2 = 90°$$

The figure below shows that K_I yields a nonlinear curve, which has extreme points at $\theta = 0$ and $\theta = \pi/2$ corresponding to the maximum and minimum K_I values, respectively. It has been assumed that the crack extends along its own plane.

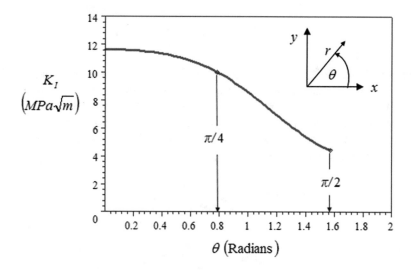

Despite that the stress intensity factor is one of the most important parameters in the fracture mechanics, determining its value at a particular loading mode related to specimen configuration is rather an experimental difficulty due to the ASTM requirements for validity.

This example problem only includes the logical mathematical procedure in a complex domain for calculating the stress intensity factor K_I. The ASTM validity assessment is not included since the determination of the plane-strain fracture toughness, K_{IC}, is not required. Therefore, the above procedure is acceptable from a theoretical perspective.

4.6.2 Point Forces

Although there are many reports and solutions to elasticity problems that can be found in the literature, it is pertinent to highlight a complex variable method for solving the problem of concentrated forces on the face of a rivet hole containing a crack in an infinite plate [7, 32, 39, 43, 49, 51]. Figure 4.16 schematically shows a rivet hole under uniform pressure (Fig. 4.16a) and two cracks emanating from a rivet hole (Fig. 4.16b) due to concentrated forces.

Consider the infinite plate containing a single hole bounded by the contour C and assume that the origin of the Cartesian coordinates lies in the center of the hole. If the resultant force $F = X + iY$ acts over C (Fig. 4.16a), then Eq. (4.114) gives the analytic complex potentials as

$$\gamma(z) = -\frac{X + iY}{2\pi(1 + \kappa)} \log(z) + \gamma^*(z)$$

(4.139)

$$\psi(z) = \frac{\kappa(X - iY)}{2\pi(1 + \kappa)} \log(z) + \psi^*(z)$$

where $\gamma^*(z) = \sum_{n=1}^{\infty} a_n z^{-n}$ and $\psi^*(z) = \sum_{n=1}^{\infty} b_n z^{-n}$ are analytic and single-valued complex potentials in the plate for large $|z|$ [24, 38].

In fact, $\gamma^*(z)$ and $\psi^*(z)$ can be eliminated from Eq. (4.139) since $\log(z)$ dominates the complex potentials at large distances from the hole. This implies that the moment is zero when the contour C surrounds the point $z = 0$ and consequently, the resultant force is simply $X + iY$ [24]. Thus,

$$\gamma(z) = -\frac{X + iY}{2\pi(1 + \kappa)} \log(z)$$

(4.140)

$$\psi(z) = \frac{\kappa(X - iY)}{2\pi(1 + \kappa)} \log(z)$$

However, if the resultant force $X + iY$ and the moment M act at a point z_o, then the induced complex potentials are [24]

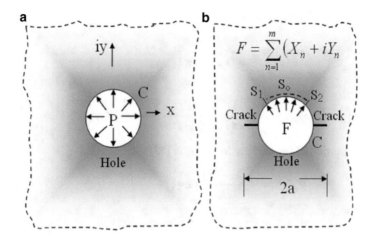

Fig. 4.16 (a) Rivet hole under uniform force and (b) a crack emanating from a rivet hole under concentrated forces on the upper face in an infinite plate

$$\gamma\,(z) = -\frac{X + iY}{2\pi\,(1 + \kappa)}\,\log\,(z - z_o)$$

(4.141)

$$\psi\,(z) = \frac{\kappa\,(X - iY)}{2\pi\,(1 + \kappa)}\,\log\,(z - z_o) + \frac{\kappa\,(X + iY)}{2\pi\,(1 + \kappa)}\,\frac{\bar{z}_o}{z - z_o}$$

$$+\,\frac{iM}{2\pi\,(z - z_o)}$$

Consider the case shown in Fig. 4.16b for a single force acting on the hole at point S_o. Thus, the applicable complex potentials described by Muskhelishvili [38] for a region mapped outside the unit circle take the form

$$\gamma\,(z) = -\frac{X + iY}{2\pi\,(1 + \kappa)}\,\log\,(z) + \gamma_o\,(z)$$

(4.142)

$$\psi\,(z) = \frac{\kappa\,(X - iY)}{2\pi\,(1 + \kappa)}\,\log\,(z) + \psi_o\,(z)$$

For a point force on the unit circle, the boundary condition on the contour C is

$$h\,(z) = \gamma\,(z) + \overline{\gamma'\,(z)} + \overline{\psi\,(z)}$$

(1.143)

Substituting Eq. (4.142) into (4.143) yields

$$h_o = \frac{(X - iY)}{2\pi}\,\log\,(z) + \frac{(X - iY)}{2\pi\,(1 + \kappa)}\,\frac{(1 + z^2)}{(1 - z^2)}$$

(4.144)

Assume that the Cauchy integrals are

$$\gamma_o(z) = -\frac{1}{2\pi i} \int_C \frac{h_o d\zeta}{\zeta - z}$$

(4.145)

$$\psi_o(z) = -\frac{1}{2\pi i} \int_C \frac{h_o d\zeta}{\zeta - z} - \frac{\zeta\left(\zeta^2 + 1\right)}{\zeta^2 - 1} \gamma_o'(\zeta)$$

Then, the solution of the Cauchy integrals is [32]

$$\gamma_o(z) = -\frac{(X - iY)}{2\pi} [\log(\zeta - z) - \log(\zeta)]$$

$$\psi_o(z) = \frac{(X - iY)}{2\pi} [\log(\zeta - z) - \log(\zeta)]$$

(4.146)

$$-\frac{\zeta\left(\zeta^2 + 1\right)}{\zeta^2 - 1} \frac{(X - iY)}{2\pi} \left[\frac{1}{\zeta - z} + \frac{1}{\zeta}\right]$$

Combining Eqs. (4.142) and (4.146) gives

$$\gamma(z) = -\frac{X + iY}{2\pi} \left[\log(\zeta - z) + \frac{\kappa}{\kappa + 1} \log(\zeta)\right]$$

$$\psi(z) = \frac{\kappa(X - iY)}{2\pi(1 + \kappa)} \log(\zeta) + \frac{(X - iY)}{2\pi} [\log(\zeta - z) - \log(\zeta)]$$

(4.147)

$$-\frac{\zeta\left(\zeta^2 + 1\right)}{\zeta^2 - 1} \frac{(X - iY)}{2\pi} \left[\frac{1}{\zeta - z} + \frac{1}{\zeta}\right]$$

Figure 4.16b illustrates that there exists multiple forces having real and imaginary parts. Hence, the total complex force and its conjugate take the form

$$F = \sum_{n=1}^{m} (X_n + iY_n)$$

(a)

$$\overline{F} = \sum_{n=1}^{m} (X_n - iY_n)$$

(b)

Substituting Eqs. (a) and (b) into (1.47) yields potential functions

$$\gamma(z) = -\frac{F}{2\pi} \left[\log(\zeta - z) + \frac{\kappa}{\kappa + 1} \log(\zeta)\right]$$

$$\psi(z) = \frac{\kappa \overline{F}}{2\pi(1 + \kappa)} \log(\zeta) + \frac{\overline{F}}{2\pi} [\log(\zeta - z) - \log(\zeta)]$$

(4.148)

$$-\frac{\zeta\left(\zeta^2 + 1\right)}{\zeta^2 - 1} \frac{\overline{F}}{2\pi} \left[\frac{1}{\zeta - z} + \frac{1}{\zeta}\right]$$

Fig. 4.17 Traction forces
acting on the elliptical crack
surface in the z-plane

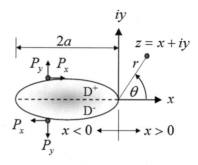

where F and $\overline{F}$ are total forces per unit thickness. Now the complex stress intensity factor (SIF) equation at the crack tips can be defined as [23, 32]

$$K_I - iK_{II} = 2\sqrt{\frac{\pi}{a}}\,\gamma'(z) \quad \text{for } |z| = \pm 1 \tag{4.149}$$

From Eq. (4.148),

$$K_I - lK_{II} = F\sqrt{\frac{1}{\pi a}}\left[\frac{1}{z - \zeta} - \frac{\kappa}{(\kappa + 1)\zeta}\right] \tag{4.150}$$

where K_I and K_{II} are the real and imaginary parts given by Eq. (4.150), respectively.

4.6.3   Complex T-Stress

Furthermore, it is appropriate to include the T-stress (T_x) in terms of complex potentials to account for the effects of stress biaxiality on the stress intensity factor K_I. The theoretical analysis that follows is described by Chen et al. [18] using the crack front position and crack back position techniques Consider the case when a stress distribution or traction forces ($P_y - iP_{xy}$) act on the crack surface. Thus, Eq. (4.89) along with the T_x acting parallel to the crack flanks can be written as

$$\begin{bmatrix} \sigma_x & \tau_{xy} \\ \tau_{xy} & \sigma_y \end{bmatrix} = \frac{K_I}{\sqrt{2\pi r}}\begin{bmatrix} f_x(\theta) & f_{xy}(\theta) \\ f_{xy}(\theta) & f_y(\theta) \end{bmatrix} \tag{4.151}$$

$$+ \frac{K_{II}}{\sqrt{2\pi r}}\begin{bmatrix} g_x(\theta) & g_{xy}(\theta) \\ g_{xy}(\theta) & g_{xy}(\theta) \end{bmatrix} + \begin{bmatrix} T_x & P_{xy} \\ P_{xy} & P_y \end{bmatrix}$$

For mode I, assume a straight crack in a infinite plate subjected to infinite biaxial stresses, and let the origin of the Cartesian coordinates be located at the crack tip as shown in Fig. 4.17.

Evaluating Eq. (4.151) at $\theta = 0$ and $r = x$ as $r \to 0$ yields

$$\sigma_x = \frac{K_I}{\sqrt{2\pi r}} + T_x \text{ and } \sigma_y = \frac{K_I}{\sqrt{2\pi r}} + P_y \qquad (4.152)$$

and subtracting the expressions in Eq. (4.152) gives the T-stress in the crack front as

$$T_x = \lim_{r \to 0} \left(\sigma_x - \sigma_y\right) + P_y \text{ or } T_x = -\lim_{r \to 0} \left(\sigma_y - \sigma_x\right) + P_y \qquad (4.153)$$

For mode II, the upper and lower stresses due to the traction forces acting on the crack surfaces (Fig. 4.14) are evaluated at $\theta = \pm\pi$ and $r = -x$ as $r \to 0$ so that

$$\sigma_x^+ = -\frac{2K_{II}}{\sqrt{2\pi r}} + T_x \text{ and } \sigma_y^+ = P_y \text{ at } \theta = +\pi \qquad (4.154a)$$

$$\sigma_x^- = \frac{2K_{II}}{\sqrt{2\pi r}} + T_x \text{ and } \sigma_y^- = P_y \text{ at } \theta = -\pi \qquad (4.154b)$$

Combining Eqs. (1.95) and (4.153) yields the T-stress in terms of complex potentials at the crack front [18]

$$T_x = -\text{Re}\left[\sigma_y - \sigma_x + 2i\tau_{xy}\right]_{z=r} + P_y$$
$$T_x = -2\text{Re}\left[\bar{z}\gamma''(z) + \psi'(z)\right]_{z=r} + P_y \qquad (4.155)$$

Similarly, Eqs. (1.95) and (4.154) yields the T-stress for the crack back position

$$T_x = \lim_{r \to 0}\left[\left(\sigma_x^+ + \sigma_y^+\right) + \left(\sigma_x^- + \sigma_y^-\right)\right]_{z=r} - P_y$$
$$T_x = 2\text{Re}\left[\gamma'(z)^+ + \overline{\gamma'(z)^-}\right]_{z=-x} - P_y \qquad (4.156)$$
$$T_x = 4\text{Re}\left[\gamma'(z)^+\right]_{z=-x} - P_y$$

In summary, the T-stress expressions, Eqs. (4.155) and (4.156), can be evaluated in the interval $-\infty < z < \infty$ when $y = 0$. If mode I prevails, then T_x is determined by Eq. (4.89). It should be mentioned that the influence of shear stresses due to dislocations ahead of the crack tip has been excluded in the above analytical analysis in order to keep the mathematics as simple as possible.

4.7 Problems

4.1. (a) Calculate K_I using the singularity and non-singularity stresses for a single-edge cracked plate having $a = 4\,\text{mm}$, $x = 0.1$, and $L/w \geq 1.5$ and subjected to $300\,\text{MPa}$ in tension. (b) Plot the stress ratio T_x/σ and the stress biaxiality ratio β as functions of $x = a/w$. Here, w is the width of the plate.

4.2. Using the information given in problem 4.1, calculate the stress intensity factor when $T_x = 0$.

4.3. Assume that a single-edge crack in a plate is loaded in tension. Derive the dominant near crack tip stresses in rectangular coordinates using the Westergaard stress function

$$Z(z) = \frac{K_I}{\sqrt{2\pi z}} \quad \text{where} \quad K_I = \sigma\sqrt{\pi a}$$

4.4. Consider a unit circle with its center at the origin and let the function $F(z) = P(z) + iQ(z)$ be holomorphic inside the contour C. Also let the function $F(\zeta)$ take definite boundary values where $\zeta = e^{i\delta}$ gives the points on C. If the boundary condition is

$$F(\zeta) + \overline{F(\zeta)} = f(\delta)$$

determine the Cauchy integral formula and the integral for $F(z)$.

4.5. Show that the fundamental combination of stresses can be defined in terms of Cauchy integral formula:

$$\sigma_x + \sigma_y = \frac{1}{\pi i}\int_C \frac{hd\zeta}{(\zeta - z)^2} + \frac{1}{\pi i}\int_C \frac{\overline{h}d\zeta}{(\zeta - z)^2} - \frac{1}{\pi i}\int_C \frac{h(\zeta)\,d\zeta}{\zeta^2}$$

$$\sigma_y - \sigma_x + 2i\tau_{xy} = \frac{1}{\pi i}\int_C \frac{hd\zeta}{(\zeta - z)^3} + \frac{1}{\pi i}\int_C \frac{\overline{h}d\zeta}{(\zeta - z)^2} + \frac{1}{\pi iz^2}\int_C \frac{hd\zeta}{(\zeta - z)^2}$$

$$-\frac{1}{\pi iz^2}\int_C \frac{hd\zeta}{\zeta^2}$$

4.6. Suppose that a plate containing a single unstressed crack (Fig. 4.15) is deformed by unknown stresses at infinity and assume that the complex potential, $f'(z) = \sigma_{yy} - i\tau_{xy}$, represents the stress distributions across the crack contour C. (a) Determine the stress distribution on $y = 0$ outside the crack and (b) the upper and lower boundary functions for $|x| \leq a$.

4.7. Consider an infinite plate subjected to a remote tensile stress (S) normal to the direction of a central crack. If the complex potentials for this type of crack are

$$\gamma(z) = \frac{S}{4}\left[2\sqrt{z^2 - a^2} - z\right]$$

$$\chi'(z) = \psi(z) = \frac{S}{2}\left[z - \frac{a^2}{\sqrt{z^2 - a^2}}\right]$$

then determine the crack tip stresses using the Westergaard stress function $Z(z) = 2\gamma'(z)$ when $x >> a$.

4.8. Consider the elliptical crack shown below and assume that the crack is in the z-plane where $z = \zeta + a$ and $p(z) = p(\zeta)$. Derive the stress equations using the given crack geometry and the Westergaard complex method. Make assumptions if needed.

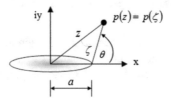

4.9. Assume that an infinite plate contains a through-central crack along the x-axis. If the plate is subjected to a remote or infinite stress loading condition, $\sigma_y = \sigma_y^\infty = S, \sigma_x^\infty = \tau_{xy}^\infty = 0$, then use the Sanford [18] complex equation

$$2\gamma'(z) = Z(z) - \psi^*(z)$$

Here, $\gamma'(z)$ may be defined by a Cauchy integral, $Z(z)$ is the Westergaard stress function, $\psi^*(z)$ is a complex polynomial, and z is the complex variable define as $z = x + iy$. Based on this information, determine a function for the stress intensity factor K_I and expand $\psi^*(z)$ when $n = 0$ and 1. Let $z = a$ be half the total crack length.

4.10. Derive the stresses defined by Eq. (4.13) using the Westergaard functions and the complex stress equations given below

$$Z' = \frac{K_I}{\sqrt{2\pi(z-a)}} \quad \text{and } Z'' = \frac{K_I}{\sqrt{4\pi(z-a)^3}}$$

and

$$\sigma_x = \text{Re}\,Z' - y\,\text{Im}\,Z''$$
$$\sigma_x = \text{Re}\,Z' + y\,\text{Im}\,Z''$$
$$\tau_{xy} = -y\,\text{Re}\,Z''$$

References

1. M. Adda-Bedia, Brittle fracture dynamics with arbitrary paths III. The branching instability under general loading. J. Mech. Phys. Solids **53**, 227–248 (2005)
2. M. Adda-Bedia, R. Arias, Brittle fracture dynamics with arbitrary paths I. Kinking of a dynamic crack in general antiplane loading. J. Mech. Phys. Solids **51**, 1287–1304 (2003)
3. A.R. Akisanya, N.A. Fleck, Brittle fracture of adhesive joints. Int. J. Fract. **58**, 93–114 (1992)
4. A.R. Akisanya, N.A. Fleck, Analysis of a Wavy Crack in Sandwich Specimens. Int. J. Fract. **55**, 29–45 (1992)
5. M.H. Aliabadi, D.P. Rooke, *Numerical Fracture Mechanics*, chap. 2 (Computational Mechanics Publications, Kluwer Academic Publishers, Boston, 1992)
6. T.L. Anderson, *Fracture Mechanics: Fundamentals and Applications* (Taylor & Francis Group, LLC, CRC Press, Inc., Boston, 1991), p. 49
7. A. Atre, W.S. Johnson, J.C. Newman Jr., *Assessment of Residual Stresses and Hole Quality on the Fatigue Behavior of Aircraft Structural Joints. Volume 3: Finite Element Simulation of Riveting Process and Fatigue Lives*, DOT/FAA/AR-07/56, V3, National Technical Information Service (NTIS), Springfield, VA (March 2009), pp. 1–166
8. A. Azhdari, M. Obata, S. Nemat-Nasser, Alternative solution methods for crack problems in plane anisotropic elasticity, with examples. Int. J. Solids Struct. **37**, 6433–6478 (2000)
9. T.L. Becker Jr., R.M. Cannon, R.O. Ritchie, Finite crack kinking and T-stresses in functionally graded materials. Int. J. Solids Struct. **38**, 5545–5563 (2001)
10. K. Bertram Broberg, *Cracks and Fracture* (Elsevier Inc., New York, 1999)
11. D. Broek, *Elementary Engineering Fracture Mechanics*, chap. 3, 4th edn. (Kluwer Academic Publisher, Boston, 1986)
12. E. Bouchbinder, J. Mathiessen, I. Procaccia, Stress field around arbitrarily shaped cracks in two-dimensional elastic materials. Phys. Rev. E **69**(2), 026127-1–026127-7 (2004)
13. A.F. Bower, *Applied Mechanics of Solids* (CRC Press, Taylor and Francis Group, LLC, New York, 2010), pp. 278–280, 572–573
14. J.R. Bowler, Thin-skin eddy-current inversion for the determination of crack shapes, in *Inverse Problems*, vol. 18, No. 6 (IOP Publishing, Bristol, 2002), pp. 1891–1905
15. J.R. Bowler, N. Bowler, Evaluation of the magnetic field near a crack with application to magnetic particle inspection. J. Phys. D: Appl. Phys. **35**, 2237–2242 (2002)
16. A. Carabineanu, N. Peride, E. Rapeanu, E.M. Craciun, Mathematical modelling of the interface crack. A new improved numerical method. Comput. Mater. Sci. **46**, 677–681 (2009)
17. B. Chen, The effect of the T-stress on crack path selection in adhesively bonded joints. Int. J. Adhes. Adhes. **21**(5), 357–368 (2001)
18. Y.Z. Chen, Z.X. Wang, X.Y. Lin, Crack front position and crack back position techniques for evaluating the T-stress at crack tip using functions of a complex variable. J. Mech. Mater. Struct. **3**(9), 1659–1673 (2008)
19. B. Cotterell, J.R. Rice, Slightly curved or kinked cracks. Int. J. Fract. **16**, 155–169 (1980)
20. E.-M. Craciun, Antiplane fracture analysis of a crack in a monoclinic composite. An. St. Univ. Ovidius Constanta **11**(1), 69–82 (2003)
21. J.W. Dally, W.F. Riley, *Experimental Stress Analysis*, 3rd edn. (McGraw-Hill, Inc., New York, 1991)
22. M. Denda, Y.F. Dong, Complex variable approach to the BEM for multiple crack problems. Comput. Meth. Appl. Mech. Eng. **141**, 247–264 (1997)
23. M. Denda, I. Kosaka, Dislocation and point-force-based approach to the special Green's function BEM for elliptic hole and crack problems in two dimensions. Int. J. Numer. Meth. Eng. **40**, 2857–2889 (1997)
24. A.H. England, *Complex Variable Methods in Elasticity*, chap. 3, vol. 56 (Dover Publications, Inc., Mineola, NY, 2003), pp. 77–81

25. T. Fett, *A Compendium of T-Stress Solutions*. Forschungszentrum Karlsruhe, Technik and Umwelt, Wissenschaftliche Berichte, FZKA 6057 (1998), pp. 1–72, www.ubka.uni-karlsruhe. de
26. T. Fett, *Stress Intensity Factors and T-stress for Cracked Circular Disks*. Forschungszentrum Karlsruhe, Technik and Umwelt, Wissenschaftliche Berichte, FZKA 6484 (2000), pp. 1–68, www.ubka.uni-karlsruhe.de
27. T. Fett, Stress intensity factors and T-stress for interiorly cracked circular disks under various boundary conditions. Eng. Fract. Mech. **68**(9), 1119–1136 (2001)
28. T. Fett, Stress intensity factors and T-stress for single and double-edge cracked circular disks under mixed boundary conditions . Eng. Fract. Mech. **69**(1), 69–83 (2002)
29. T. Fett, D. Munz, G. Thun, Fracture toughness testing on bars under opposite cylinder loading, in *ASME, International Gas Turbine and Institute, Turbo Expo IGTI*, vol. 4A (2002), pp. 97–102
30. L.L.Fischer, G.E. Beltz, The effect of crack blunting on the competition between dislocation nucleation and cleavage. J. Mech. Phys. Solids **49**, 635–654 (2001)
31. N.A. Fleck, J.W. Hutchinson, Z. Suo, Crack path selection in a brittle adhesive layer. Int. J. Solids Struct. **27**(13), 1683–1703 (1991)
32. S. Hawanale, A.K. Bakare, S.P. Chavan, A novel method for the determination of stress intensity factor for a crack emanating from riveted hole. J. Inst. Eng. (India) Mech. Eng. Division **86**, 160–163 (2005)
33. K. Hellan, *Introduction to Fracture Mechanics* (McGraw-Hill Book Company, New York, 1984)
34. J.H. Kim, G.H. Paulino, T-stress, mixed-mode stress intensity factors and crack initiation angles in fuctionally graded materials: a unified approach using the interaction integral method. Comput. Meth. Appl. Mech. Eng. **192**, 1463–1494 (2003)
35. M.T. Kirk, R.H. Dodds Jr., T.L. Anderson, An application technique for predicting size effects on cleavage fracture toughness (J_c) using elastic T-stress, in *Fracture Mechanics*, vol. 24, ASTM STP 1207 (1994), pp. 62–86
36. S.G. Larsson, A.J. Carlsson, Influence of non-singular stress terms and specimen geometry on small-scale yielding at crack tips in elastic–plastic materials. J. Mech. Phys. Solids **21**, 263–277 (1973)
37. P.S. Leevers, J.C. Radon, Inherent biaxiality in various fracture specimen geometries. Int. J. **19**, 311–325 (1982)
38. N.I. Muskhelishvili, *Some Basic Problems of the Mathematical Theory of Elasticity*, 4th Russian edition (1954) & 2nd English edition (1977), Translated from the Russian Language by J.R.M. Radok, chaps. 7, 12–15,18 (Noordhoff International Publishing, Layden, 1977)
39. J.C. Newman Jr., C.E. Harris, R.S. Piascik, D.S. Dawicke, *Methodology for Predicting the Onset of Widespread Fatigue Damage in Lap-Splice Joints*, NASA/TM-1998-208975, NASA Langley Research Center Hampton, VA (December 1998), pp. 1–20
40. N.P. O'Dowd, C.F. Shih, Family of crack-tip fields characterized by a triaxiality parameter: I structure of fields. J. Mech. Phys. Solids **39**, 989–1015 (1991)
41. N.P. O'Dowd, C.F. Shih, Family of crack-tip fields characterized by a triaxiality parameter: II fracture applications. J. Mech. Phys. Solids **40**, 939–963 (1992)
42. J. Qu, Wang, *Elastic T-Stress Solutions of Embedded Elliptical Cracks Subjected to Uniaxial and Biaxial Loadings*. ASTM STP, vol. 1480 (ASTM, *Philadelphia*, 2007), pp. 295–308
43. A. Rahman, J. Bakuckas Jr., C. Bigelow, P. Tan, *Boundary Correction Factors for Elliptic Surface Cracks Emanating from Countersunk Rivet Holes Under Tension, Bending, and Wedge Loading Conditions*, DOT/FAA/AR-98/37, National Technical Information Service (NTIS), Springfield, VA, (March 1999), pp. 1–70
44. M.H. Sadd, *Elasticity: Theory, Applications and Numerics*, chap. 10, 2nd edn. (Elsevier Inc., New York, 2009)
45. A.S. Saada, *Elasticity Theory and Applications*, 2nd edn. (J. Ross Publishing, Inc., New York, 2009), p. 764

46. A. Saxena, *Nonlinear Fracture Mechanics for Engineers* (CRC Press, LLC, New York, 1998), p. 221
47. A.H. Sherry, C.C. France, L. Edwards, Compendium of T-stress solutions for two and three dimensional cracked geometries. Int. J. **18**, 141–155 (1995)
48. S.P. Timoshenko, J.N. Goodier, *Theory of Elasticity*, 3rd edn. (McGraw-Hill, New York, 1970)
49. M.P. Szolwinski, G. Harish, P.A. McVeigh, T.N. Farris, *Experimental Study of Fretting Crack Nucleation in Aerospace Alloys with Emphasis on Life Prediction*, Fretting Fatigue: Current Technologies and Practices, ASTM STP 1367 (1999)
50. J. Weertman, *Dislocation Based Fracture Mechanics* (World Scientific, New Jersey, 1996), pp. 6–18, 166
51. R.B. Waterhouse, Avoidance of fretting fatigue failures, in *Fretting Fatigue* (Applied Science Publishers, London, 1981), pp. 221–240
52. M.L. Williams, On the stress distribution at the base of a stationary crack. J. Appl. Mech. **24**, 109–114 (1957)

Crack Tip Plasticity

5

5.1 Introduction

This chapter includes some theoretical aspects of linear-elastic fracture mechanics (LEFM) when a crack tip undergoes some plasticity [1–10, 12–24]. For a small-scale yielding (*SSY*) approximation, the plastic zone size (r) is used to define the classical singularity in the elastic field equations derived in Chap. 4 as $r^{-1/2}$. Some models for the plastic zone size (r) are considered herein since r can be related to an effective crack length ($a_e = a + r$) for determining the effective stress intensity factor (K_{eff}) for modes *I*, *II*, or *III*. Generally, mode *I* is the most common load. However, if plasticity ahead of the crack tip is significantly large, then elastic-plastic fracture mechanics (EPFM) is used to characterize the fracture resistance of a ductile material.

As a result of plastic deformation ahead of the crack tip, the plastic zone develops and the crack tip blunts to an extent, while the crack faces separate, making the crack mouth opening displacement (CMOD), the crack opening displacement (COD), the crack tip opening displacement (CTOD), or the crack tip opening angle (CTOA) a measurable parameter related to fracture resistance [32]. In particular, during the plastic deformation process, the plastic region ahead of the crack tip is characterized by determining its size. Usually, the crack tip opening displacement ($\delta_t = CTOD$) [68] and the *J*-integral [52] are the two most common parameters used in EPFM for determining the fracture toughness of an elastic-plastic material. In fact, EPFM considers a substantial plastic deformation under quasi-static conditions, and it is analyzed using the large-scale yielding (*LSY*) approach for which r in elastic-plastic (ductile) materials is significantly large ($r \geq a, w, B$) and failure is typically governed by the critical crack tip opening displacement ($\delta_{tc} = CTOD$) and the path-independent *J*-integral (J_{IC}) at the onset of crack growth under mode *I* loading. Recall that the *J*-integral is a line integral around the crack tip that measures fracture toughness in terms of nonlinear energy release rate for crack growth. The *J*-integral can also be used for other loading modes.

© Springer International Publishing Switzerland 2017
N. Perez, *Fracture Mechanics*, DOI 10.1007/978-3-319-24999-5_5

5.2   Crack Tip Stress State

Consider mode I loading under quasi-static conditions in the field of linear-elastic fracture mechanics (LEFM), where the crack opening stress derived in Chap. 4 takes the form $\sigma_y = \left(K_I / \sqrt{2\pi r} \right) f_I (\theta)$. This equation describes the magnitude of the stress $(K_I / \sqrt{2\pi r})$ and its distribution ahead of the crack tip at (r, θ). Recall that $f_I (\theta)$ is a trigonometric function, r is the plastic zone size, and theoretically $\sigma_y \to \infty$ as $r \to 0$. In practice, σ_y is limited to the yield strength (σ_{ys}) of the material, r is very small (SSY = small-scale yielding) compared with the specimen dimensions ($r << a, w, B$), and fracture is governed by the critical stress intensity factor $K_C = G_C / E$ for plane stress and $K_{IC} = \left(1 - v^2 \right) G_{IC} / E$ for plane-strain conditions, where G_C or G_{IC} is the critical strain energy release rate or the driving force for crack propagation at $a_c > a$, where a_c is the critical length and a is allowed to grow to a maximum so that $a = a_c$.

Most engineering metallic materials are subjected to an irreversible plastic deformation. If plastic deformation occurs, then the elastic stresses are limited by yielding since stress singularity cannot occur, but stress relaxation takes place within the plastic zone. This plastic deformation occurs in a small region and it is called the crack tip plastic zone.

On the other hand, a large-scale yielding (LSY) corresponds to a large plastic zone, which occurs in ductile materials due to $r >> a$. As a consequence of plastic deformation ahead of the crack tip, the linear-elastic fracture mechanics (LEFM) theory is limited to $r << a$; otherwise, elastic-plastic fracture mechanics (EPFM) theory controls the fracture process since $(r \geq a)$.

Plane Strain

1. Large thickness B, $\epsilon_z \simeq 0$ in an internal region and $\sigma_z = v \left(\sigma_x + \sigma_y \right)$. This means that the material is constrained in the z-direction due to a sufficiently large thickness and the absence of strain in this axis. In fact, the stress in the z-direction develops due to the Poisson's effect.
2. Yielding is suppressed due to the kinematics constraints from the surrounding elastic material. Kinematics describes the motion of points.
3. Plastic deformation is associated with the hinge mechanism (internal necking in Fig. 5.1).
4. The plastic zone size is small in the midsection of the plate (Fig. 5.1a). This condition implies that the plastic zone must be smaller than the crack length.

Plane Stress

1. The thickness B is small, $\sigma_z = 0$, and $\epsilon_z \neq 0$ through the whole thickness. Consequently, a biaxial state of stress may result.
2. If $\sigma_y \geq \sigma_x > 0$ (Tresca criterion), then yielding occurs by a cumulative slip mechanism (Fig. 5.1b).

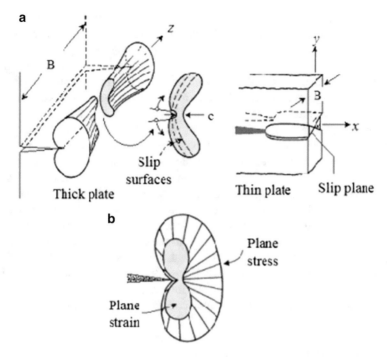

Fig. 5.1 Schematic plastic zone shapes. (**a**) 3-dimensional plastic zone and the plastic hinge model [25] and (**b**) plastic zone surfaces [13]

3. The height of the yielded zone is limited due to the slip mechanism.
4. The total motion has a necking effect in front of the crack as it opens.

5.3   The Plastic Zone Shape

Figure 5.1 shows schematic plastic zones for plane stress (thin plate) and plane-strain (thick plate) conditions. Some requirements for plane conditions were introduced in Chap. 1, but they are included henceforth from a different perspective. Also depicted in Fig. 5.1 is the hinge model for plasticity ahead of a crack tip [9, 16, 25–34, 36, 44, 49].

In this chapter, a few models for the configuration or the shape of the crack tip plasticity are included. It is essential to have a thorough knowledge of the shape and size of the plastic zone in order to compare theoretical and experimental results for plane stress and plane-strain conditions. Furthermore, the formation of the plastic zone depends on the material properties, specimen or structural element configuration, and loading conditions [50–65, 67–70].

Most solid materials develop plastic strains when the yield strength is exceeded in the region near a crack tip. Thus, the amount of plastic deformation is restricted by the surrounding material, which remains elastic during loading.

Theoretically, linear-elastic stress analysis of sharp cracks predicts infinite stresses ($\sigma_{ij} \to \infty$) at the crack tip. In fact, inelastic deformation, such as plasticity in metals and crazing in polymers, leads to relaxation of crack tip stresses caused by the yielding phenomenon at the crack tip [2]. As a result, a plastic zone is formed containing microstructural defects such as dislocations and voids. Consequently, the local stresses are limited to the yield strength of the material. This implies that the elastic stress analysis becomes increasingly inaccurate as the inelastic region at the crack tip becomes sufficiently large and, consequently, linear-elastic fracture mechanics (LEFM) is no longer useful for predicting the field equations [2].

The size of the plastic zone (r) can be estimated when moderate crack tip yielding occurs. Thus, the introduction of the plastic zone size, as a correction parameter that accounts for plasticity effects adjacent to the crack tip, is vital in determining the effective crack length ($a_e = a + r$) and the effective stress intensity factor, $K_{eff} = f(\sigma, a_e)$. The plastic zone, which develops in materials subjected to local yielding at the crack tip, can be determined for plane conditions, that is, plane strain for maximum constraint on relatively thick components and plane stress for variable constraint due to thickness effects of thin solid bodies.

For convenience, Eqs. (3.27) and (3.28) are included in this chapter as a starting point for determining the plastic zone size equation. Thus,

$$\sigma_y = \sigma \sqrt{\frac{a}{2r}} \tag{5.1}$$

$$K_I = \sigma \sqrt{\pi a} \tag{5.2}$$

Setting $\sigma_y = \sigma_{ys}$ in Eq. (5.1) means that plasticity exists adjacent to the crack tip. Notice that the double subscript in this local stress has being changed to one just for convenience. Combining these equations yields the plastic zone size as [28]

$$r = \frac{a}{2} \left(\frac{\sigma}{\sigma_{ys}} \right)^2 \tag{5.3}$$

Here, σ is the applied stress (MPa), σ_{ys} is the yield strength (MPa), and a is the starting crack length. It has been reported [28] that yielding at the crack tip causes the crack to behave as if it is larger than the actual size and, therefore, the tensile stress reaches a finite value.

Furthermore, the stress σ_y produces plastic work when $\sigma_{ys} \leq \sigma_y \leq \sigma_u$ due to an external applied stress σ, and it is transferred into strain energy density for plastic deformation to occur. This implies that the plastic zone size reaches a maximum magnitude when $\sigma_y = \sigma_u$, where σ_u is the ultimate tensile strength. Consequently, the effective crack size, $a_e = a + r$, becomes the fictitious crack length, which extends through the plastic zone due to initiation and coalescence of voids. This may repeat and continue until the actual crack reaches a critical value at the onset of crack propagation for fracture or separation.

5.4   Irwin's Approximation

Irwin [28] has shown that the effect on the plastic zone is to artificially extend the crack by a distance r_1 (Fig. 5.2) known as Irwin's plastic zone correction under small-scale yielding (*SSY*) for plates with thickness B.

The crack model in Fig. 5.2 can be modeled as a two-dimensional plate surface containing a crack as a plane with a particular domain in the (x,y) Cartesian coordinates. Accordingly, r and θ become the polar coordinates with the origin at the crack tip on the same plane.

The elastic stress distribution shown in Fig. 5.2 indicates that $\sigma_y \rightarrow \infty$ as $r \rightarrow 0$. Actually, σ_y is limited to σ_{ys} as shown by the elastic-plastic stress distribution. This means that $\sigma_y \rightarrow \infty$ occurs mathematically, not physically.

In order to account for the changes due to the artificial crack extension or virtual crack length and to visualize the plastic zone as a cylinder, the crack length a can be replaced by a_e in Eq. (5.3). Moreover, the virtual crack length defined by a_e is referred to as the effective crack length.

The conditions of equilibrium for a stationary crack tip include internal and external forces per unit length [5,69]. In such a case, the areas related to the shedding loads P_s and P_{ys} due to equal yielding; that is, $AP_s = AP_{ys}$ when the plastic zone size is $r \ll a$. This is schematically depicted in Fig. 5.2 for *SSY*. In fact, Irwin's

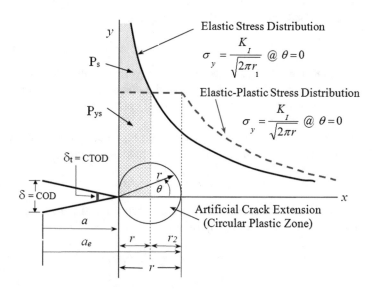

Fig. 5.2 Crack tip plastic zone model for mode I loading

approximation is an elastic-plastic fracture mechanics approach since $r > 0$ and $r << a$, but the elastic fracture mechanics part controls the fracture process ahead of the crack tip. These loads are the equilibrium forces per unit length defined by [16, 28]

$$P_s = B \int_o^{r_1} \left(\sigma - \lambda \sigma_{ys} \right) dx \tag{5.4}$$

$$P_{ys} = B \int_o^{r_2} \lambda \sigma_{ys} dx \tag{5.5}$$

where $\lambda = 1$ for plane stress and $\lambda = \sqrt{3}$ for plane strain [28].

For equilibrium conditions, the force balance $\sum \left(P_s + P_{ys} \right) = 0$ leads to the determination of the plastic zone size. Hence,

$$\int_o^{r_1} \left(\sigma - \lambda \sigma_{ys} \right) dx - \int_o^{r_2} \lambda \sigma_{ys} dx = 0 \tag{5.6}$$

Inserting Eq. (3.1) into (5.6) and integrating yields

$$\int_o^{r_1} \left(\frac{K_I}{\sqrt{2\pi x}} - \lambda \sigma_{ys} \right) dx - \int_o^{r_2} \lambda \sigma_{ys} dx = 0 \tag{a}$$

$$\frac{2r_1 K_I}{\sqrt{2\pi r_1}} - \lambda \sigma_{ys} \left(r_1 + r_2 \right) = 0 \tag{b}$$

$$2r_1 \sigma_y - \lambda \sigma_{ys} \left(r_1 + r_2 \right) = 0 \tag{5.7}$$

When yielding occurs, the boundary between the elastic and the plastic regions is limited to the yield strength (σ_{ys}) of the solid material. Thus,

$$\sigma_y = \lambda \sigma_{ys} \tag{5.8}$$

Inserting Eq. (5.8) into (5.7) gives $2r_1 = r_1 + r_2$ which implies that $r_1 = r_2$ and $r = r_1 + r_2$ as shown in Fig. 5.2. Recall that the effective crack length is defined as $a_e = a + r$, and it is referred to as the virtual crack length proposed by Irwin [16, 28].

Replacing the crack length a for $a_e = a + r$ into Eq. (3.28) provides the effective stress intensity factor for a stationary crack under monotonic mode I loading as

$$K_I = \sigma \sqrt{\pi \left(a + r \right)} = \sigma \sqrt{\pi a_e} \tag{5.9}$$

This K_I equation is the corrected stress intensity factor due to finite specimen size and plasticity. Now, inserting Eq. (5.3) into (5.9) yields

$$K_I = \sigma \sqrt{\pi a \left[1 + \frac{1}{2} \left(\frac{\sigma}{\sigma_{ys}} \right)^2 \right]} \tag{5.10}$$

Furthermore, the plastic zone size for plane conditions can easily be determined by combining Eqs. (3.1) and (5.8). Thus, Irwin's plastic zone expression is derived as

$$r = \frac{1}{2\pi} \left(\frac{K_I}{\lambda \sigma_{ys}} \right)^2 = \frac{a}{2} \left(\frac{\sigma}{\lambda \sigma_{ys}} \right)^2 \tag{5.11}$$

5.5 Dugdale's Approximation

Dugdale [19] proposed a strip yield model (cohesive zone model) for thin elastic perfectly plastic material to characterize the plastic zone under mode I plane stress conditions. Barenblatt [4] considered a slightly different approach to yield similar results.

Consider Fig. 5.3 which shows the plastic zones in the form of narrow strips extending a distance r each and carrying a closure stress equals to the yield stress σ_{ys} to prevent the crack from opening [5, 19]. The phenomenon of crack closure is caused by internal stresses since they tend to close the crack in the region where $a < x < c$.

In addition, the term $\delta_t = 2\mu_y$ in Fig. 5.3 is the so-called the crack tip opening displacement (CTOD) being twice the displacement (μ_y) in the y-direction perpendicular to the applied stress. Details on CTOD are included in a later section.

The Dugdale's model introduces an effective crack length (a_e) by adding the wedge-like plastic zone size (r). For half crack length due to symmetry, $a_e = a + r$ which makes the crack fictitiously longer than the real crack size. Consequently, this treatment induces an effective stress intensity factor $K_e > K_I$.

The central crack problem in Fig. 5.3 can be modeled using Westergaard complex function of the form [60]

$$Z(z) = \frac{P}{\pi} \frac{\sqrt{a^2 - x^2}}{(z - x) \sqrt{z^2 - a^2}} \tag{5.12}$$

Now the internal stress along the crack plane in the complex plane $z = x + iy$ with the system boundary at $y = 0$ is

$$\sigma_y = \text{Re}\, Z(z) = \frac{P}{\pi} \frac{\sqrt{a^2 - x^2}}{(z - x) \sqrt{z^2 - a^2}} \tag{5.13}$$

Fig. 5.3 Dugdale strip yield model [19] for non-strain hardening solids under plane stress conditions: Dugdale's strip yield model for a wedge crack loaded in tension. Taken from [9]

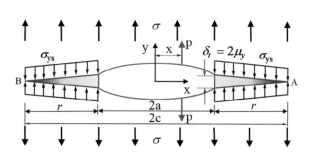

The stress intensity factor due to wedge internal forces is derived using Eqs. (3.1) and (5.14). For point A in Fig. 5.3,

$$K_A = \lim_{r \to 0} \sigma_y \sqrt{2\pi r} \quad \text{and} \quad r = (x - a) \tag{5.14}$$

$$K_A = \lim_{z \to a} \frac{P}{\pi} \frac{\sqrt{(a+x)(a-x)}\sqrt{2\pi (z-a)}}{(z-x)\sqrt{(z+a)(z-a)}} \tag{a}$$

$$K_A = \lim_{z \to a} \frac{P}{\pi} \frac{\sqrt{(a+x)(a-x)}\sqrt{2\pi}}{(z-x)\sqrt{(z+a)}} \tag{b}$$

$$K_A = \frac{P}{\pi} \frac{\sqrt{(a+x)(a-x)}\sqrt{2\pi}}{(a-x)\sqrt{(a+a)}} \tag{c}$$

Simplifying Eq. (c) yields

$$K_A = \frac{P}{\sqrt{\pi a}}\sqrt{\frac{a+x}{a-x}} \tag{5.15}$$

Similarly, the stress intensity factor at point B is

$$K_B = \frac{P}{\sqrt{\pi a}}\sqrt{\frac{a-x}{a+x}} \tag{5.16}$$

Furthermore, assume that stress singularities disappear when the following equality is true $K_\sigma = -K_I$, where K_σ is the applied stress intensity factor and K_I is due to yielding ahead of the crack tip [9]. The total stress intensity factor, $K_I = K_A + K_B$, is derived in a practical mathematical form by using a differential load per unit thickness, $dP = \sigma_{ys}dx$. Thus,

$$dK_A = \frac{dP}{\sqrt{\pi a}}\sqrt{\frac{a+x}{a-x}} = \frac{\sigma_{ys}}{\sqrt{\pi a}}\sqrt{\frac{a+x}{a-x}}dx \tag{a}$$

$$dK_B = \frac{dP}{\sqrt{\pi a}}\sqrt{\frac{a-x}{a+x}} = \frac{\sigma_{ys}}{\sqrt{\pi a}}\sqrt{\frac{a-x}{a+x}}dx \tag{b}$$

and

$$dK_I = \frac{\sigma_{ys}}{\sqrt{\pi a}}\int_a^{a+r} \left(\sqrt{\frac{a+x}{a-x}} + \sqrt{\frac{a-x}{a+x}} \right) dx \tag{5.17}$$

$$K_I = 2\sigma_{ys}\sqrt{\frac{a}{\pi}}\int_a^{a+r} \frac{dx}{\sqrt{a^2 - x^2}} \tag{5.18}$$

$$K_I = \frac{2\sigma_{ys}}{\pi}\sqrt{\pi (a+r)}\arccos\frac{a}{a+r} \tag{5.19}$$

Equating Eqs. (5.9) and (5.19) yields

$$\frac{\pi \sigma}{2\sigma_{ys}} = \arccos \frac{a}{a+r} \tag{5.20}$$

Let $y = \pi\sigma/2\sigma_{ys}$ so that

$$\frac{a}{a+r} = \cos y \tag{5.21}$$

$$r = a\,(\sec y - 1) \tag{5.22}$$

Expanding the trigonometric function Eq. (5.22) gives

$$\sec y = 1 + \frac{y^2}{2!} + \frac{y^4}{4!} + \frac{y^6}{6!} + \dots \simeq 1 + \frac{y^2}{2} \tag{5.23}$$

Inserting Eq. (5.23) into (5.22) gives

$$r = \frac{ay^2}{2} = \frac{a}{2}\left(\frac{\pi\sigma}{2\sigma_{ys}}\right)^2 \tag{5.24}$$

Substituting Eq. (5.24) into (5.9) gives the corrected stress intensity factor due to plasticity at the crack tip

$$K_I = \sigma \sqrt{\pi a \left[1 + \frac{1}{2}\left(\frac{\pi\sigma}{2\sigma_{ys}}\right)^2\right]} \tag{5.25}$$

This expression, Eq. (5.25), is similar to Irwin's expression, Eq. (5.10). Figure 5.4 compares the normalized stress intensity factors as per Irwin's and Dugdale's plastic zone models.

The curves significantly differ as $\sigma/\sigma_{ys} \to 1$; however, similarities occur at $0 < \sigma/\sigma_{ys} < 0.2$. This strongly suggests that both Irwin's and Dugdale's method should be used very carefully at large stress ratios because of their differences in normalized stress intensity factors.

Essentially, if $r \ll a$ plasticity corrections are not necessary. If $r > a$, linear-elastic fracture mechanics (LEFM) is a doubtful approach for solving engineering problems using brittle or elastic solids, and, therefore, the most attractive approach is the elastic-plastic fracture mechanics (EPFM), which will be dealt with in a later section.

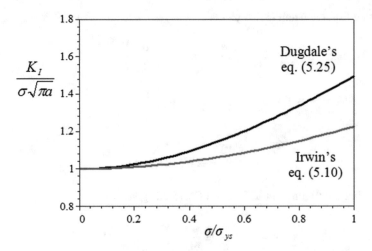

Fig. 5.4 Normalized stress intensity factor as a function of stress ratio for plane stress condition

Comparing Irwin's and Dugdale's strip yield model, also known as Dugdale's approximation, can easily be done by combining Eqs. (5.11) and (5.24) for plane stress conditions. Thus,

$$r\,[Irwin] = 0.41r\,[Dugdale] \tag{5.26}$$

Furthermore, Dugdale's strip yield model assumes that the stress singularity predicted by linear-elastic fracture mechanics vanishes because the elastic stress $\sigma_y \rightarrow \sigma_{ys}$ near the crack tip, inducing plastic deformation. Actually, Dugdale's original work on crack tip plasticity is a convenient tool for characterizing thin sheets in mode I, where plastic deformation ahead of the crack tip is confined within an infinitely thin strip placed along the crack plane.

Example 5.1. *Plot $r/a = f\left(\sigma/\sigma_{ys}\right)$ as per Irwin's and Dugdale's plastic zone size, Eqs. (5.11) and (5.24), respectively. Compare.*

Solution. *Irwin's Eq. (5.11):* $\quad r/a = 0.5\left[\dfrac{\sigma}{(1)\sigma_{ys}}\right]^2 = 0.5\left(\sigma/\sigma_{ys}\right)^2$ *for plane*

stress and $r/a = 0.5\left[\dfrac{\sigma}{(\sqrt{3})\sigma_{ys}}\right]^2 = 0.16667\left(\sigma/\sigma_{ys}\right)^2$ *for plane strain Dugdale's*

Eq. (5.24): $r/a = 0.5\left[\dfrac{\pi\sigma}{2\sigma_{ys}}\right]^2 = 1.2337\left(\sigma/\sigma_{ys}\right)^2$ *for plane stress*

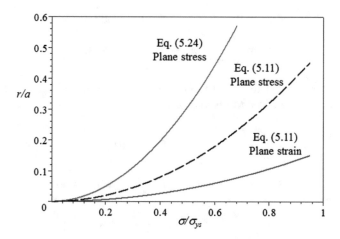

5.6   Yield Criteria

It is well known that significant crack tip plasticity occurs in elastic-plastic materials. The Von Mises and the Tresca yield criteria are commonly used to derive analytical closed-form solutions for the plastic zone sizes, which in turn give the crack plastic zone shapes.

5.6.1   Von Mises Yield Criterion

This criterion is deduced from the maximum distortion energy theory in which the state of stress is referred to as the principal stress directions. The Von Mises yield criterion is

$$(\sigma_1 - \sigma_2)^2 + (\sigma_2 - \sigma_3)^2 + (\sigma_3 - \sigma_1)^2 = 2\sigma_{ys}^2 \tag{5.27}$$

and the principal stress σ_1 are σ_2 defined by

$$\sigma_1, \sigma_2 = \frac{\sigma_x + \sigma_y}{2} \pm \sqrt{\left(\frac{\sigma_x - \sigma_y}{2}\right)^2 + \tau_{xy}^2} \tag{5.28}$$

Substituting Eq. (4.13) into (5.28) yields the principle stresses defined by

$$\sigma_1 = \frac{K_I}{\sqrt{2\pi r}} \cos\frac{\theta}{2}\left(1 + \sin\frac{\theta}{2}\right) \tag{5.29}$$

$$\sigma_2 = \frac{K_I}{\sqrt{2\pi r}} \cos\frac{\theta}{2}\left(1 - \sin\frac{\theta}{2}\right) \tag{5.30}$$

$$\sigma_3 = 0 \qquad \text{For plane stress} \qquad (5.31)$$

$$\sigma_3 = \frac{2vK_I}{\sqrt{2\pi r}} \cos\frac{\theta}{2} \qquad \text{For plane strain} \qquad (5.32)$$

Substituting Eq. (5.29) through (5.32) into (5.27) and manipulating the resultant expressions yields the Von Mises yield criterion as

$$\frac{K_I^2}{2\pi r}\left[\frac{3}{2}\sin^2\theta + h\left(1 + \cos\theta\right)\right] = 2\sigma_{ys}^2 \qquad (5.33)$$

from which the plastic zone size takes the following analytical form

$$r = \frac{1}{4\pi}\left(\frac{K_I}{\sigma_{ys}}\right)^2\left[\frac{3}{2}\sin^2\theta + h\left(1 + \cos\theta\right)\right] \qquad (5.34)$$

Here, $h = 1$ for plane stress and $h = (1 - 2v)^2$ for plane strain. Substituting Eq. (3.29) into (5.34) along with $\theta = 0$ gives the plastic zone as

$$r = \frac{h}{2\pi}\left(\frac{K_I}{\sigma_{ys}}\right)^2 = \frac{ha}{2}\left(\frac{\alpha\sigma}{\sigma_{ys}}\right)^2 \qquad (5.35)$$

This equation resembles Eqs. (5.3) and (5.11) for plane stress condition.

5.6.2   Tresca Yield Criterion

This criterion is based on the maximum shear stress theory, which predicts that yielding occurs when the maximum shear stress reaches half value of the yield stress in a uniaxial tension test. This is known as the Tresca yield criterion. Thus,

$$\tau_{max} = \frac{1}{2}\sigma_{ys} \qquad (5.36)$$

According to Mohr's circle theory, the maximum shear stress for plane strain is the largest of the following τ_{max} equations

$$\tau_{max} = \frac{1}{2}\left(\sigma_1 - \sigma_2\right) \qquad (5.37)$$

$$\tau_{max} = \frac{1}{2}\left(\sigma_1 - \sigma_3\right) \qquad (5.38)$$

Henceforward, σ_1 is algebraically the largest and σ_3 algebraically the smallest principal-stress components. Combining Eqs. (5.36) and (5.37) yields the maximum shear stress or Tresca yield criterion as

$$\sigma_1 = \sigma_{ys} \quad \text{for plane stress} \tag{5.39}$$

$$\sigma_1 - \sigma_2 = \sigma_{ys} \quad \text{for plane strain} \tag{5.40}$$

$$\sigma_1 - \sigma_3 = \sigma_{ys} \quad \text{for plane strain} \tag{5.41}$$

Substituting the stresses given in Eq. (5.29) through (5.32) into (5.39) through (5.41) yields the plastic zone size as

$$r = \frac{1}{2\pi}\left(\frac{K_I}{\sigma_{ys}}\right)^2 \left[\left(\cos\frac{\theta}{2}\right)\left(1 + \sin\frac{\theta}{2}\right)\right]^2 \quad \text{from Eq. (5.39)} \tag{5.42}$$

$$r = \frac{1}{2\pi}\left(\frac{K_I}{\sigma_{ys}}\right)^2 \left(2\cos\frac{\theta}{2}\sin\frac{\theta}{2}\right)^2 \quad \text{from Eq. (5.40)} \tag{5.43}$$

$$r = \frac{1}{2\pi}\left(\frac{K_I}{\sigma_{ys}}\right)^2 \cos^2\frac{\theta}{2}\left(1 - 2v + \sin\frac{\theta}{2}\right)^2 \quad \text{from Eq. (5.41)} \tag{5.44}$$

The shapes of the normalized plastic zone size $[2\pi r\,(\sigma_{ys}/K_I)^2]$, as per Eqs. (5.34) and (5.43), are plotted in polar coordinates as depicted in Fig. 5.5 for plane stress and plane-strain conditions. As expected, the former condition gives a larger plot than the latter one.

In addition, McClintock and Irwin [43] used the Von Mises yield criterion to determine the plastic zone shapes for mode *II* and *III* loading. Figure 5.6 shows these

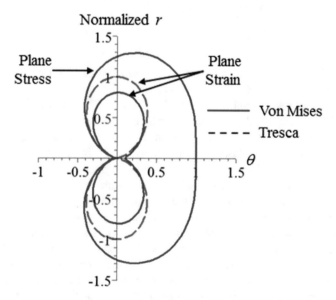

Fig. 5.5 Normalized plastic zone shapes for mode I loading according to (a) Von Mises (*blue continuous line*) and (b) Tresca yield criteria (*dashed line*)

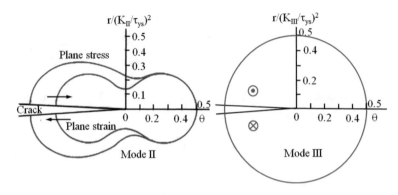

Fig. 5.6 Plastic zone shapes for modes II and III [43]

authors' analyses. Nevertheless, the preceding analytical and theoretical results were limited to the yield strength. This analytical procedure led to an error on the plastic zone size expressions due to the exclusion of the extra load that a material has to carry outside the plastic zone boundaries [9]. It is apparent that the mathematical models for plotting the plastic zone size do not include the main effect of strain hardening during plastic deformation ahead of the crack tip.

Measurements of the plastic zone can be accomplished by using techniques like:

1. Surface replicas	4. Photoelastic coatings
2. Moiré patterns	5. Etching
3. X-ray diffraction	6. Microhardness

Figures 5.7 and 5.8 illustrate experimental results obtainable by using relaxation methods [31]. For instance, Fig. 5.8 compares experimental and theoretical normalized results from several authors [6, 24, 29, 54, 64]. The data scatter in this figure is due to different theoretical procedures used by these authors.

However, difficulties do arise when analyzing the outcome of experiments because the elastic and plastic strains cannot easily be distinguished, and the measurements are usually restricted to specimen surfaces. These difficulties may be avoided, to an extent, by using the Hahn–Rosengren etching technique [14, 39], which requires a proper polycrystalline material and an etching solution so that dislocations and slip band would be etched in all grains. This way the area of plastic yielding can be delineated with fewer difficulties. Nonetheless, Hahn and Rosenfield [24] affirmed that the above theoretical mathematical approaches do not provide satisfactory description of the plastic zone shape. Therefore, none of the existing theories appear suitable for predicting the plastic zone shape and size at $\theta = 0$. Figure 5.9 shows real plastic zone shapes obtained by using the Hahn and Rosenfield etching technique [24].

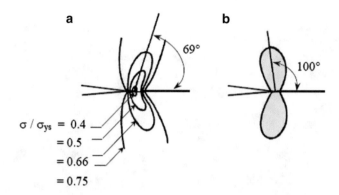

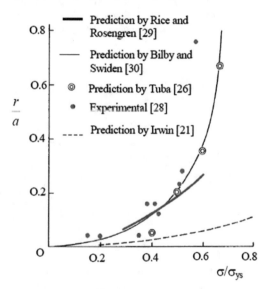

Fig. 5.7 Plastic zones in mode I according to (**a**) Tuba [64] and (**b**) Rice [51]

Fig. 5.8 Effects of normalized stress on the normalized plastic zone size [24]

Additionally, Theocaris and Andrianopoulos [63] used the Von Mises yield, $r = f(\theta)$, and the Strain Energy Density Factor, $S = f(\theta)$, criteria for developing the plastic zone shapes under mixed-mode I and II conditions. These theoretical results are depicted in Fig. 5.10, in which the plastic zone shape for the Von Mises yield criterion is larger than the one for the Strain Energy Density Factor (S). Furthermore, these shapes get enlarged and rotated as the inclined angle β increases. Denote that the plastic zone shapes depicted in Fig. 5.10 are similar at different crack incline angles.

Moreover, inherent local mixed-mode interaction, during plastic deformation ahead of the crack tip, is associated with the local normal stress ($\sigma_y \rightarrow \sigma_{ys}$) becoming a finite stress instead of a singularity stress. As a result, the plastic zone reaches a maximum size prior to incremental crack advance (growth).

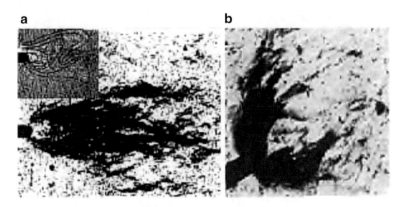

Fig. 5.9 (a) Interference pattern with strain contours (*top-left corner*) and the corresponding plastic zone revealed by etching a 3Si-steel specimen with a thickness of 0.4 mm [24] and (**b**) shape region of high shear in plane stress plastic zone of an Al-Cu-Mg alloy [8]

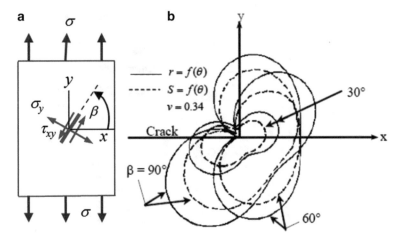

Fig. 5.10 Mixed-mode I and II interaction of a through-thickness inclined crack in a plate exhibiting several plastic zone shapes [63]. (**a**) Plate, (**b**) plastic zones

Example 5.2. *A thin steel plate having a 8-mm through the thickness single-edge crack size is designed to hold a 50 kN static tension load. The steel properties are $K_{IC} = 71\,MPa\sqrt{m}$, $\sigma_{ys} = 1896\,MPa$, and $v = 0.30$. The plate is 4 m long, 60 mm wide, and 3 mm thick. Assume that the steel plate is part of a structure that operates at room temperature and controlled environment, and that the nominal applied stress is a Von Mises stress. Determine (**a**) the safety factor (S_F) based on the fracture mechanics approach, (**b**) the critical crack length and the crack extension.*

For comparison purposes, calculate (c) the applied stress using the above safety factor and (d) the Von Mises plastic zone size using the yielding criterion (strength of materials) approach. Explain.

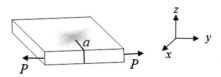

Solution. (a) *Calculate the nominal stress:*

$$\sigma = \frac{P}{A} = \frac{P}{Bw} = \frac{50{,}000 \text{ N}}{(3 \times 10^{-3} \text{ m}) (60 \times 10^{-3} \text{ m})} \simeq 278 \text{ MPa}$$

For $a/w = 8/60 = 0.13$, the geometry correction factor is

$$\alpha = 1.12 - 0.23 \left(\frac{a}{w}\right) + 10.55 \left(\frac{a}{w}\right)^2 - 21.71 \left(\frac{a}{w}\right)^3 + 30.38 \left(\frac{a}{w}\right)^4$$

$$\alpha = 1.12 - 0.23 \, (0.13) + 10.55 \, (0.13)^2 - 21.71 \, (0.13)^3 + 30.38 \, (0.13)^4 = 1.23$$

From Eq. (3.29),

$$K_I = \alpha\sigma \sqrt{\pi a} = (1.23) \, (278 \text{ MPa}) \sqrt{\pi \, (8 \times 10^{-3} \text{ m})} \simeq 54 \text{ MPa}\sqrt{\text{m}}$$

Thus, the safety factor is

$$S_F = \frac{K_{IC}}{K_I} = \frac{71 \text{ MPa}\sqrt{\text{m}}}{54 \text{ MPa}\sqrt{\text{m}}} \simeq 1.31$$

This safety factor is actually very low for designing against fracture. Therefore, crack propagation or sudden fracture may be expected to occur at small overloads.

(b) *The critical crack length is*

$$K_{IC} = \alpha\sigma \sqrt{\pi a_c}$$

$$a_c = \frac{1}{\pi} \left(\frac{K_{IC}}{\alpha\sigma}\right)^2 = \frac{1}{\pi} \left[\frac{71 \text{ MPa}\sqrt{\text{m}}}{(1.23) \, (278 \text{ MPa})}\right]^2 \simeq 14 \text{ mm}$$

Therefore, the crack extended is $\Delta a = a_c - a = 14 \text{ mm} - 8 \text{ mm} = 6 \text{ mm}$

(c) *The applied stress as per the yielding criterion is*

$$\sigma = \sigma_{ys}/S_F = (1896 \text{ MPa}) \, /1.31 \simeq 1447 \text{ MPa}$$

which is much greater than 278 MPa. Therefore, fracture mechanics is the most restrict approach for designing against fracture because the specimen contains a crack (defect). On the other hand, the yielding criterion assumes that there are no defects prior to loading the specimen.

(d) *Using Eq. (5.35) along the crack plane* $(\theta = 0)$ *the plastic zone size becomes*

$$r = \frac{h}{2\pi}\left(\frac{K_I}{\sigma_{ys}}\right)^2$$

$$r = \frac{(1-2v)^2}{2\pi}\left(\frac{K_I}{\sigma_{ys}}\right)^2 = \frac{[1-2\,(0.3)]^2}{2\pi}\left(\frac{54\ \text{MPa}\sqrt{\text{m}}}{1896\ \text{MPa}}\right)^2$$

$$r = 20.66\ \mu\text{m}$$

Therefore, the steel is brittle as deduced from the calculated small plastic zone size. The figure below shows the profile of the plastic zone size as a function of the yield strength. Denote the nonlinear behavior.

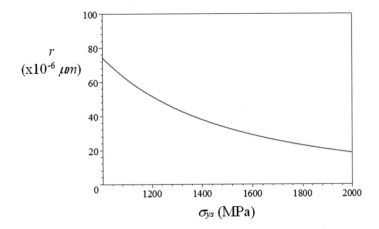

Conclusively, the plastic zone size, $r = f(\sigma)_{ys}$, *exhibits a nonlinear behavior as the yield strength increases. However, the inherent local mixed-mode interaction during plastic deformation ahead of the crack tip must be associated with the local normal stress,* $\sigma_y \to \sigma_{ys}$, *which becomes a finite stress. As a result, the plastic zone reaches a maximum size prior to an incremental crack advance. In other words, LEFM requires a sharp crack subjected to a remote opening stress, inducing local infinite stress* $(\sigma_y \to \infty)$ *at the crack tip. This means that the actual plastic zone size is entirely finite, and consequently, the local opening stress is totally finite as well since* $\sigma_y \to \sigma_{ys}$.

5.7   Fracture Parameters

The most common fracture resistance parameters used nowadays for characterizing stable crack growth in thin and ductile metallic materials under quasi-static loading are

δ_m = The crack mouth opening displacement (CMOD) [2]
δ_t = The crack tip opening displacement (CTOD) [2, 68]
δ_5 = The crack opening displacement (COD) [32]
ψ_t = The crack tip opening angle (CTOA) [17]

These parameters require special procedures for determining their critical values as measure of fracture toughness of most common specimens, such as bending $SE(B)$, compact $C(T)$, and middle-crack $M(T)$ geometries having specific dimensions in order to control the required low constraint. The critical values are determined from force (P) versus crack extension (Δa) plots. For instance, the ASTM E2472 standard method provides relevant details on test procedures for low constraint and fatigue pre-cracked $C(T)$ and $M(T)$ specimen geometries with low constraint conditions, which are assured by having specimens with $a/B \geq 4$ and $b/B \geq 4$ [46]. Here, a is the original crack length, B is the specimen thickness, b is the specimen ligament ($b = w - a$), and w is the specimen width.

In general, experimental force (P) versus crack extension (Δa) or P versus displacement (μ) plots are needed for assessing fracture behavior, followed by measurements of the CMOD, CTOD, COD, and CTOA parameters, which, in turn, can be implemented in finite element analysis to predict crack growth behavior and their critical values [34, 41, 45]. Figure 5.11 illustrates the locations of δ_m, δ_t, δ_5, and ψ_t.

Fig. 5.11 Definition of CMOD, δ_5 CTOD, and CTOA

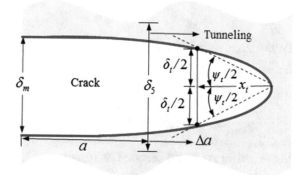

5.8   Crack Tip Opening Displacement

So far, the characterization of cracked brittle materials has been restricted to the linear-elastic fracture criterion (LEFM), which treats localized plasticity as a small deformed area at a small-scale yielding (*SSY*); otherwise, LEFM invalidates the applicability of closed-form analytical solutions associated with a material resistance to crack growth at a large-scale yielding. In the latter case, therefore, the physical sense of K_{IC}, acting as the controlling fracture (critical) parameter, is lost to a great extent due to large plasticity ahead of the crack tip.

The aforementioned plasticity ahead of the crack front is fundamentally and quantitatively governed by elastic-plastic fracture mechanics (EPFM) through models that include the crack tip opening displacement (*CTOD* $\rightarrow$ δ_t-model), the crack tip opening angle (*CTOA* $\rightarrow$ ψ_t-model), and the *J*-integral (*J*-model) as parameters. Subsequently, these models provide the fracture criteria at critical states for determining fracture toughness, which is determined at the onset of crack growth.

In particular, Wells [68] first proposed the CTOD as a fracture criterion when $\delta_t \geq \delta_{tc}$, Dawicke–Sutton [17] idealized the CTOA criterion when $\psi_t \geq \psi_{tc}$, and Rice [51] developed the *J*-integral as energy fracture criterion, say, $J_I \geq J_{IC}$ for mode *I*. Both CTOD and CTOA are a measure of fracture toughness of solid materials that undergo plane stress-strain transition and elastic-plastic or fully plastic behavior as in large structures (ships, pressure vessels) [64]. On the other hand, the *J*-model treats the presence of relatively large deformations as an integral part of analytical formulations that comprehensively describe the state of plastic stresses and strains [14, 21].

According to the ASTM E1290 Standard, the critical CTOD (δ_{tc}) is used when the K_{IC} requirements are not met [64]. From Fig. 5.3b, δ_t is defined as twice the displacement of the crack flanks in the y-direction [9]. Hence,

$$\delta_t = 2\mu_y \tag{5.45}$$

Furthermore, definitions of δ_t for two cases are defined as [9, 53]

$$3\delta_t = \frac{4\sigma}{E}\sqrt{a^2 - x^2} \qquad \text{(uncorrected)} \tag{5.46}$$

$$\delta_t = \frac{4\sigma}{E}\sqrt{(a + r)^2 - x^2} \qquad \text{(corrected)} \tag{5.47}$$

If $x = a$, then Eq. (5.47) becomes [9, 10, 24]

$$\delta_t = \frac{4\sigma}{E}\sqrt{(a + r)^2 - a^2} \simeq \frac{4\sigma}{E}\sqrt{2ar} \tag{5.48}$$

Inserting Eqs. (5.11) and (5.24) independently into (5.48) under plane stress conditions yields the crack tip opening displacement as

$$\delta_t = \frac{4a\sigma^2}{\pi E \sigma_{ys}} = \frac{4K_I^2}{\pi \lambda E \sigma_{ys}} \qquad \text{(Irwin)} \tag{5.49}$$

$$\delta_t = \frac{2\pi a\sigma^2}{E \sigma_{ys}} = \frac{2K_I^2}{E \sigma_{ys}} \qquad \text{(Dugdale)} \tag{5.50}$$

These two equations can be related as

$$\delta_t^{Irwin} = \frac{2}{\pi} \delta_t^{Dugdale} \tag{5.51}$$

Alternatively, Burdekin [12] and Rice [51] independently developed δ_t closed-form solutions based on Dugdale's strip yield model [19] and the definition of Eq. (5.45) for plane stress and plane-strain conditions, respectively. Hence,

$$\delta_t = \frac{8a\sigma_{ys}}{\pi E} \ln \left[\sec \left(\frac{\pi \sigma}{2\sigma_{ys}} \right) \right] \qquad \text{(Burdekin)} \tag{5.52}$$

$$\delta_t = \frac{2(\kappa + 1)(1 + v) a\sigma_{ys}}{\pi E} \ln \left[\sec \left(\frac{\pi \sigma}{2\sigma_{ys}} \right) \right] \qquad \text{(Rice)} \tag{5.53}$$

and from Eqs. (4.65) and (4.66),

$$\kappa = \frac{3 - v}{1 + v} \qquad \text{for plane stress} \tag{5.54}$$

$$\kappa = 3 - 4v \qquad \text{for plane strain} \tag{5.55}$$

Expanding the natural logarithmic function in Eqs. (5.52) and (5.53) using Taylor's series, $\ln x \simeq x - 1$ for $0 < x < 2$, along with the approximation given by Eq. (5.23) yields

$$x = 1 + \frac{y^2}{2} \tag{5.56}$$

$$\ln x \simeq x - 1 = 1 + \frac{y^2}{2} - 1 \tag{a}$$

$$\ln x \simeq= \frac{y^2}{2} \tag{5.57}$$

$$\ln \left[\sec \left(\frac{\pi \sigma}{2\sigma_{ys}} \right) \right] \simeq \frac{1}{2} \left(\frac{\pi \sigma}{2\sigma_{ys}} \right)^2 \tag{5.58}$$

Thus, Eqs. (5.52) and (5.53) become the elastic CTOD

Fig. 5.12 Comparison of crack tip opening displacement for a material having $E = 207\,\mathrm{GPa}$, $\sigma_{ys} = 700\,\mathrm{MPa}$, $K_{IC} = 60\,\mathrm{MPa}\sqrt{\mathrm{m}}$, and $v = 1/3$. Legend: Dugdale = curve 1, Irwin = curves 2 and 4, Burdekin = curve 3, and Rice = curves 5 and 6

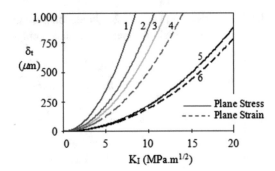

$$\delta_{te} = \frac{\pi a \sigma^2}{E \sigma_{ys}} = \frac{K_I^2}{E \sigma_{ys}} \qquad \text{(Burdekin)} \qquad (5.59)$$

$$\delta_{te} = \frac{(\kappa + 1)\,(1 + v)\,a\sigma^2}{4 E \sigma_{ys}} = \frac{(\kappa + 1)\,(1 + v)\,K_I^2}{4\pi E \sigma_{ys}} \quad \text{(Rice)} \qquad (5.60)$$

The preceding procedure provides a selection of mathematical models, which define (δ_t). In fact, the applicability of these models for characterizing fracture of thin sheets is of great importance since the dimensional requirements are not too strict as in the K_I-criterion for thick materials (ASTM E399 standard).

Figure 5.12 shows the relationship of $\delta_t = f(K_I)$ as per the above models for plane stress and plane-strain conditions.

Denote that the δ_t curves do not agree with each other due to different assumptions used by each cited author to develop a δ_t-model. However, Burdekin's model, Eq. (5.59), agrees with the work done by Broek and Vlieger [10], Robinson and Tetelman [56], and Bowles [7].

Considerable work has been confined to the δ_t-model since it has the advantage of measuring fracture toughness for elastic materials and it is very sensitive to variations in temperature, loading rate, specimen thickness, and thermomechanical processing [63].

In general, yielding in the vicinity of a crack tip is related to complex dislocation networks, such as dislocation pileup. In this regard, reference should be made to the classical work of Weertman [66], Puttick [50], Stroh [62], and Rogers [57], who analyzed and studied several dislocation models for the nucleation of cracks.

Returning to the δ_t-model, the crack tip plastic deformation in a thin sheet is not restricted to lateral contraction which causes localized thinning at the crack tip. Thereby, crack tip blunting causes an extensive increase in the crack tip radius (ρ) in the order of sheet thickness (B), and thus, the plastic zone size (r) may be estimated as $r \simeq B$ [65].

In addition, ASTM E1820 standard includes the basic and resistance curve procedures for determining the CTOD for some specimen configurations. According to the LEFM approach, the hinge model is schematically depicted in Fig. 5.13 for a SE(B) specimen under quasi-static mode I loading [2].

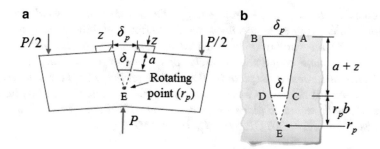

Fig. 5.13 Hinge model for plastic crack mouth opening displacement (CMOD) in a SE(B) specimen, after reference [2]. (**a**) SE(T) specimen, (**b**) similar triangles

This, then, induces the elastic CTOD (δ_{te}) to be defined by

$$\delta_{te} = \frac{K_I^2}{m\sigma_{ys}E'} \tag{5.61}$$

Using similar triangles with shared parts shown in Fig. 5.13 gives [2]

$$\frac{\delta_p}{r_pb + a + z} = \frac{\delta_{tp}}{r_pb} \tag{5.62}$$

Solving for δ_{tp}, the plastic CTOD, yields

$$\delta_{tp} = \frac{r_pb\delta_p}{r_pb + a + z} \tag{5.63}$$

where $m = 1$ and $E' = E$ for plane stress, $m = 2$ and $E' = E/\left(1 - v^2\right)$ for plane strain, v is the Poisson's ratio, r_p is the plastic rotational factor, $\delta = \delta_e + \delta_p$ is the CMOD, and z is the knife-edge height.

For $SE(B)$ and $C(T)$ specimens, the δ_{tp} expressions along with the ligament $b = w - a$ are, respectively [2],

$$\delta_{tp} = \frac{0.44\,(w - a)\,\delta_p}{0.44\,(w - a) + a + z} \rightarrow SE(B) \tag{5.64}$$

$$\delta_{tp} = 0.4 \left\{ 1 + 2\left[\frac{1}{2} + \frac{a}{b} + \left(\frac{a}{b}\right)^2\right]^{1/2} - 2\left(\frac{1}{2} + \frac{a}{b}\right)\right\} \rightarrow C(T) \tag{5.65}$$

According to the ASTM E1820 standard, the total CTOD expression for any point on the load-displacement curve, $P = f\,(\mu)$, becomes

$$\delta_t = \frac{\left(1 - v^2\right)}{2}\frac{K_I^2}{\sigma_{ys}E} + \frac{r_p\,(w - a)\,\delta_p}{r_p\,(w - a) + a + z} \tag{5.66}$$

Example 5.3. *If the critical strain energy release rate and the yield strength of a 13-mm thick C(T) steel specimen are 32 kJ/m² and 1500 MPa, respectively, determine* **(a)** *the validity of the fracture mechanics tension test as per ASTM E399 for the plate containing a single-edge crack of 10 mm long at fracture,* **(b)** *the fracture stress if the plate is 20 mm wide,* **(c)** *the critical crack tip opening displacement,* **(d)** *the displacement, and* **(e)** *the plastic zone size and* **(f)** *interpret the results with regard to plane-strain condition. Given data for steel:* $v = 1/3$ *(Poisson's ratio),* $E = 207$ *GPa, and* $\sigma_{ys} = 1500$ *MPa.*

Solution. *Given data:*

$$G_{IC} = 32 \text{ kJ/m}^2 = 32 \times 10^{-3} \text{ MPa m}, \, \sigma_{ys} = 1500 \text{ MPa}$$
$$E = 207{,}000 \text{ MPa}, \, v = 1/3, \, a = 10 \text{ mm}, \, w = 20 \text{ mm}$$
$$a/w = 0.5, \, B = 13 \text{ mm}$$

(a) *Using Eq. (3.5) yields the critical stress intensity factor*

$$G_{IC} = \frac{K_{IC}^2}{E'} = \frac{\left(1 - v^2\right) K_{IC}^2}{E}$$

$$K_{IC} = \sqrt{\frac{EG_{IC}}{\left(1 - v^2\right)}} = \sqrt{\frac{(207{,}000 \text{ MPa}) (32 \times 10^{-3} \text{ MPa m})}{1 - 1/9}}$$

$$K_{IC} = 86.33 \text{ MPa}\sqrt{m}$$

The ASTM E399 minimum size requirements Eq. (3.30), are,

$$a_{min}, B_{min} \geq 2.5 \left(\frac{K_{IC}}{\sigma_{ys}}\right)^2 = 2.5 \left(\frac{86.33 \text{ MPa}\sqrt{m}}{1500 \text{ MPa}}\right)^2 = 8.28 \text{ mm}$$

Therefore, the test is valid because $a, B \, (10, 13) > a_{min}, B_{min} \, (8.28)$ mm *and* $a/w = 0.5$ *is within the* $0.2 \leq a/w \leq 1$ *valid range. Proceed with the required calculations.*

(b) *The fracture stress along with* $\alpha = 9.6591$ *since* $a/w = 0.5$ *(consult Table 3.1) and* $K_{IC} = \alpha \sigma_f \sqrt{\pi a}$ *is*

$$\sigma_f = \frac{K_{IC}}{\alpha \sqrt{\pi a}} = \frac{71 \text{ MPa}\sqrt{m}}{(9.6591) \sqrt{\pi} (10 \times 10^{-3} \text{ m})} = 41.47 \text{ MPa} < \sigma_{ys}$$

(c) *The crack tip opening displacement is calculated using Rice's equation, Eq. (5.60), with* $\kappa = 3 - 4v = 5/3$ *and* $(\kappa + 1)(1 + v) = 32/9$

$$\delta_{tc} = \frac{(\kappa + 1)(1 + v) K_{IC}^2}{4\pi E \sigma_{ys}}$$

$$\delta_{tc} = \frac{32}{36\pi} \frac{\left(86.33 \text{ MPa}\sqrt{m}\right)^2}{(207{,}000 \text{ MPa}) (1500 \text{ MPa})} = 6.80 \text{ } \mu\text{m}$$

(d) *From Eq. (5.45), the displacement is*

$$\mu_y = \delta_{tc}/2 = (6.80 \ \mu m) \, /2 = 3.40 \ \mu m$$

(e) *The plastic zone can be calculated using Eq. (5.35) along with the constant* $h = (1 - 2v)^2 = 1/9$

$$r = \frac{h}{2\pi} \left(\frac{K_{IC}}{\sigma_{ys}} \right)^2 = \frac{1}{18\pi} \left(\frac{86.33 \ \text{MPa}\sqrt{m}}{1500 \ \text{MPa}} \right)^2 = 58.57 \ \mu m$$

(f) *The above results suggest that the plate met the ASTM E399 size requirements because* $(a, B) > (a_{min}, B_{min})$ *and* $0.2 \le a/w = 0.5 \le 1$. *It can be assumed that the C(T) specimen breaks in a brittle manner because both the plastic zone size (r) and the crack tip opening displacement* δ_{tc} *are very small. These results suggest that ASTM E399 standard test is most suitable for brittle material in question.*

Example 5.4. *A single-edge bend, SE(B), specimen made out of a hypothetical ductile metallic material having mechanical properties like* $\sigma_{ys} = 400 \ MPa$, $K_{IC} = 60 \ MPa\sqrt{m}$, $v = 1/3$ *(Poisson's ratio) and* $E = 72 \ GPa$ *was tested according to the ASTM E1820 specifications for the crack tip opening displacement approach. The specimen configuration and the schematic load-displacement plot are, respectively,*

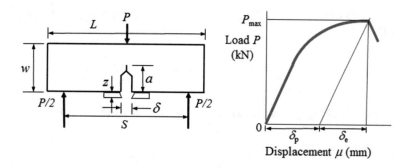

Assume that the material's microstructural homogeneity prevails. The elastic (δ_e) *and plastic* (δ_p) *opening displacements at the crack mouth are graphically defined in the load-displacement plot. The specimen dimensions are*

$$B = 20 \ \text{mm} \qquad a = 8 \ \text{mm}$$

$$S = 150 \ \text{mm} \qquad \delta_p = 0.4 \ \text{mm}$$

$$w = 40 \ \text{mm} \qquad z = 1.5 \ \text{mm}$$

Calculate the critical crack tip opening displacement, CTOD δ_{tc}, when the maximum load is 38 kN. Recall that δ_{tc} is a genetic term for the resistance to crack growth, namely, fracture toughness of the material, and that the δ_t is done materials that exhibit plastic deformation prior to failure.

Solution. *For the elastic behavior, the geometry correction factor for the SE(B) specimen configuration along with $x = a/w = 8/40 = 0.2$ is*

$$\alpha = \frac{3x^{1/2}\left\{1.99 - x(1-x)\left[2.15 - 3.93x + 2.7x^2\right]\right\}}{2(1+2x)(1-x)^{3/2}}$$

$$\alpha = \frac{3(0.2)^{1/2}\left\{1.99 - (0.2)(1-0.2)\left[2.15 - 3.93 * 0.2 + 2.7 * 0.2^2\right]\right\}}{(2)(1+2*0.2)(1-0.2)^{3/2}}$$

$$\alpha = 1.1749$$

According to the ASTM E1820 standard, the elastic stress intensity factor becomes

$$K_I = \frac{\alpha PS}{Bw^{3/2}}$$

$$K_I = \frac{(1.1749)(38000\ \text{N})\left(150 * 10^{-3}\ \text{m}\right)}{(20 * 10^{-3}\ \text{m})(40 * 10^{-3}\ \text{m})^{3/2}}$$

$$K_I = 41.86\ \text{MPa}\sqrt{\text{m}} < K_{IC}$$

Adding Eqs. (5.61) and (5.64) gives the total CTOD δ_t expression in the form given below [2]

$$\delta_{tc} = \delta_{te} + \delta_{tp} = \frac{K_I^2}{m\sigma_{ys}E'} + \frac{r_p(w-a)\delta_p}{r_p(w-a)+a+z}$$

The elastic CTOD (δ_{te}) for plane strain along with $E' = E/(1-v^2)$ is given by

$$\delta_{te} = \frac{K_I^2}{m\sigma_{ys}E'} = \frac{(1-1/9)(41.86)^2}{(2)(400)(72000)}$$

$$\delta_{te} = 2.7041 \times 10^{-2}\ \text{mm} \simeq 0.027041\ \text{mm}$$

For the SE(B) specimen, the plastic rotational factor is $r_p = 0.44$ and the plastic CTOD δ_{tp} becomes

$$\delta_{tp} = \frac{0.44(w-a)\delta_p}{0.44(w-a)+a+z} = \frac{(0.44)(40-8)(0.4)}{(0.44)(40-8)+8+1.5}$$

$$\delta_{tp} = 0.23885\ \text{mm}$$

Thus, the critical CTOD is

$$\delta_{tc} = \delta_{te} + \delta_{tp} = 0.027041 \text{ mm} + 0.23885 \text{ mm}$$
$$\delta_{tc} = 0.26589 \text{ mm} \simeq 0.27 \text{ mm}$$

Therefore, $\delta_{tp} > \delta_{te}$ as expected.

Example 5.5. *Apply the effective crack length a_e on Eq. (5.60) to calculate the elastic CTOD for plane conditions. Let the Poisson's ratio be $1/3$, $\sigma_{ys} = 400 \text{ MPa}$, $K_I = 30 \text{ MPa}\sqrt{m}$, $a = 4 \text{ mm}$, and $E = 72 \text{ GPa}$. Assume a single-edge crack under tension loading.*

Solution. *Irwin's effective crack length, $a_e = a + r$, approach on Rice's CTOD, Eq. (5.60), becomes*

$$\delta_{te} = \frac{(\kappa + 1)(1 + v) a_e \sigma^2}{4E\sigma_{ys}} = \frac{(\kappa + 1)(1 + v) \sigma^2}{4E\sigma_{ys}} (a + r) \tag{a}$$

where the plastic zone size and the plane condition factor are

$$r = \frac{1}{2\pi} \left(\frac{K_I}{\sigma_{ys}} \right)^2 \quad \& \quad \kappa = \frac{3 - v}{1 + v} \tag{b}$$

Useful expressions:

$$K_I = \sigma \sqrt{\pi a} \quad \& \quad K_I^2 = \pi a \sigma^2 \tag{c}$$

$$\sigma = \frac{K_I}{\sqrt{\pi a}} \quad \& \quad \sigma^2 = \frac{K_I^2}{\pi a} \tag{d}$$

Thus, Eq. (a) becomes

$$\delta_{te} = \frac{(\kappa + 1)(1 + v) \sigma^2}{4E\sigma_{ys}} \left[a + \frac{1}{2\pi} \left(\frac{K_I}{\sigma_{ys}} \right)^2 \right]$$

$$\delta_{te} = \frac{(\kappa + 1)(1 + v) K_I^2}{4\pi a E\sigma_{ys}} \left[a + \frac{1}{2\pi} \left(\frac{K_I}{\sigma_{ys}} \right)^2 \right]$$

Substituting the given values gives

$$\delta_{te} = 0.03 \text{ mm}$$

For plane strain,

$$r = \frac{1}{6\pi}\left(\frac{K_I}{\sigma_{ys}}\right)^2 \quad \& \quad \kappa = 3 - 4v$$

$$\delta_{te} = \frac{(\kappa + 1)(1 + v)K_I^2}{4\pi a E \sigma_{ys}}\left[a + \frac{1}{6\pi}\left(\frac{K_I}{\sigma_{ys}}\right)^2\right]$$

Thus,

$$\delta_{te} = 0.02 \text{ mm}$$

These results are small, implying that $\delta_{tp} \gg \delta_{te}$.

5.8.1   Crack Opening Displacement (COD) δ_5

The crack opening displacement, COD δ_5, is a force-induced separation between two points as depicted in Fig. 5.14 and described in the ASTM E2472 standard for performing experiments using C(T) or M(T) specimens. The δ_5 parameter is measured using a 5-mm COD transducer (COD clip gage) located at the fatigue pre-crack tip. This simply implies that the clip gage is place 2.5 mm above and below the initial fatigue crack tip [46]. In fact, COD δ_{25} (25-mm separation) has been used for characterizing asphalt concrete samples [61].

The main purpose of the δ_5 measurements is to determine stable crack extension (Δa). Suitable data includes $P = f(\delta_5)$ and $\delta_5 = f(\Delta a)$, where P is the applied load. Here, $P = f(\delta_5)$ and $\delta_5 = f(\Delta a)$ are nonlinear functions to be determined numerically using experimental data for a particular specimen geometry. Figure 5.15 schematically depicts a possible $P = f(\delta_5)$ profile for a ductile metallic material [46].

Furthermore, fracture mechanics tests can be performed using the δ_5 definition under quasi-static or cyclic loading modes at temperatures, provided that the δ_5 gage is designed for this purpose. Besides the C(T) or M(T) specimens, other specimen sizes and geometries should be considered for evaluating possible effects of specimen dimensions using the δ_5 method. Abundant experimental and predicted results using the δ_5 method can be found elsewhere [33, 46].

Fig. 5.14 The 5-mm crack tip opening displacement (δ_5 COD) method for a stable crack growth

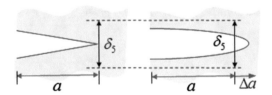

Fig. 5.15 Schematic
$P = f(\delta_5)$ profile

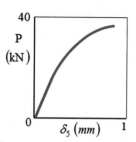

5.8.2 Crack Tip Opening Angle (CTOA)

Once crack growth occurs in a thin ductile material, the crack front may exhibit
a slight curvature within the specimen thickness. This deformation phenomenon is
schematically depicted in Fig. 5.11, and it implies that the crack length measured on
the specimen surface is slightly smaller than that in the interior. This observable
plastic deformation is known as crack tunneling, where the crack front bows
out forming a curvature along the crack front. The tunneling crack front shape
can be analyzed in an interrupted load manner followed by fatigue cracking in
order to resharpen the crack [32, 47]. This process is usually characterized by
a two-dimensional [45] or three-dimensional [34] analysis along with the proper
constraints (plane stress and plane-strain conditions).

Figure 5.11 also depicts a schematic top view of the crack front and possible
location for experimental measurements of the local angle (ψ_t) and the local
constraint. This means that ψ_t can be determined at different locations along the
crack front boundary as a function of distance x_t behind the crack tip [18, 34].
The local constraints along the crack front curvature go from high constraint at the
crack midpoint (high triaxial plane-strain condition) to lower constraint at the crack
surface (high stress condition) [34]. Actual crack faces show a zigzag pattern [71].

The above schematic crack tunneling model suggests that the crack front
curvature has significant effects on fracture toughness due to the varying crack
extension (Δa) in this region. Nowadays, δ_t or ψ_t criterion is widely used to
characterize the crack initiation in ductile thin plates and thin sheets.

Simple trigonometry (Fig. 5.11) gives a straightforward relationship between δ_t
and ψ_t in a closed form. Thus,

$$\tan(\psi_t/2) = \frac{\delta_t/2}{x_t} \tag{5.67}$$

$$\delta_t = 2x_t \tan(\psi_t/2) \tag{5.68}$$

$$\psi_t = 2\tan^{-1}\left(\frac{\delta_t}{2x_t}\right) \tag{5.69}$$

where x_t is a common distance behind the crack tip for ψ_t and δ_t. The δ_t and ψ_t relationship can also be derive using the J-integral approach for strain hardening materials in the limit of small-scale yielding as [2, 22, 38]

$$J = m\sigma_{ys}\delta_t = m\sigma_F\delta_t \tag{5.70}$$

where $m = 1$ for plane stress and $m = 2$ for plane-strain conditions [22] and σ_{ys} is the monotonic yield strength, which can be replaced by the flow stress, $\sigma_F = 0.5\left(\sigma_{ys} + \sigma_t\right)$. Moreover, the strip yield criterion for nonhardening materials also assumes $m = 1$ under plane stress conditions [2]. Fundamentally, strain hardening is a basic metallurgical property of a metal since a flow stress is required for plastic deformation.

The derivative dJ/da is the slope of the J_R-curve (Fig. 3.12), and it can be defined here using Eq. (5.70). Thus [22],

$$\frac{dJ}{da} = \frac{dJ}{d\delta_t}\frac{d\delta_t}{da} = m\sigma_{ys}\frac{d\delta_t}{da} \tag{5.71}$$

Solving Eq. (5.71) for $d\delta_t/da$ gives the definition of ψ_t [22, 42]

$$\psi_t = \frac{d\delta_t}{da} = \frac{dJ/da}{m\sigma_{ys}} \tag{5.72}$$

For mode I loading, the J-integral defined by Eq. (3.63) and its derivative with respect to crack extension Δa are

$$J_I = C_1\left(\Delta a\right)^{C_2} \tag{5.73}$$

$$\frac{dJ_I}{d\left(\Delta a\right)} = C_1C_2\left(\Delta a\right)^{C_2-1} \tag{5.74}$$

Then ψ_t becomes

$$\psi_t = \frac{d\delta_t}{d\left(\Delta a\right)} = \frac{dJ_I/d\left(\Delta a\right)}{m\sigma_{ys}} \tag{5.75}$$

$$\psi_t = \frac{C_1C_2\left(\Delta a\right)^{C_2-1}}{m\sigma_{ys}} \tag{5.76}$$

Recall that C_1 and C_2 are least-squares curve fitting coefficients. Measurements of CTOA for a particular specimen may exhibit constant CTOA values from crack growth initiation to failure [46]. Significant details on the CTOA criterion and CTOA measurements can be found elsewhere [37, 38, 46] using microtopography for characterizing ductile fracture behavior. Figure 5.16a depicts a schematic $\psi_t = f\left(\Delta a\right)$ curve which represents a typical ψ_t behavior.

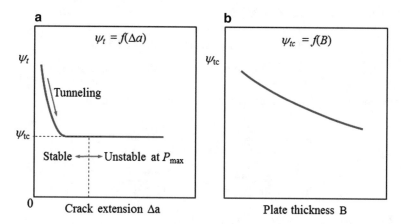

Fig. 5.16 Crack tip opening angle (CTOA) profiles. (**a**) CTOA vs. crack extension and (**b**) critical CTOA vs. plate thickness

Initially, ψ_t decreases with increasing crack extension (Δa), apparently due to crack tunneling during the initial stable crack tip tearing, followed by a constant ψ_t value referred to as the critical CTOA; that is, $\psi_t = \psi_{tc}$. At some maximum load (vertical dashed line), a transition from stable to unstable crack growth takes place at ψ_{tc} [17, 34, 40, 41].

Figure 5.16b schematically shows the effect of plate thickness (B) on ψ_{tc}. Hence, $\psi_{tc} = f(B)$ decreases with increasing thickness up to a limit, not yet defined [40, 41]. Furthermore, ductile fracture can also be characterized using the tearing modulus as per Paris et al. [48]. For mode I loading, the tearing modulus is defined as

$$T_I = \frac{E}{\sigma_{ys}^2} \frac{dJ}{da} \tag{5.77}$$

Combining Eqs. (5.71), (5.72), and (5.77) yields [55]

$$T_I = m\sigma_{ys} \frac{E}{\sigma_{ys}^2} \frac{d\delta_t}{da} \tag{5.78}$$

$$T_I = \left(\frac{mE}{\sigma_{ys}}\right)\psi_t \tag{5.79}$$

Example 5.6. *A single-edge cracked thin sheet made out of a hypothetical ductile Al-alloy is subjected to a quasi-static tension loading mode. Assume that the crack extension (Δa) and the ψ_{tc} (CTOA) values are $\Delta a_1 = 3\,mm$ and $\psi_{tc1} = 15°$ and $\Delta a_2 = 6\,mm$ and $\psi_{tc2} = 14°$. Calculate (**a**) δ_{tc} and x_t, (**b**) J_I, and (**c**) the power-law coefficients. Assume plane stress conditions.*

Solution. **(a)** *The critical COTD are*

$$\delta_{tc1} = 2\,(1\text{ mm}) \tan\,(15\pi/180) = 0.54\text{ mm}$$

$$\delta_{tc2} = 2\,(1\text{ mm}) \tan\,(14\pi/180) = 0.50\text{ mm}$$

The distance x_t behind the crack tip for ψ_t measurements are

$$x_{t1} = \frac{\delta_{tc1}}{2\tan\,(\psi_{tc1}/2)} = \frac{0.54\text{ mm}}{2\tan\,(15°)} = 1.01\text{ mm}$$

$$x_{t2} = \frac{\delta_{tc2}}{2\tan\,(\psi_{tc2}/2)} = \frac{0.50\text{ mm}}{2\tan\,(14°)} = 1.00\text{ mm}$$

(b) *The J-integral values along with $m = 1$ are*

$$J_I = m\sigma_{ys}\delta_{tc}$$

$$J_{I1} = (1)\,(500\text{ MPa})\,(0.54 \times 10^{-3}\text{ m}) = 0.27\text{ MJ/m}^2$$

$$J_{I2} = (1)\,(500\text{ MPa})\,(0.50 \times 10^{-3}\text{ m}) = 0.25\text{ MJ/m}^2$$

(c) *The J_I power law*

$$J_I = C_1\,(\Delta a)^{C_2}$$

$$\ln\,(J_I) = \ln\,(C_1) + C_2 \ln\,(\Delta a)$$

Then,

$$C_2 = \frac{\ln\,(J_{I2}/J_{I1})}{\ln\,(\Delta a_2/\Delta a_1)} = \frac{\ln\,(0.27/0.25)}{\ln\,(6/3)} = 0.11103$$

$$C_1 = \frac{J_{I1}}{(\Delta a_1)^{C_2}} = \frac{0.27\text{ MJ/m}^2}{(3 \times 10^{-3}\text{ m})^{0.11}} = 0.51154\text{ MJ/m}^{2.11103}$$

Hence, the power-law equation becomes

$$J_I = (0.51154)\,(\Delta a)^{0.11103}$$

for $3\text{ mm} \leq \Delta a \leq 6\text{ mm}$.

5.9 Problems

5.1. Use the inequality $K_{IC} \geq K_I$ as a criterion for crack instability where K_I is defined by Irwin's plastic zone corrected expression for a finite size, to determine if a steel pressure vessel is susceptible to explode under $\sigma = 200\,\text{MPa}$ hoop stress. The vessel contains an internal circular crack perpendicular to the hoop stress. If the properties of the steel are $K_{IC} = 60\,\text{MPa}\sqrt{m}$ and $\sigma_{ys} = 700\,\text{MPa}$, and the crack size is $a = 20\,\text{mm}$, (a) determine the ASTM E399 thickness requirement and the minimum thickness to be used to prevent explosion. (b) Will crack propagation occur at 200 MPa? (c) Plot $B = f\left(\sigma/\sigma_{ys}\right)$ for $a = 10$, 20, and 30 mm, and (d) will the pressure vessel explode when the crack size is 30 mm? Why or why not? (e) When will the pressure vessel explode? [Solution: (a) $B_{min} = 8.28\,\text{mm}$; (b) no, it will not because $K_I < K_{IC}$; (d) no; and (e) it explodes when $a_c = 67.91\,\text{mm}$].

5.2. A project was carried out to measure the elastic strain energy release rate as a function of normalized stress $\left(\sigma/\sigma_{ys}\right)$ of large plates made out a hypothetical brittle solid. All specimens had a single-edge crack of 3-mm long. Plot the given data and do regression analysis on this data set. Determine (a)the maximum allowable σ/σ_{ys} ratio for $G_{IC} = 30\,\text{kPa\,m}$ and (b) K_{IC} in $\text{MPa}\sqrt{m}$. Given data: $\nu = 0.3$, $\sigma_{ys} = 900\,\text{MPa}$, $E = 207\,\text{GPa}$,

σ/σ_{ys}	0	0.10	0.20	0.30	0.40	0.50	0.60	0.70	0.80	0.90
G_{IC}	0	0.40	1.70	1.90	7.00	12.00	19.00	26.00	36.00	48.00

5.3. Calculate the critical crack length of Problem 5.2. [Solution: $a_c = 3\,\text{mm}$].

5.4. A large brittle plate containing a central crack 4-mm long is subjected to a tensile stress of 800 MPa. The material has $K_{IC} = 80\,\text{MPa}\sqrt{m}$, $\sigma_{ys} = 1200\,\text{MPa}$ and $\nu = 0.30$. Calculate (a) the applied K_I, (b) the plastic zone size using the Von Mises yield criterion and prove that $r = r_{max}$ when $\theta = \theta_o$. Consider all calculations for plane stress and plane-strain conditions, and (c) draw the entire plastic zone contour where the crack tip is the origin of the coordinates.

5.5. Use the data given in Example 3.4 for a pressure vessel containing a semielliptical crack (Fig. 3.6) to calculate Irwin's and Dugdale's (a) plastic zones, (b) using Kabayashi's finite size correction factor and plasticity correction factor. (c) Compare results and determine the percent error against each case. (d) Is it necessary to include a plastic correction factor? Explain. [Solution: (a) $r\,(Irwin) = 0.54\,\text{mm}$, and $r\,(Dugdale) = 1.33\,\text{mm}$, (b) $K_I\,(Irwin) = 5.63\,\text{MPa}\sqrt{m}$ and $K_I\,(Dugdale) = 6.22\,\text{MPa}\sqrt{m}$].

5.6. A 50-mm thick pressure vessel is to support a hoop stress of 300 MPa at room temperature under no action of corrosive agents. Assume that a semielliptical crack

(Fig. 3.6) is likely to develop on the inner surface with the major axis $2c = 40\,\text{mm}$ and semiminor axis $a = 10\,\text{mm}$. A 300-M steel, which is normally used for airplane landing gear, is to be considered. Will crack propagation occur at $300\,\text{MPa}$ hoop stress? Make sure you include the Irwin's plastic zone correction in your calculations. Is it necessary to do such a plastic correction? Use the data below and select the suitable tempered steel.

300-M Steel	σ_{ys} (MPa)	K_{IC} $(\text{MPa}\sqrt{\text{m}})$
650 °C Tempering	1070	152
300 °C Tempering	1740	65

5.7. If localized plasticity is to be considered, explain the physical meaning of the following inequality $\pi a \sigma^2 / E > \delta_t \sigma_{ys}$.

5.8. Show that $r = \delta_t / (2\pi\epsilon_{ys})$ where r is the plastic zone size due to dislocation networks within the plastic zone area ahead of the crack tip.

5.9. Show that $\delta_t / \epsilon_{ys} = (K_I / \sigma_{ys})^2$ and give a reasonable interpretation of this equality.

5.10. A large strong plate containing a through-the-thickness central crack of $2a_c = 20\,\text{mm}$ has $E = 207\,\text{MPa}$, $\sigma_{ys} = 1275\,\text{MPa}$ and $\delta_c = 9.47\,\mu\text{m}$ at service temperature. Determine (a) the plane-strain fracture toughness, (b) the design stress intensity factor for a safety factor (S_F) of 2, (c) the critical fracture stress, and (d) the design service stress.

5.11. Predict δ_t for a glass using $\delta = \left(4\sigma\sqrt{a^2 - x^2}\right)/E$.

5.12. Derive an expression for δ_t using a Von Mises material. Compare it with that of a Tresca material under plane-strain conditions.

5.13. A material has $E = 70\,\text{MPa}$, $\sigma_{ys} = 500\,\text{MPa}$ and $\nu = 1/3$. It has to be used as a plate in a large structure. Nondestructive evaluation detects a central crack of $50\,\text{mm}$ long. If the displacement at fracture is $0.007\,\text{mm}$ and the plate width is three times the thickness, calculate (a) the crack tip opening displacement, (b) the plane-strain fracture toughness, (c) the plane-strain energy release rate, and (d) the plate thickness, and (e) what is the safety factor being indirectly included in this elastic-plastic fracture mechanics approach? Assume plane-strain conditions as per Eq. (5.49) and a fracture load of 200 kN. [Solution: (a) $\delta_c = 0.014\,\text{mm}$, (b) $K_{IC} = 41.61\,\text{MPa}\sqrt{\text{m}}$, (c) $\sigma_c = 148.48\text{MPa}$, (d) $B = 21.19\,\text{mm}$, and (e) $S_F = 5.43$].

5.14. Repeat Problem 5.13 using Eq. (5.60). Compare results.

5.15. A hypothetical large metallic plate containing a 10-mm central crack is 30-mm wide and 5-mm thick and mechanically loaded in tension. This plate has $E = 69\,\text{GPa}$, $\sigma_{ys} = 500\,\text{MPa}$, $v = 1/3$ and $\epsilon = 0.3\%$ for plane stress strain. Determine (a) δ_t, (b) G_I and (c) σ as per Irwin, Dugdale, Burdekin, and Rice equations. Compare results. [Solution: (a) $\delta_t = 0.015\,\text{mm}$, (b) $G_I = 7.5\,\text{kJ/m}^2$ and (c) σ (Irwin) $= 161\,\text{MPa}$, σ (Dugdale) $= 128\,\text{MPa}$, σ (Burdekin) $= 182\,\text{MPa}$, σ (Rice) $= 322\,\text{MPa}$].

5.16. Determine (a) the critical crack tip opening displacement (δ_c), (b) the plastic zone size (r) and (c) the fracture stress (σ_f) for a large aluminum alloy plate containing a central crack of 5-mm long. Use the following available data and assume plane-strain conditions: $K_{IC} = 25\,\text{MPa}\sqrt{\text{m}}$, $\sigma_{ys} = 500\,\text{MPa}$, and $E = 70\,\text{MPa}$.

5.17. Show that $\delta \simeq \delta_t\sqrt{1 + (E/8a\sigma)^2\,\delta_t^2}$ for plane stress conditions. Plot $\delta = f(\delta_t)$ for various σ and fixed a value.

5.18. If the plane-strain fracture toughness (K_{IC}) and the yield strength (σ_{ys}) of a 12-mm thick $C(T)$ steel specimen are $71\,\text{MPa}\,\text{m}^{1/2}$ and $1896\,\text{MPa}$, respectively, determine (a) the strain energy release rate (G_{IC}) and the validity of the fracture mechanics tension test as per ASTM E399 for the plate containing a single-edge crack of 10 mm long at fracture, (b) the fracture stress if the plate is 20 mm wide, (c) the critical crack tip opening displacement, and (d) the plastic zone size and (e) interpret the results with regard to plane-strain condition. Use a Poisson's ratio of 1/3 and assume that the elastic modulus of the steel 207 GPa.

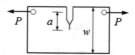

5.19. Assume that isotropic solid material having a single-edge crack is subjected to a remote tensile stress at room temperature. Let the properties of the material be $\sigma_{ys} = 500\,\text{MPa}$, Poisson's ratio $v = 1/3$, and $E = 72\,\text{GPa}$. Let the applied stress intensity factor for mode I loading be $K_I = 20\,\text{MPa}\sqrt{\text{m}}$. Excluding microstructural details and microscale defects, use the Tresca yielding criterion to derive (a) an expression for the critical plastic zone angle (θ_c) and its magnitude when minimum principle stresses are equal ($\sigma_2 = \sigma_3$) and (b) determine when $\sigma_{min} = \sigma_2$ and $\sigma_{min} = \sigma_3$ by knowing the value of θ_c, (c) the plastic zone size at θ_c, and $K_I = 20\,\text{MPa}\sqrt{\text{m}}$. The Tresca yielding criterion is based on the maximum shear stress reaching a critical or failure level. Hence, the definition of the maximum shear stress for this criterion is $\tau_{max} = 0.5\,(\sigma_{max} - \sigma_{min}) = 0.5\sigma_{ys}$, where σ_{max} and σ_{min} are principle stresses and σ_{ys} is the monotonic tensile yield strength of a solid material. Let $\sigma_{max} = \sigma_1$ and $\sigma_{min} = \sigma_2$ or $\sigma_{min} = \sigma_3$.

5.20. Consider a ductile steel plate containing a 50-mm through-thickness central crack subjected to a remote tensile stress of 40 MPa. If the yield strength of the steel is 300 MPa, then calculate as per LEFM, Irwin's approximation and the Dugdale's strip yield criterion. Compare results. Repeat all calculations for a 290 MPa remote stress. Explain.

5.21. A single-edge SE(B) specimen with $B = 20$ mm, $w = 40$ mm is used to determine the critical CTOD δ_{tc} and J_{IC}. See Example 5.4 for the specimen configuration and the load-displacement plot. Assume plane-strain conditions and let

$P_{\max} = 38\,kN;\ \sigma_{ys} = 800\,\text{MPa};\ E = 207\,\text{GPa}$
$K_{IC} = 60\,\text{MPa}\sqrt{\text{m}};\ v = 0.3$
$B = 20\,\text{mm};\ w = 40\,\text{mm};\ \delta_p = 1.00\,\text{mm}$
$S = 150\,\text{mm};\ z = 1.5\,\text{mm};\ a = 12\,\text{mm}$

References

1. M.H. Aliabadi, D.P. Rooke, *Numerical Fracture Mechanics*, Chap. 2 (Computational Mechanics Publications, Kluwer Academic, Boston, 1992)
2. T.L. Anderson, *Fracture Mechanics*, Chap. 3 and 7, 3rd edn. (CRC Press Taylor & Francis, New York, 2005), pp. 17, 61
3. ASTM E1290, *Standard Method for Crack tip Opening Displacement (CTOD) Fracture Toughens Measurement* (Materials Committee, 1998), p. 846
4. G.I. Barenblatt, The mathematical theory of equilibrium of cracks in brittle fracture. Adv. Appl. Mech. **7**, 55–129 (1962)
5. J.M. Barsom, A.T. Rolfe, *Fracture and Fatigue Control in Structures; Applications of Fracture Mechanics*, 3rd edn. (Butterworth-Heinemann, Woburn/ASTM, West Conshohocken, 1999), pp. 51–56, 61
6. B.A. Bilby, K.H. Swiden, Representation of plasticity at notches by linear dislocation arrays. Proc. R. Soc. **A285**, 22–30 (1965)
7. C.Q. Bowles, Army Mat. And Mech. Research Center, Watertown, Massachusetts, Report AMMRC Cr 70-23, 1970
8. D. Broek, A study on ductile fracture, Nat. Aerospace Inst. Amsterdam, Report TR 71021, 1971
9. D. Broek, *Elementary Engineering Fracture Mechanics*, Chap. 4, 4th edn. (Kluwer Academic, Boston, 1986)
10. D. Broek, H. Vlieger, National Aerospace Lab. TR 74032, 1974
11. W.F. Brown Jr., J.E. Srawley, *Plane Strain Crack Toughness Testing of High Strength Metallic Materials*. ASTM STP, vol. 410 (American Society for Testing and Materials, Philadelphia, 1966), pp. 1–65
12. F.M. Burdekin, D.E. Stone, The crack opening displacement approach to fracture mechanics in yielding materials. J. Strain Anal. Eng. Des. **1**(2), 145–153 (1966)
13. R. Chona, M.S. Thesis, University of Maryland, College Park, August 1985
14. J.A. Collins, *Failure of Materials in Mechanical Design: Analysis, Prediction, Prevention* (Wiley, New York, 1981)
15. T.H. Courtney, *Mechanical Behavior of Materials* (McGraw-Hill, New York, 1990)
16. W.W. Dally, W.F. Riley, *Experimental Stress Analysis*, Chap. 4, 3rd edn. (McGraw-Hill, New York, 1991)
17. D.S. Dawicke, M.A. Sutton, CTOA and crack-tunneling measurements in thin sheet 2024-T3 aluminum alloy. Exp. Mech. **34**(4), 357–368 (1994)

18. D.S. Dawicke, J.C. Newman, Jr., C.A. Bigelow, Three-dimensional CTOA and constraint effects during stable tearing in thin sheet material, in *STM STP*, vol. 1256, ed. by W.G. Reuter, J.H. Underwood, J.C. Newman Jr. (American Society for Testing and Materials, Philadelphia PA, 1995), pp. 223–242

19. D.S. Dugdale, Yielding of steel sheets containing slits. J. Mech. Phys. Solids **8**, 100–104 (1960)

20. J.D. Eshelby, The determination of the elastic field of an ellipsoidal inclusion and related problems. Proc. R. Soc. Lond. Ser. A **241**, 376–396 (1957)

21. H.O. Fuchs, R.I. Stephens, *Metal Fatigue in Engineering* (Wiley, New York, 1980)

22. A.S, Gullerud, R.H. Dodds Jr., R.W. Hampton, D.S. Dawicke, Three-dimensional modeling of ductile crack growth in thin sheet metals: computational aspects and validation. Eng. Fract. Mech. **63**, 347–734 (1999)

23. G.T. Hahn, A.R. Rosenfield, Local yielding and extension of a crack under plane stress. Acta Metall. **13**, 293–306 (1965)

24. G.T. Hahn, A.R. Rosenfield, Plastic flow in the locale on notches and cracks in Fe-3Si steel under condition approaching plane strain. Report to ship structure committee SSC-191, U.S. Coast Guard, Washington, DC, November 1968

25. K. Hellan, *Introduction to Fracture Mechanics* (McGraw-Hill, New York, 1984), p. 16

26. R.W. Hertzberg, *Deformation and Fracture Mechanics of Engineering Materials*, 3rd edn. (Wiley, New York, 1989).

27. G.R. Irwin, Disccusion of paper "dynamic stress distribution surrounding a running crack - a photoelastic analysis," by A. Wells and D. Post. Proc. SESA **XVI**(1), 93–96 (1958)

28. G.R. Irwin, Fracture I, in *Handbuch der Physik VI*, ed. by S. Flugge (Springer, New York, 1958)

29. G.R. Irwin, Plastic zone near a crack and fracture toughness, in *Proceedings of 7th Sagamore Conference* (1960), p. IV-63

30. G.R. Irwin, Plastic zone near a crack and fracture toughness, in *Proceedings of the Seventh Sagamore Ordnance Materials Research Conference*, Session VI, Syracuse University Research Institute, Syracuse, New York (1960), pp. 63–78

31. J.A. Jacobs, Relaxation methods applied to the problem of plastic flow. Phil. Mag. F **41**, 349–358 (1950)

32. M.A. James, J.C. Newman Jr., Three dimensional analysis of crack-tip-opening angles and δ_5-resistance curves for 2024-T351 aluminum alloy, in *Fatigue and Fracture Mechanics: 32nd Volume, ASTM STP*, vol. 1406, ed. by R. Chona (American Society for Testing and Materials, West Conshohocken, PA, 2002), pp. 279–297

33. M.A. James, J.C. Newman Jr., Three-dimensional analyses of crack-tip-opening angles and δ_5-resistance curves for 2024-T351 aluminum alloy, in *ASTM STP*, vol. 1406 (2002), pp. 279–297. This reference is cited in [46]

34. W.M. Johnston, M.A. James, A relationship between constraint and the critical crack tip opening angle, NASA/CR-2009-215930, National Aeronautics and Space Administration, Langley Research Center, Hampton, Virginia, December 2009

35. M.F. Kanninen, C.H. Popelar, *Advanced Fracture Mechanics*, Chapter 5 (Oxford University Press, New York, 1985)

36. J.F. Lancaster, *Metallurgy of Welding*, 5th edn. (Chapman & Hall, New York, 1993)

37. W.R. Lloyd, Microtopography for ductile fracture process characterization part 1: theory and methodology. Eng. Fract. Mech. **70**, 387–401 (2003)

38. W.R. Lloyd, Microtopography for ductile fracture process characterization part 2: application for CTOA analysis. Eng. Fract. Mech. **70**, 403–415 (2003)

39. J.R. Low, The fracture of metals. Prog. Mater. Sci. **12**(1), 3–96 (1963)

40. S. Mahmoud, K. Lease, The effect of specimen thickness on the experimental characterization of critical crack-tip-opening angle in 2024-T351 aluminum alloy. Eng. Fract. Mech. **70**, 443–456 (2003)

41. S. Mahmoud, K. Lease, Two-dimensional and three-dimensional finite element analysis of critical crack-tip-opening angle in 2024-T351 aluminum alloy at four thicknesses. Eng. Fract. Mech. **71**, 1379–1391 (2004)

42. F.A. McClintock, in *Physics of Strength and Plasticity*, ed. by A.S. Argon (MIT Press, Cambridge, MA, 1969), pp. 307–326
43. F.A. McClintock, G.R. Irwin, Plasticity aspects of fracture mechanics, in *ASTM STP*, vol. 381 (1965), pp. 84–113
44. M.A. Meyers, K.K. Chawla, *Mechanical Behavior of Materials* (Prentice Hall, Upper Saddle River, 1999), pp. 347–351
45. J.C. Newman Jr., U. Zerbst, Fundamentals and applications of the crack-tip-opening-angle (CTOA), engineering fracture mechanics. Special issue of engineering fracture mechanics from a workshop, vol. 70 (2003), pp. 367–577
46. J.C. Newman Jr., M.A. James, U. Zerbst, A review of the CTOA & CTOD fracture criterion. . Eng. Fract. Mech. **70**, 371–385 (2003)
47. J.C. Newman Jr., B.C. Booth, K.N. Shivakumar, An elastic-plastic finite-element analysis of the J-resistance curve using a CTOD criterion, in *Fracture Mechanics: Eighteenth Symposium, ASTM STP*, vol. 945 ed. by D.T. Read, R.P. Reed (American Society for Testing and Materials, Philadelphia, PA, 1988), pp. 665–685
48. P.C. Paris, H. Tada, A. Zahoor, H. Ernst, Instability of the tearing model of elastic-plastic crack growth, in *ASTM STP*, vol. 668 (1979), p. 5.36
49. V.Z. Parton, E.M. Morozov, *Mechanics of Elastic-Plastic Fracture*, 2nd edn. (Hemisphere, New York, 1989)
50. K.E. Puttick, Ductile fracture in metals. Philos. Mag. **4**, 964–969 (1959)
51. J.R. Rice, in *Proceedings of 1st International Conference on Fracture*, Sendai, 1965, vol. I, ed. by T. Yokobori, et al. (Japanese Society for strength and Fracture of Materials, Tokyo, 1966), p. 283
52. J.R. Rice, A path independent integral and the approximate analysis of strain concentration by notches and cracks. J. Appl. Mech. **35**, 379–386 (1968)
53. J.R. Rice, in *Fracture an Advanced Treatise*, vol. II, ed. by H. Liebowitz (Academic Press, New York, 1968), p. 191
54. J.R. Rice, G.F. Rosengren, Plane strain deformation near a crack tip in a power-law hardening material. J. Mech. Phys. Solids **16**, 1 (1968)
55. R.O. Ritchie, A.W. Thompson, On macroscopic and microscopic analyses for crack initiation and crack growth toughness in ductile alloys. Metall. Trans. A, **16A**, 233–248 (1985)
56. J.N. Robinson, A.S. Tetelman, Un. Cal. Los Angeles Report Eng. 7360, 1973
57. H.C. Rogers, The tensile fracture of ductile metals. Trans. Metall. Soc. AIME **218**, 498–506 (1960)
58. A.R. Rosenfield, P.K. Dai, G.T. Hahn, Crack extension and crack propagation under plane stress, in Proceedings of the 1st International Conference on Fracture, Sendai, vol. 1, ed. by T. Yokobori, T. Kawasaki, J.L. Swedlow (Japanese Society for Strength and Fracture of Materials, Tokyo, 1966), pp. 223–258;
    Second Progress Report Project SR-164 "Local Strain Measurement" to the Ship Structure Committee, Department of the Navy, Bureau of Ships Contract NObs 92383. Washington, DC. National Academy of Sciences, National Research Council, Originator's Report Numder SSC-172 (March 1966), pp. 1–38
59. R.J. Sanford, A critical re-examination of the westergaard method for solving opening-mode crack problems. Mech. Res. Commun. **6**(5), 289–294 (1979)
60. R.J. Sanford, *Principles of Fracture Mechanics*, Chap. 3 (Pearson Education, Upper Saddle River, NJ, 2003)
61. S.H. Song, M.P. Wagoner, G.H. Paulino, W.G. Buttlar, δ_{25} Crack opening displacement in cohesive zone models: experimental and simulations in asphalt concrete. Fatigue Fract. Eng. Mater. Struct. **31**, 850–856 (2008)
62. A.N. Stroh, A theory of fracture of metals. Adv. Phys. **6**, 418–465 (1957)
63. P.S. Theocaris, N.P. Andrianopoulos, The mises elastic-plastic boundary as the core region in fracture criteria. Eng. Fract. Mech. **16**(3), 425–432 (1982)
64. I.S. Tuba, A method of elastic-plastic plane stress and plane strain analysis. J. Strain Anal. **1**, 115–122 (1966)

65. J.H. Underwood, D.P. Kendall, Measurement of plastic strain distribution in the region of a crack tip. Exp. Mech. **9**, 296–304 (1969)
66. J. Weertman, *Dislocation Based Fracture Mechanics* (World Scientific, New Jersey, 1996), p. 21
67. G.H. Wellman, S.T. Rolfe, R.H. Dodds, Three dimensional elastic-plastic finite element analysis of three point bend specimens. Weld. Res. Counc. Bull. **299**, 15–25 (1984)
68. A.A. Wells, Unstable crack propagation in metals: cleavage and fast fracture, in *Proceedings of the Crack Propagation Symposium*, vol. 1, Paper 84, Cranfield, 1961
69. H.M. Westergaard, Bearing pressures and cracks. J. Appl. Mech. **G1**, A49–A53 (1939)
70. M.L. Williams, On the stress distribution at the base of a stationary crack. J. Appl. Mech. **24**, 109–114 (1957)
71. U. Zerbst, M. Heinimann, C.D. Donne, D. Steglich, Fracture and damage mechanics modelling of thin-walled structures - an overview. Eng. Fract. Mech. **76**, 5–43 (2009)

The Energy Principle

6

6.1   Introduction

In this chapter, the elastic behavior of solids containing cracks is examined using the energy principle approach, which includes all forms of energy since loading develops mechanical work, energy absorption around the crack tip, and energy dissipation as heat. Williams [25] and Broek [3] undertook this energy approach as the primary form of energy (mechanical work) being considered. If work is done, then crack growth occurs and elastic energy is released. With the exception of pure brittle solids, engineering materials undergo some form of plastic deformation at the crack tip due to an applied external stress. Such plastic deformation is an irreversible process or plastic flow.

Theoretically, the action of a remote external force on a body containing at least one crack is to cause crack growth and to disturb the material potential energy. This action introduces mechanical energy that balances out with the potential energy of the body, inducing energy dissipation as the body deforms ahead of the crack tip. Some of this input mechanical energy is absorbed by the body as strain energy, and some is dissipated as heat during mechanical deformation. In practice, the potential energy (stored energy) of the body changes upon the introduction of a crack and upon the action of the external force under a quasi-static or dynamic loading mode.

This chapter includes an energy balance for characterizing the fracture of elastic materials containing cracks. The goal is to minimize the energy in relation to the crack length in order to derive the critical stress for crack propagation.

In a general sense, plastic flow is referred to as a permanent and non-recoverable deformation in most common solid materials. On the other hand, viscous flow describes the mechanical behavior of plastics, such as noncrystalline polymers, and it is temperature and time dependent. The term viscoelasticity is also used to indicate viscous flow.

© Springer International Publishing Switzerland 2017
N. Perez, *Fracture Mechanics*, DOI 10.1007/978-3-319-24999-5_6

6.2   Energy Balance

The focus of this section is to determine the energy balance and the crack driving force for a slow crack growth event due to the action of an external quasi-static load. The crack may be embedded, on the surface or through-the-thickness. Consider a body with a boundary contour Γ as shown in Fig. 6.1 subjected to an energy input increment (dU) due to an external loading [3, 7, 25].

From Fig. 6.1, the shaded area represents a very small process zone surrounded by an elastic continuum; S_c and S_c^+ are the existing and new crack surface [7]. The energy change in a loaded plate occurs due to the displacements arising from the fracture area change $dA = B\partial a$ for a constant thickness B and variable crack length a. Thus, the input energy change (dU_1) is divided into the change in dissipated energy (dU_2) as heat which arises due to the irreversible process during plastic or viscous flow, the change in stored energy or total potential elastic energy (dU_3), and the change in kinetic energy (dU_4) of the system [25].

For an isothermal case, dU_2 is transferred across the contour Γ of the system, and for an adiabatic case, dU_2 is not transferred and the system temperature rises. Consequently, the conservation of energy change due to the displacements arising from the fracture area change dA can be defined as [25]

$$\frac{dU_1}{dA} - \frac{dU_2}{dA} = \frac{dU_3}{dA} + \frac{dU_4}{dA} \qquad (6.1)$$

For a growing crack, dU_2/dA is the energy dissipated in propagating fracture over an increment of area dA which is referred to as the fracture resistance R. On the other hand, $dU_1/dA - dU_2/dA$ can be defined as the net energy input. In fact, the energy dissipated can be treated as the strain energy release rate during fracture [25].

Fig. 6.1 Cracked body with energy changes

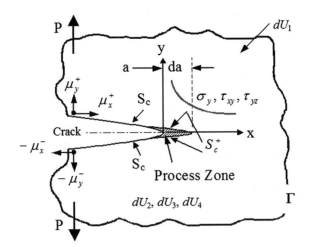

The energy dissipated is the strain energy release rate during fracture. On the other hand, the input energy is the supplied energy to a crack tip for inducing crack growth and possible plastic deformation ahead of the crack tip. The input energy must balance out with the amount of energy dissipated during the formation of new crack surfaces, which are related to crack extension.

Another important definition is the energy release rate G_i, where the subscript stands for mode of loading; that is, $i = I, II, III$ or a combination of loading modes described in Fig. 3.1. Thus [25],

$$G_i = \frac{dU_1}{dA} - \frac{dU_3}{dA} \tag{6.2a}$$

$$G_i = \frac{dU_1}{Bda} - \frac{dU_3}{Bda} \tag{6.2b}$$

Combining Eqs. (6.1) and (6.2) along with $R = dU_2/dA$ and $U_4 = 0$ for a stationary body, the strain energy release rate criterion for crack tip instability is [25]

$$G_{iC} \geq R \tag{6.3}$$

Here, R is the materials fracture resistance. If the cracked plate shown in Fig. 6.1 is subjected to an external load P and the crack grows very slowly, then the load points undergo a relative displacement $d\mu$ perpendicular to the crack plane, and the crack length extends an amount da. Consequently, the work done responsible for such an increment in displacement and crack length is defined by the input energy gradient. Thus,

$$\frac{dU_1}{da} = P\frac{d\mu}{da} \tag{6.4}$$

Nonetheless, G_i is the energy release rate for crack growth, commonly known as the strain energy release rate. Assuming that the plastic zone or the size and shape of the energy dissipation zone remains nearly constant during brittle fracture, then the strain energy release rate reaches a critical value (G_{iC}), also known as **fracture toughness**. Thus, $G_{iI} = G_{iC}$ at fracture and it is universally accepted as a material property in linear-elastic fracture mechanics.

Logically, fracture toughness in terms of energy is normally converted into a critical stress intensity factor (K_{iC}) used in designing applications. Actually, mode I is the most studied and used type of loading for doing fracture mechanics testings. Thus, $G_{iC} \rightarrow G_{IC}$ and $K_{iC} \rightarrow K_{IC}$.

Generally, G_{iC} in Eq. (6.3) depends on the applied load and specimen geometry, while R depends on the fracture properties of the materials as well as the specimen geometry. Fundamentally, G_{iC} is the energy of the mechanics of solids containing discontinuities known as flaws or cracks. Anyhow, if $G_{iC} = R$, then G_{iC} is a material property.

6.3   Linear Compliance

Assume that the solid body is elastic and its elastic mechanical behavior can be characterized by a linear load-displacement relationship, $P = f(\mu)$, as depicted in Fig. 6.2 for which the slope is $0 < dP/da < \infty$ [7]. The slope is referred to as the stiffness and the inverse of stiffness is called the compliance.

Consider mode I (tension) loading shown in Fig. 6.2, where the curve AB is a possible load-displacement trajectory $P(\mu, a)$ for a moving crack [7].

The stored energy due to tension loading can be defined as the area under the curve (Area OAE) [25]

$$U_3 = \frac{1}{2}P\mu \tag{6.5}$$

from which

$$\frac{dU_3}{da} = \frac{P}{2}\frac{d\mu}{da} + \frac{\mu}{2}\frac{dP}{da} \tag{6.6}$$

Inserting Eqs. (6.4) and (6.6) into (6.2b) gives

$$G_I = \frac{1}{2B}\left(P\frac{d\mu}{da} - \mu\frac{dP}{da}\right) \tag{6.7}$$

$$G_I = \frac{1}{B}\frac{dU_3}{da} \quad @ \, P = \text{Constant} \tag{6.7a}$$

For essentially elastic response, the linear compliance is the inverse of the slope of the load-line OA in Fig. 6.2, from which the displacement takes the form [25]

$$\mu = PC \tag{6.8}$$

Fig. 6.2 Schematic linear load-displacement for a growing crack. OA and OB are the loading and unloading lines, respectively, and the curve AB is a possible trajectory $P(\mu, a)$ for a moving crack during unloading. After reference [7]

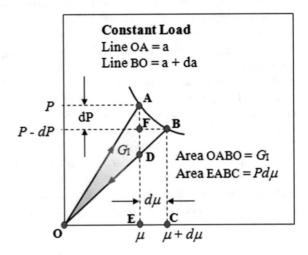

from which

$$\frac{d\mu}{da} = P\frac{dC}{da} + C\frac{dP}{da} \tag{6.9}$$

Substituting Eqs. (6.8) and (6.9) into (6.7) yields the crack driving force as

$$G_I = \frac{P^2}{2B}\frac{dC}{da} \tag{6.10}$$

$$G_I = \frac{1}{2B}\left(\frac{\mu}{C}\right)^2\frac{dC}{da} \tag{6.11}$$

$$G_I = \frac{U_3}{BC}\frac{dC}{da} \tag{6.12}$$

Moreover, the G_I expression, Eq. (6.7), can also be derived using the segments and areas given in Fig. 6.2 [7]. For instance,

- $OA =$ Initial loading line
- $AB =$ Unloading line since

$$P \to (P - dP) \text{ as } \mu \to (\mu + d\mu) \tag{a}$$

Consequently, the crack area changes from A to $A + dA$ or the crack grows from a to $a + da$.
- Area $OAE =$ Stored energy at fracture

$$OAE = P\mu/2 \tag{b}$$

- Area $OBC =$ Stored energy after fracture

$$OBC = (P - dP)(\mu + d\mu)/2 \tag{c}$$

- Area $ABCE =$ Work done along with external fixed load P

$$ABCE = \int_{\mu}^{\mu+d\mu} Pd\mu = Pd\mu \tag{d}$$

- Area $OAB = OAE + ABCE - OBC =$ Release of elastic energy $= G_I dA = G_I Bda$. Thus, the strain energy release rate equation becomes

$$OAB = \frac{1}{2}P\mu + Pd\mu - \frac{1}{2}(P - dP)(\mu + d\mu)$$

$$G_I = \frac{1}{2B}\left(P\frac{d\mu}{da} - \mu\frac{dP}{da}\right) \tag{6.7}$$

Furthermore, this expression, Eq. (6.7), can take two different definitions based on special cases. For instance, when the applied load P is constant, G_I Eq. (6.7) gives

$$G_I = \frac{P}{2B} \frac{d\mu}{da} \qquad (a)$$

On the other hand, if the displacement is constant, then

$$G_I = -\frac{\mu}{2B} \frac{dP}{da} \quad @ \; \mu = \text{Constant}$$

6.4   Nonlinear Compliance

The possible load lines for a nonlinear behavior is shown in Fig. 6.3 [3, 7, 25]. The analysis is carried out using a nonlinear compliance expression of the form [25]

$$\mu = (PC_n)^{1/n} \qquad (6.13)$$

where n = strain hardening exponent

The strain energy release rate is defined by [25]

$$G_I = \frac{1}{B} \left[P\frac{d\mu}{da} - \frac{d}{da}\left(\frac{P\mu}{1+n}\right) \right] \qquad (6.14)$$

$$G_I = \frac{1}{(1+n)B}\left(nP\frac{d\mu}{da} - \mu\frac{dP}{da} \right) \qquad (6.15)$$

Combining Eqs. (6.13) and (6.15) yields

$$G_I = \frac{1}{(1+n)B} P^{(1+n)/n} C_n^{(1-n)/n} \frac{dC_n}{da} \qquad (6.16)$$

Fig. 6.3 Possible load trajectory for a growing crack in a nonlinear-elastic-plastic material

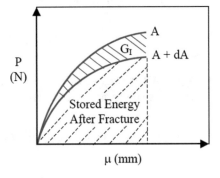

P
(N)

G_I

A

A + dA

Stored Energy
After Fracture

μ (mm)

$$G_I = \frac{1}{(1+n)\,B}\,\frac{\mu^{1+n}}{C_n^2}\,\frac{dC_n}{da} \tag{6.17}$$

$$G_I = \frac{U_3}{BC_n}\,\frac{dC_n}{da} \tag{6.18}$$

A remarkable observation is that if $n = 1$, then Eq. (6.16) yields Eq. (6.10).

6.5 Traction Forces

Consider two-dimensional problems in rectangular system x, y, where $z = x_3$ has a unit length, and it is treated here as a coordinate. Assume that the path of the crack tip is along x-axis and that the origin is at the crack tip (Fig. 6.1) for a pure elastic state under quasi-static loading. This implies that the crack area along with a unit length in the x_3-axis becomes $\Delta A = \Delta a x_3 \rightarrow \Delta a$.

By definition, the traction exerted through S_c^+ from the continuum on the process zone is $-\vec{T} = \{T_i\}$, where i denotes the type of stress per stressing mode I, II, or III.

According to Hellan [7], the magnitudes of the traction forces on the upper and lower crack sides are $T_i = -\sigma_{yi}$ and $T_i = +\sigma_{yi}$ or $T_i = -\sigma_{vi}\,(+\sigma_{vi})$, respectively. The corresponding displacements are $\mu_i = \mu_i^+$ and $\mu_i = \mu_i^-$ or $\mu_i = \mu_i^+\,(\mu_i^-)$.

Furthermore, the dominant stresses along the crack line can be determined at $\theta = 0$ and the displacements along the crack sides at $\theta = \pm\pi$. This implies that the crack tip is located at $(x, y) = (\geq 0, 0)$ for the principle stresses and $(x, y) = (\leq 0, 0)$ for the displacements.

Accordingly, the translation of the fields at the crack front leads to $\Delta a \rightarrow 0$, which is required to determine the limit of the crack driving force. This can clearly be appreciated by considering a two-dimensional analysis of a pure elastic solid body subjected to a quasi-static loading as depicted in Fig. 6.1. Thus, the strain energy release rate on the elastic continuum over the crack surface S_c^+ can be defined by [7]

$$G_i = \lim_{\Delta A \rightarrow 0} \int_{S_c^+} \left(\int_{(a)}^{(b)} T_i\,d\mu_i \right) ds \quad \text{for } \Delta A \rightarrow 0 \tag{6.19}$$

$$G_i = \lim_{\Delta a \rightarrow 0} \int_0^{\Delta a} \left[\int_{(a)}^{(b)} \sigma_{yi}\,d\left(\mu_i^+ - \mu_i^-\right) \right] dx \tag{6.20}$$

$$G_i = \frac{1}{2\Delta a} \int_0^{\Delta a} \sigma_{yi}^{(a)} \left(\mu_i^+ - \mu_i^-\right)^{(b)} dx \tag{6.21}$$

where ds = Differential surface area
$d\mu_i$ = Displacement increments at ds
ΔA = Change in crack area

(a) = Representation of a stage for plastic process

(b) = Representation of a stage for crack extension

In order to solve Eq. (6.21), the stresses and the displacements for the three modes of loading (as depicted in Fig. 3.1) being derived in Chap. 4 can be defined in a general form by replacing the plastic zone size r for x in the stress equations for each stress mode and r for $\Delta a - x$ in the displacement equations when crack extension occurs.

Moreover, if stable crack growth occurs along the crack plane, then the displacements are related as follows: $\mu_y^+ = -\mu_y^-$ and $\mu_y^+ - \mu_y^- = 2\mu_y^+$.

For the symmetric mode I at $y = 0$ and $\mu_y^{\pm} = -\mu_\theta$,

$$\sigma_y = \sigma_\theta = \frac{K_I}{\sqrt{2\pi x}} \qquad \text{for } x \geq 0;\ \theta = 0 \qquad (6.22)$$

$$\mu_y^{\pm} = \pm \frac{(\kappa + 1)\,(1 + \nu)\,K_I}{E} \sqrt{-\frac{x}{2\pi}} \quad \text{for } x \leq 0;\ \theta = \pm \pi \quad (6.23)$$

$$\sigma_y\left(\mu_y^+ - \mu_y^-\right) = \frac{(\kappa + 1)\,(1 + \nu)\,K_I^2}{\pi E} \sqrt{\frac{\Delta a - x}{2\pi}} \qquad (6.24)$$

For the antisymmetric mode II at $y = 0$ and $\mu_x^{\pm} = -\mu_r$,

$$\tau_{xy} = \tau_{r\theta} = \frac{K_{II}}{\sqrt{2\pi x}} \qquad \text{for } x \geq 0;\ \theta = 0 \qquad (6.25)$$

$$\mu_x^{\pm} = \pm \frac{(\kappa + 1)\,(1 + \nu)\,K_{II}}{E} \sqrt{-\frac{x}{2\pi}} \quad \text{for } x \leq 0;\ \theta = \pm \pi \quad (6.26)$$

$$\tau_{xy}\left(\mu_x^+ - \mu_x^-\right) = \frac{(\kappa + 1)\,(1 + \nu)\,K_{II}^2}{\pi E} \sqrt{\frac{\Delta a - x}{2\pi}} \qquad (6.27)$$

For the antisymmetric mode III at $y = 0$,

$$\tau_{yz} = \tau_{z\theta} = \frac{K_{III}}{\sqrt{2\pi x}} \qquad \text{for } x \geq 0;\ \theta = 0 \qquad (6.28)$$

$$\mu_z^{\pm} = \pm \frac{4\,(1 + \nu)\,K_{III}}{E} \sqrt{-\frac{x}{2\pi}} \qquad \text{for } x \leq 0;\ \theta = \pm \pi \quad (6.29)$$

$$\tau_{yz}\left(\mu_z^+ - \mu_z^-\right) = \frac{4\,(1 + \nu)\,K_{III}^2}{\pi E} \sqrt{\frac{\Delta a - x}{2\pi}} \qquad (6.30)$$

From Chap. 4, the constant κ is defined by

$$\kappa = \frac{3 - \nu}{1 + \nu} \quad \text{for plane stress} \qquad (6.31)$$

$$\kappa = 3 - 4\nu \quad \text{for plane strain} \qquad (6.32)$$

Substituting Eqs. (6.24), (6.27), and (6.30) into (6.21) yields the crack driving force for a mixed-mode interaction:

$$G_i = \frac{1}{2\Delta a} \int_o^{\Delta a} \left[\sigma_y \left(\mu_y^+ - \mu_y^- \right) + \tau_{xy} \left(\mu_x^+ - \mu_x^- \right) + \tau_{yz} \left(\mu_z^+ - \mu_z^- \right) \right] dx \qquad (a)$$

$$G_i = \frac{2(1+v)}{\pi \Delta a E} \left[\frac{\kappa + 1}{4} \left(K_I^2 + K_{II}^2 \right) + K_{III}^2 \right] \int_o^{\Delta a} \sqrt{\frac{\Delta a - x}{x}} \, dx \qquad (b)$$

The integral can be solved by letting $x = \Delta a \sin^2 \alpha$ for $0 \le \alpha \le \pi/2$ so that $dx = 2\Delta a \sin \alpha \cos \alpha \cdot d\alpha$. Thus,

$$\int_o^{\pi/2} \sqrt{\frac{\Delta a - x}{2\pi}} \, dx = 2\Delta a \int_o^{\pi/2} \cos^2 \alpha \cdot d\alpha$$

$$= \Delta a \left[\frac{\alpha}{2} + \frac{\sin 2\alpha}{4} \right]_0^{\pi/2} \qquad (c)$$

$$= \frac{\pi \Delta a}{2}$$

and

$$G_i = \frac{(1+v)}{E} \left[\frac{\kappa + 1}{4} \left(K_I^2 + K_{II}^2 \right) + K_{III}^2 \right] \qquad (6.33)$$

Algebraic manipulation of Eq. (6.33) along with Eqs. (6.31) and (6.32) gives G_i for crack motion on its tangent plane

$$G_i = \frac{K_I^2}{E'} + \frac{K_{II}^2}{E'} + \frac{(1+v) K_{III}^2}{E} \qquad (6.34)$$

where $E' = E$ for plane stress
$E' = E/ \left(1 - v^2 \right)$ for plane strain.

For convenience, the effect of Poisson's ratio on the strain energy release rate for mode I loading is shown in Fig. 6.4.

Thus far it has been assumed that the material is homogeneous, isotropic, and linear elastic and that the crack extends in a self-similar manner (along its own

Fig. 6.4 Variation of the energy release rate $E'G_I = f(K_I)$

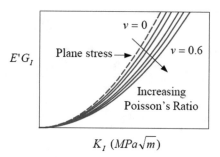

plane). Based on these assumptions, the strain energy release rate schematic profile depicted in Fig. 6.4 was produced using Eq. (6.34) for mode I. Evidently, $E'G_I = f(K_I)$ is fundamentally a simple relation, but gives a nonlinear profile dependent on the Poisson's ratio v.

6.6   Load and Displacement Control

Assume that the slender ($a >> h$) double cantilever beam (*DCB*) shown in Fig. 6.5 is loaded in tension with no rotation at the end of the beam. If the displacement and linear compliance equations are [5,7,25]

$$\mu = \frac{2Pa^3}{3E'I} \tag{6.35}$$

$$C = \frac{\mu}{P} = \frac{2a^3}{3E'I} \tag{6.36}$$

where a = Crack length
I = Moment of inertia = $Bh^3/12$

For a **constant load** condition, the gradients of the compliance, displacement, and load gradient are, respectively,

$$\frac{dC}{da} = \frac{2a^2}{E'I} \tag{6.37}$$

$$\frac{d\mu}{da} = \frac{2Pa^2}{E'I} \tag{6.38}$$

$$\frac{dP}{da} = 0 \tag{6.39}$$

Fig. 6.5 Slender double cantilever beam (DCB) with $a >> h$

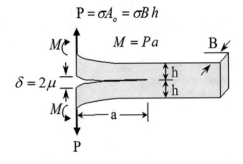

Substituting Eqs. (6.38) and (6.39) into (6.7) gives

$$G_I = \frac{P}{2B} \frac{d\mu}{da} \tag{6.40}$$

$$G_I = \frac{P^2 a^2}{BE'I} \quad \text{for } P = \text{constant} \tag{6.41}$$

If Eq. (6.10) is used instead of (6.7), it yields the same result. Thus,

$$G_I = \frac{P^2}{2B} \frac{dC}{da} = \frac{P^2 a^2}{BE'I} \quad \text{for } P = \text{constant} \tag{6.42}$$

As a result, $G_I \propto a^2$ (proportional) which implies that G_I increases rapidly as a increases due to the availability of strain energy. Crack propagation occurs when $G_I = G_{IC}$ and $a = a_c$. In addition, crack instability occurs under load-control if $dP/da < 0$ when $G_I \propto a^2$. This is clearly demonstrated by solving Eq. (6.42) for P and deriving the load gradient

$$P = \frac{\sqrt{G_I BE'I}}{a} \tag{6.43}$$

$$\frac{dP}{da} = -\frac{\sqrt{G_I BE'I}}{a^2} < 0 \tag{6.44}$$

Consider the slender DCB under **constant displacement**. In this case, the load and load gradient expressions are

$$P = \frac{\mu}{C} = \frac{3\mu E'I}{2a^3} \tag{6.45}$$

$$\frac{dP}{da} = -\frac{9\mu E'I}{2a^4} \tag{6.46}$$

$$\frac{d\mu}{da} = 0 \tag{6.47}$$

Insert Eqs. (6.46) and (6.47) into (6.7) to get

$$G_I = -\frac{\mu}{2B} \frac{dP}{da} = \frac{9\mu^2 E'I}{4Ba^4} \quad \text{For } \mu = \text{Constant} \tag{6.48}$$

This result suggests that $G_I \propto a^{-4}$ and G_I decreases as a increases. Therefore, crack stability occurs if $d\mu/da > 0$ and $dG_I/da < 0$ when $G_I \propto a^{-4}$; that is,

$$\mu = \frac{2a^2}{3} \sqrt{\frac{4G_I B}{9E'I}} \tag{6.49}$$

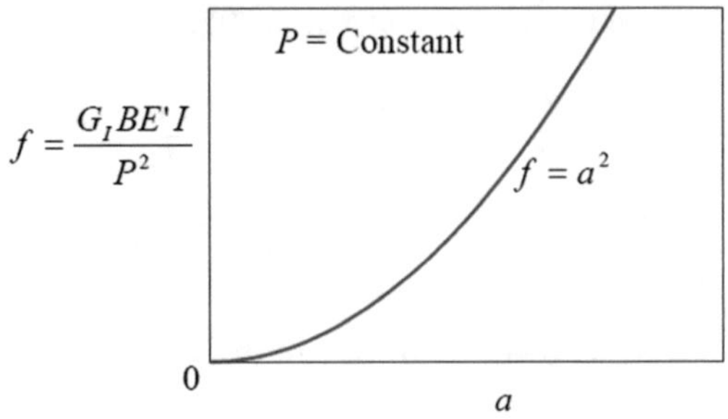

Fig. 6.6 Normalized crack driving force as a function of crack length under load control system

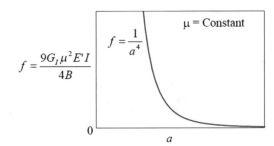

$$f = \frac{G_I BE'I}{P^2}$$

Fig. 6.7 Normalized crack driving force as a function of crack length under displacement control system

$$f = \frac{9G_I \mu^2 E'I}{4B}$$

$$\frac{d\mu}{da} = \frac{4a}{3}\sqrt{\frac{4G_I B}{9E'I}} > 0 \tag{6.50}$$

$$\frac{dG_I}{da} = -\frac{9\mu^2 E'I}{Ba^3} < 0 \tag{6.51}$$

The theoretical trend of the behavior of the strain energy release rate as per Eqs. (6.42) and (6.48) is shown in Figs. 6.6 and 6.7, respectively.

To this point, the energy principles being described in terms of the strain energy release rate provide significant supplementary insight into the fracture mechanics field. For instance, the derived equations for G_I at constant load P and constant displacement μ_y are the fundamental mathematical expressions for generating an important foundation for analytical methods in fracture mechanics. The importance of the energy principle methods is further explained below in relation to Eqs. (6.42) and (6.48).

It is evident that G_I strongly depends on the crack length (a) at constant load condition. According to Eq. (6.42), G_I has a nonlinear behavior since $G_I \propto a^2$. This implies that any small amount of crack growth induces a steep increment on G_I. This is actually depicted in Fig. 6.6.

On the other hand, G_I is also strongly dependent on the crack length (a) at constant displacement condition. In this case, $G_I \propto a^{-4}$ and G_I exhibits a nonlinear behavior in the opposite sense. Therefore, G_I drastically decreases as a increases as shown in Fig. 6.7.

Therefore, there is a significant difference between the two energy methods since the strain energy release rate has been shown to depend on the exponent of the crack length; $G_I \propto a^2$ and $G_I \propto a^{-4}$. As shown in Figs. 6.6 and 6.7, they are essentially opposite in trend. However, understanding the differences between these two analytical results is extremely important when characterizing crack growth and determining fracture toughness of a particular material.

Fundamentally, the strain energy release rate is the crack tip energy release rate during crack growth, and it quantifies the rate of change of the potential energy of the cracked elastic solid undergoing elastic deformation. If the solid is elastic perfectly plastic, then the rate of change of the potential energy is inherently a mixed-mode interaction.

Example 6.1. *Suppose that three specimens made of 7075-T651 Al-alloy ($E = 70\,GPa$, $\sigma_{ys} = 850\,MPa$ and $v = 0.30$) that have identical dimensions were loaded in tension and exhibited linear behavior. The data shown below was obtained at room temperature.*

No.	a (mm)	B (mm)	P (kN)	μ_y (mm)	Fracture
1	25.00	25	145		Yes
2	25.00	25	100	0.2500	No
3	25.50	25	100	0.2525	No

Determine the plane-strain fracture toughness K_{IC}.

Solution. *The compliance for specimens 2 and 3 and the compliance gradient are*

$$C_2 = \frac{\mu_{y2}}{P_2} = \frac{0.2500 \times 10^{-3}\,\text{m}}{100 \times 10^3\,\text{N}} = 2.500 \times 10^{-9}\,\text{m/N}$$

$$C_3 = \frac{\mu_{y3}}{P_3} = \frac{0.2525 \times 10^{-3}\,\text{m}}{100 \times 10^3\,\text{N}} = 2.525 \times 10^{-9}\,\text{m/N}$$

$$\frac{dC}{da} \simeq \frac{C_3 - C_2}{a_3 - a_2} = \frac{2.525 \times 10^{-9}\,\text{m/N} - 2.500 \times 10^{-9}\,\text{m/N}}{(25.50 - 25.00) \times 10^3}$$

$$\frac{dC}{da} \simeq 5.00 \times 10^{-8}\,\text{N}^{-1}$$

From Eq. (6.42),

$$G_{IC} = \frac{P_f^2}{2B}\frac{dC}{da}$$

$$G_{IC} = \frac{\left(145 \times 10^3\,\text{N}\right)^2 \left(5.00 \times 10^{-8}\,\text{N}^{-1}\right)}{2\left(25 \times 10^{-3}\,\text{m}\right)}$$

$$G_{IC} = 21{,}025\,\text{N/m} \simeq 0.021025\,\text{MN/m}$$

For pure mode I loading, Eq. (6.34) yields

$$G_{IC} = \frac{K_{IC}^2}{E'} = \frac{(1 - v^2)\, K_{IC}^2}{E} \tag{6.34}$$

$$K_{IC} = \sqrt{\frac{G_{IC} E}{(1 - v^2)}}$$

$$K_{IC} = \sqrt{\frac{(0.021025 \,\mathrm{MN/m}) \left(70 \times 10^3 \,\mathrm{MN/m^2}\right)}{1 - 0.30^2}}$$

$$K_{IC} = 40.22 \,\mathrm{MPa}\sqrt{\mathrm{m}}$$

According to the ASTM E399 standard thickness requirement is

$$B_{ASTM} \geq 2.5 \left(\frac{K_{IC}}{\sigma_{ys}}\right)^2 = 5.60 \,\mathrm{mm}$$

Therefore, this result is valid because $B > B_{ASTM}$.

6.7   Crack Resistance Curves

For plane stress conditions, the Griffith energy criterion for crack growth $(R = G_i)$ was modified by Irwin [11] when he proposed that crack instability should occur when

$$\frac{dG_i}{da} = \frac{dR}{da} \tag{6.52}$$

The shape of the crack resistance curve (R-curve) is horizontal for pure brittle materials since the surface energy is an invariant material property and R is independent of crack size [1, 3]. For pure mode I, the Griffith instability criterion on the R-curve is shown in Fig. 6.8a as a dashed horizontal line for which $G = G_{IC} = R$ [2]. The characterization of this fracture criterion anticipates that R increases when the plastic zone at a small-scale yielding (SSY) increases and strain hardens. However, local material separation should occur due to void initiation and coalescenceat high strains and stresses. In this case, the plastic zone must reach a critical size for crack growth to occur since sufficient energy must be available; otherwise, the crack would be stationary. Hence, both G_I and R increase with increasing stress level, and subsequently, unstable crack growth occurs when the crack acquires a critical value shown in Fig. 6.8b at a stability point where $K_I = K_{IC}$ at a particular load P_4.

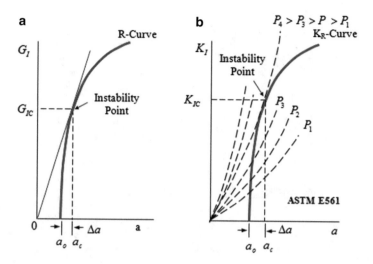

Fig. 6.8 Schematic resistance curves showing the instability point for graphically defining fracture toughness using (**a**) the strain energy release rate and (**b**) the stress intensity factor for mode I loading

Accordingly, the ASTM E561 Standard Practice includes the R-curve to provide a toughness diagram in the form of applied $K_I = f(\Delta a)$ so that crack instability (onset of unstable fracture process) occurs when $K_I = K_{IC}$ and $a = a_c$ for a specific applied load P_4 (Fig. 6.8b). This implies that the resistance to fracture of a metallic solid containing an initial crack size (a) may be characterized by an R-curve, provided that crack growth or extension develops slowly and stably.

Furthermore, the R-curve can be constructed by drawing secant lines on a load-displacement curve as shown in Figure 6.9a [2]. The slope of the secant lines as a measure of the compliance C and the crack lengths a_i are determined. Subsequently, follow the sequence in Fig. 6.9.

Plot the compliance C as a function a/w, where $a = a_o + \Delta a + r$ is the effective crack length. This is depicted in Fig. 6.9b. Plot $\alpha = f(a/w)$ (Fig. 6.9c). Calculate K_I and G_I, and plot G_I as a function of crack length a. The resultant plot is the R-curve for G_I as shown in Fig. 6.9d.

The point of intersection between the secant line and the R-curve defines the critical strain energy release rate (G_{IC}), which in turn is related to the critical stress intensity factor or plane-strain fracture toughness (K_{IC}).

Thus far, the above compliance approach has been based on elastic deformation ahead of the crack tip. For most materials, the stresses induced at the crack tip cause initially elastic deformation, and subsequently, plastic deformation induces further energy dissipation confined to a small plastic zone. The main goal so far is to introduce analytical methods for determining the energy associate with crack growth and the critical energy at fracture or at the onset of slow crack growth.

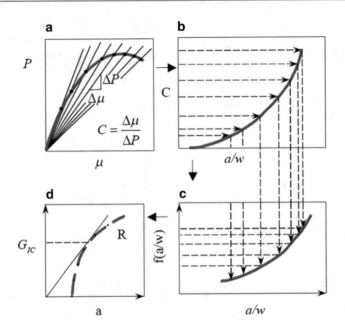

Fig. 6.9 Schematic (**a**) load-displacement, (**b**) compliance-normalized crack length, (**c**) geometry correction factor and (**d**) resistance diagrams for mode I loading

6.8   The J-Integral

Consider the two-dimensional crack being surrounded by two arbitrary counter-clockwise contours Γ_1 and Γ_2 shown in Fig. 6.10. If a small-scale yielding prevails, then the quantities K_i and G_i, where $i = I, II$, and III represent the loading modes as defined in Chap. 4, can describe the stress state near the crack tip when the field is elastic with a relatively small plastic zone $r \ll a$; otherwise, Ki and G_i do not describe the elastic-plastic behavior of tough materials containing large plastic zones $r \geq a$ (large-scale yielding) (*LSY*). Nevertheless, the need to characterize tough solids prevails since many engineering materials are of this category.

In order to determine an energy quantity that describes the elastic-plastic behavior of tough materials, Rice [22] introduced a contour integral or line integral that encloses the crack front shown in Fig. 6.10. This integral was originally introduced by Eshelby [4] as

$$J = \int_{\Gamma} \left(W dy - \overrightarrow{T} \frac{\partial \overrightarrow{\mu}}{\partial x} ds \right) \tag{6.53}$$

where J = Effective energy release rate (*MPa.m* or *MN/m*)
   W = Elastic strain energy density or plastic loading work (J/m^3)
   $\overrightarrow{\mu}$ = Displacement vector at ds

Fig. 6.10 J-integral contours around the crack surfaces

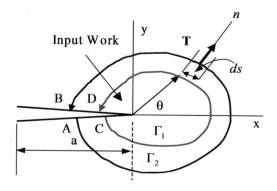

$ds = $ Differential element along the contour

$n = $ Outward unit normal to Γ

$\overrightarrow{T}\left(\partial \overrightarrow{\mu}/\partial x\right) ds = $ Input work

$s = $ Arc length

$\overrightarrow{T} = $ Tension vector (traction forces) on the body bounded by Γ

$\Gamma = $ Arbitrary counterclockwise contour

The term J in Eq. (6.53) is a line of surface integral defined around a contour Γ. It characterizes the stress–strain field around the crack front, and therefore, it must be the energy release to the crack tip during crack growth. Due to this fact, the J-integral is used as a failure criterion, and it is a measure of the fracture toughness at the onset of slow crack growth for elastic and elastic-plastic metallic materials (Fig. 3.11). The inherent characteristics of the J-integral exhibits (a) remarkable path, contour size and shape independence, and (b) an invariability in magnitude when the contour lies either inside or outside the plastic zone [15]. The former characteristic indicates that the J-integral vanishes ($J = 0$) around an arbitrary closed contour as shown by Parton and Morozov [20] using Green's formula.

The interpretation of the J-integral includes the following observations:

- The J-integral vanishes along the closed contours Γ_1 and Γ_2 because the traction forces are $\overrightarrow{T} = 0$ along the crack lower and upper surfaces and $dy = 0$ along AC and BD. Thus, Eq. (6.53) becomes $J_{\Gamma_1} - J_{\Gamma_2} = 0$. Therefore, the J-integral is path independent, and it is a measure of the straining at the crack tip that accounts for significant plastic deformation at the onset of crack initiation. The contour path can be defined arbitrarily for computational advantages [7], as it will be shown in a later section, since J is conserved. This means that the contour Γ can conveniently be defined along the plastic zone boundary so that the Von Mises plastic zone size, Eq. (5.53), can define the contour shape as shown in Fig. 5.6.
- The crack line can be included in the contour Γ_1 or Γ_2 without contributing to the value of J. For this reason, points A and B or C and D do not need to coincide.
- The J-integral along a contour around the crack is the change in potential energy (elastic energy) for a virtual crack extension da. Thus,

$$J = -\frac{1}{B}\frac{dU_3}{da} \tag{6.54}$$

- Remarkably, the J-integral can be evaluated along remote paths, where small crack tip yielding does not interfere.

With regard to Eq. (6.53), the traction forces and the strain energy density are defined as

$$\overrightarrow{T} = \sigma_{ij}n \qquad (i,j) = 1, 2, 3 \tag{6.55}$$

$$W = \int \sigma_{ij}d\epsilon_{ij} \tag{6.56}$$

In general, the elastic stresses needed in Eq. (6.56) are defined in matrix form

$$\sigma_{ij} = \begin{bmatrix} \sigma_x & \tau_{xy} & \tau_{xz} \\ \tau_{yx} & \sigma_y & \tau_{yz} \\ \tau_{zx} & \tau_{zy} & \sigma_z \end{bmatrix} \tag{6.57}$$

Thus, the strain energy density (SED) can be expressed as

$$W = \frac{1}{E'} \begin{bmatrix} \frac{1}{2}\left(\sigma_x^2 + \sigma_y^2 + \sigma_z^2\right) - v\left(\sigma_x\sigma_y + \sigma_y\sigma_z + \sigma_z\sigma_x\right) \\ + (1+v)\left(\tau_{xy}^2 + \tau_{yz}^2 + \tau_{zx}^2\right) \end{bmatrix} \tag{6.58}$$

Recall that plane stress condition requires that $\sigma_z = \tau_{zx} = \tau_{zy} = 0$ and $E' = E$. Consequently, Eq. (6.58) reduces to

$$W = \frac{1}{2E}\left[\sigma_x^2 + \sigma_y^2 - 2v\sigma_x\sigma_y + 2(1+v)\tau_{xy}^2\right] \tag{6.59}$$

For pure tension loading, W becomes

$$W = \frac{\sigma_y^2}{2E} \tag{6.60}$$

However, the elastic mixed-mode interaction described by Eq. (6.34) can be used to predict the J-integral, provided that $r \ll a$. This means that $J = G$ so that

$$J = \frac{K_I^2}{E'} + \frac{K_{II}^2}{E'} + \frac{(1+v)K_{III}^2}{E} \tag{6.61}$$

The J-integral is used as a critical parameter for determining the onset of stable crack growth and predicting fracture toughness. The fracture criterion by J_{IC} for the initiation of stable crack growth is established when the applied J-integral reaches a critical value. This is indicated below for elastic and elastic-plastic materials

$$J = J_c = G_C \qquad \text{For elastic behavior} \tag{6.62}$$

$$J = J_c \neq G_C \qquad \text{For elastic-plastic behavior} \tag{6.63}$$

$$J_I = J_{IC} = G_{IC} \qquad \text{For elastic behavior under mode I loading} \tag{6.64}$$

$$J_I = J_{IC} \neq G_{IC} \quad \text{For elastic-plastic behavior for mode I} \tag{6.65}$$

Similar fracture criteria can be used for mode *II* and *III* loading systems. In general, J_{IC} may be used as a brittle or ductile fracture toughness criterion for characterizing a material behavior that does not meet the requirements of K_{IC} as per ASTM E399 Standard Test Method, Vol. 03.01.

For a large-scale yielding case, the *J*-integral becomes the controlling factor in characterizing plastic behavior of ductile materials containing stationary cracks since *J* is the available energy per unit crack extension at the crack tip. In fact, crack blunting may occur at the crack tip and the fracture criterion used for determining the onset of crack instability depends on the amount of plasticity ahead of the crack tip. Hence, the plastic zone size is $r \geq a$.

The elastic-plastic fracture mechanics (EPFM) field or the large-scale yielding approach can be used for characterizing the behavior of an existing stationary crack tip, which undergoes significant plasticity. As a result, the critical crack tip opening displacement (δ_{tc}) or the critical *J*-integral (J_C) measures the onset of crack growth in elastic-plastic solids [22].

Using the Dugdale's strip yield model, the crack tip opening displacement (*CTOD*) δ_t indicated in Fig. 5.3 can be related to the *J*-integral, provided that the path of integration is arbitrarily chosen in the elastic regime and the contour curve Γ is taken around the yield strip or plastic zone boundary.

According to Hellan's analysis [7] for thin plates and Tresca properties, the Dugdale's model shown in Fig. 5.3 can be used as the path contour Γ needed to solve the *J*-integral. Thus, the arbitrary contour depicted in Fig. 6.10 can be shrunk to a shape similar to Dugdale's strip yield model, which is shown in Fig. 6.11 for convenience.

Let $x = a$ at the lower crack side before localized yield occurs and $x = a + r$ at the upper crack side after yielding be the limits of the *J*-integral under pure tension loading (mode I). As a result, $dy = 0$ along the crack plane and the traction force becomes $T = \sigma_y = \sigma_{ys}$. Consequently, the *J*-integral defined in Eq. (6.53) becomes [7]

$$J_I = -\int_\Gamma \vec{T} \frac{\partial \vec{\mu}}{\partial x} ds \tag{a}$$

$$J_I = -\int_a^{a+r} \sigma_{ys} \frac{\partial}{\partial x} \left(\mu_y^+ - \mu_y^- \right) dx \tag{b}$$

Fig. 6.11 Dugdale's strip yield model for COD measurements [7]

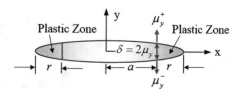

$$J_I = \int_a^{\delta_t} \sigma_{ys} d\left(\mu_y^+ - \mu_y^-\right) = \sigma_{ys}\delta_t \qquad (6.66)$$

Here, μ_y^+ and μ_y^- are the upper and lower displacements. Denote that this expression, Eq. (6.66), resembles Eq. (5.42) when $J_I = G_I$ for small-scale yielding under plane stress conditions.

It is well known that energy dissipation is confined to a small plastic zone where energy is released. Thus, the above result, Eq. (6.66), represents a simple mathematical model for determining the J-integral as the energy being released at the crack tip. So this result is used to compute the critical J-integral and the crack tip opening displacement, δ_t, at the onset of crack growth. This suggests that Eq. (6.66) may be taken as a postulate for a fracture criterion. If mode I is considered at onset of crack growth, then J_{IC} is the fracture energy taken as the fracture toughness of a ductile material, and on the other hand, δ_{tc} would also be a parameter representing fracture toughness.

Regardless of the actual mechanisms involved during crack growth, a small amount of surface energy is required to create two new crack surfaces, and subsequently, more plastic zone involves as more elastic-plastic energy is released ahead of the crack tip. This energy, then, can be quantified using Eq. (6.66) by simply measuring the crack tip opening displacement δ_t as the crack grows. At fracture, $a = a_c$, $\delta_t = \delta_{tc}$ and $J_I = J_{IC}$ is computed as expected.

6.9   The J-Integral Measurements

Most practical fracture mechanics applications are base on mode I loading. However, mode *II* and *III* may be important in certain engineering situation. In fact, mixed-mode fracture mechanics is a complex theory and it will be dealt with in details in Chap. 8.

The *J*-integral measurement is valid if unloading does not occur. In such a case, the region $J\Delta A = JB\Delta a$ shown in Fig. 6.12 represents the area between the loading curves for crack areas A and $A + \Delta A$. In fact, Fig. 6.12 illustrates load-displacement curves for determining the critical value of the *J*-integral J_{IC} (fracture toughness) at the onset of slow crack growth [7, 8] under two different conditions, and it is the nonlinear equivalent of Figure 6.2. The ASTM E1820 standard method should be consulted for experimental details.

In addition, Eq. (6.54) can be redefined according to the constant load and the constant displacement conditions indicated in Figs. 6.12a and b, respectively. Thus,

$$J = \frac{1}{B} \int_o^P \left(\frac{d\mu}{da}\right)_P dP \qquad (6.67a)$$

$$J = -\frac{1}{B} \int_o^\mu \left(\frac{d\mu}{da}\right)_\mu d\mu \qquad (6.67b)$$

Fig. 6.12 Load-displacement curves for identical specimens at (**a**) fixed load and (**b**) at fixed displacement

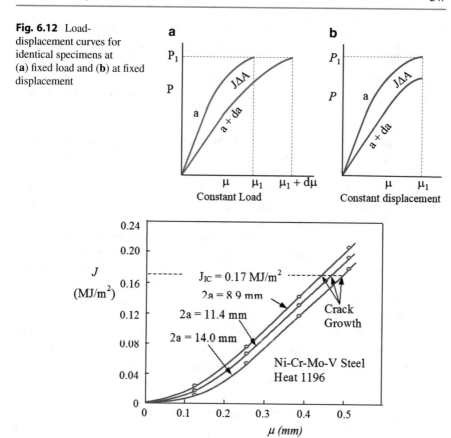

Fig. 6.13 Experimental J-integral vs. displacement [3, 12]

Obviously, either Eq. (6.67) can be solved if an expression for $d\mu/da$ is known.

With respect to Eq. (6.61), one can see that the J-integral (J) is directly related to the stress intensity for a brittle material. For a large plate containing a small crack loaded in tension (mode I), Eq. (6.61) reduces to

$$J_I = \frac{K_I^2}{E'} = \frac{\pi a \sigma^2}{E'} \qquad (6.67c)$$

Furthermore, Fig. 6.13 shows experimental data for a Ni–Cr–Mo–V steel alloy obtained using specimens containing variable original cracks [6,11]. The data represents $J = f(\mu)$ for an initial crack length $2a$, and all three curves represent mechanical behavior, which is displaced downward as $2a$ increases.

Also shown in Fig. 6.13 is the fracture toughness $J_{IC} = 0.17\,\text{MJ/m}^2$, which is the critical value of the J-integral at the onset of slow crack growth under plane-strain conditions. Denote that J is linear at $\mu \geq 0.25\,\text{mm}$. This leads to a conclusion in accordance with experimental observations that a direct measurement of the J-integral (line integral) is carried out by assuming a general plane stress state under small-scale yielding (SSY) or large-scale yielding (LSY) conditions. In principles, J-integral measurements can be carried out using any specimen configuration containing different initial crack lengths as depicted in Fig. 6.13 [3, 12]. A close control experimental setup is vital in accordance with the ASTM E1820 standard for obtaining valid data.

The primary goal in this section is to elucidate the analytical background confined to theoretical aspects of the J-integral as a strain energy release rate criterion for assessing experimental observations.

According to Rice [22] original postulate, the path-independent J-integral describes the strain field around a crack tip, and it can be used as an energy fracture criterion for the onset of crack growth. Despite that the J-integral is confined to the onset of crack growth, there remains the theoretical interest in determining the amount of crack extension prior to fracture or crack propagation using this energy approach. For most materials, the J-integral is a suitable energy method for characterizing fracture toughness by determining the crack tip opening displacement (δ_t) and the yield strength of materials. The simplest mathematical definition for J_I is described by Eq. (6.66).

In general, the J-integral method is applicable to elastic and elastic-plastic materials under plane conditions. For fracture toughness concept, Rice [22] J-integral becomes Griffith [6] strain energy release rate for elastic materials. In other words, for mode I loading $J_{IC} = G_{IC}$ for elastic behavior.

Example 6.2. *Show that J-integral vanishes in the square counterclockwise contour a-b-c-d shown in figure below. The square contour has four segments such as $\Gamma = ab - bc - cd - da$. Determine the strain energy density W and the J-integral for* **(a)** *plane stress and* **(b)** *plane-strain conditions if the applied stress is 700 MPa. Given data for a steel plate: $w = 50.80\,mm$, $\sigma_{ys} = 800\,MPa$, $v = 1/3$ and $E = 207\,GPa$.*

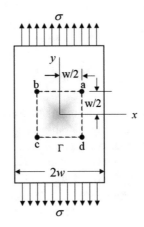

Solution. *For mode I and linear metal behavior,* $\sigma = E'\epsilon$ *(Hooke's law) and*

$$W = \int \sigma_{ij} d\epsilon_{ij} = \int E'\epsilon d\epsilon_{ij} = \frac{\sigma^2}{2E'}$$

$$\overrightarrow{T_i} = \sigma_{ij} \overrightarrow{n} = \sigma_y n$$

(a) *For plane stress condition* $(\sigma_{xx} = \sigma_{zz} = \tau_{zx} = \tau_{zy} = 0)$, *Eq. (6.53) gives*

$$J_{ab} = \int_a^b \left(Wdy - T_i \frac{\partial \mu}{\partial x} dx\right) = 0 \ \ since \ dy = 0, \ \partial \mu/dx = 0, \ ds = dx$$

$$J_{bc} = \int_b^c \left(Wdy - T_i \frac{\partial \mu}{\partial x} dy\right) = \int_b^c Wdy = \frac{(c-b)\sigma^2}{2E} \ \ since \ T_i = 0, \ ds = dy$$

$$J_{cd} = \int_c^d \left(Wdy - T_i \frac{\partial \mu}{\partial x} dx\right) = 0 \ \ since \ dy = 0, \ \partial \mu/dx = 0, \ ds = dx$$

$$J_{da} = \int_d^a \left(Wdy - T_i \frac{\partial \mu}{\partial x} dy\right) = \int_d^a Wdy = \frac{(a-d)\sigma^2}{2E} \ \ since \ T_i = 0, \ ds = dy$$

Thus,

$$J_{bc} = \frac{(-w/2 - w/2)\sigma^2}{2E} = -\frac{w\sigma^2}{2E} = -wW$$

$$J_{da} = \frac{(w/2 + w/2)\sigma^2}{2E} = \frac{w\sigma^2}{2E} = wW$$

and

$$J_\Gamma = \sum J_i = J_{ab} + J_{bc} + J_{cd} + J_{da}$$

$$J_\Gamma = 0 - \frac{w\sigma^2}{2E} + 0 + \frac{w\sigma^2}{2E} = 0$$

Therefore, the J-integral vanishes.
 If $w = 50.80\,mm$, $\sigma = \sigma_{ys} = 700\,MPa$, $v = 1/3$ *and* $E = 207\,GPa$, *then*

$$W = \frac{\sigma^2}{2E} = \frac{(700\,\text{MPa})^2}{(2)(207,000\,\text{MPa})} = 1.18\,\text{MPa}$$

$$W = 1.18\,\text{MJ/m}^3$$

$$J_{da} = -J_{bc} = \frac{w\sigma^2}{2E} = wW = \left(50.80 \times 10^{-3}\,m\right)\left(1.18\,\text{MJ/m}^3\right)$$

$$J_{da} = -J_{bc} \simeq 60\,\text{kPa}\,m = 60\,\text{kJ/m}^2$$

(b) *For plane strain,* $\sigma_z = v\left(\sigma_x + \sigma_y\right) = v\sigma$ *and*

$$W = \frac{\sigma^2}{2E'} = \frac{\left(1 - v^2\right)\sigma^2}{2E}$$

Similarly,

$$J_\Gamma = \sum J_i = J_{ab} + J_{bc} + J_{cd} + J_{da}$$

$$J_\Gamma = 0 - \frac{w\left(1 - v^2\right)\sigma^2}{2E} + 0 + \frac{w\left(1 - v^2\right)\sigma^2}{2E} = 0$$

Therefore, the J-integral vanishes.

If $w = 50.80\,mm,\ \sigma = \sigma_{ys} = 700\,MPa,\ v = 1/3$ *and* $E = 207\,GPa,$ *then*

$$W = \frac{\left(1 - v^2\right)\sigma^2}{2E} = \frac{\left(1 - 1/9\right)\left(700\,\text{MPa}\right)^2}{\left(2\right)\left(207{,}000\,\text{MPa}\right)} = 1.05\,\text{MPa}$$

$$W = 1.05\,\text{MJ/m}^3$$

$$J_{da} = -J_{bc} = \frac{w\left(1 - v^2\right)\sigma^2}{2E} = wW = \left(50.80 \times 10^{-3}\,\text{m}\right)\left(1.05\,\text{MJ/m}^3\right)$$

$$J_{da} = -J_{bc} \simeq 53.34\,\text{kPa}\,\text{m} = 53.34\,\text{kJ/m}^2$$

6.10   Tearing Modulus

Those materials (ductile) that exhibit appreciable plasticity at fracture usually show slow and stable crack growth before fracture. Stable crack growth starts at J_{IC}, but further increase of the applied stress is required to maintain the crack growing. The resistance curve in these materials is the J_R-curve, which is equivalent to the R-curve discussed previously. The crack driving force is J_I instead of G_I [8]. The criteria for stable and unstable (instability) crack growth are

$$\frac{dJ}{da} < \frac{dJ_R}{da} \quad \text{for stable crack growth} \tag{6.68}$$

$$\frac{dJ}{da} \geq \frac{dJ_R}{da} \quad \text{for unstable crack growth} \tag{6.69}$$

Paris et al. [19] have proposed a dimensionless tearing modulus (T_J) defined by multiplying Eq. (6.69) by E/σ_{ys}^2 so that

$$T_J = \frac{E}{\sigma_{ys}^2}\frac{dJ}{da} \tag{6.70}$$

Fig. 6.14 Schematic J-resistance curve describing the fracture resistance of a material

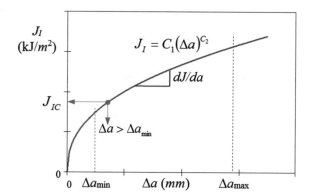

$$T_R = \frac{E}{\sigma_{ys}^2} \frac{dJ_R}{da} \tag{6.71}$$

In fact, T_J is nothing but the crack resistance for crack growth and T_R is the material resistance. Then, crack instability is reached when $T_J = T_R$ and fracture occurs if $T_J > T_R$. Here, dJ/da is the slope of the J-Δa resistance curve in the stable crack growth regimen (Fig. 6.14) designated using the following crack extension range

$$\Delta a_{min} \leq \Delta a \leq \Delta a_{max}$$

Assuming that the large-scale yielding in finite bodies does not lose its uniqueness (loss of constraint) at the crack tip [21], the J-integral for crack initiation is determined from the curve generated by a power-law function, $J_I = C_2 (\Delta a)^{C_2}$, by letting $J_I = J_{IC}$ at $\Delta a \gtrsim \Delta a_{min}$, provided that the ASTM E1820 size requirements are met [23]. Details on this type of nonlinear regression analysis is given in Chap. 3, Fig. 3.12.

Actually, Hutchinson–Paris [9] devised a procedure to validate the J-integral crack growth by stating that

1. the elastic unloading region in the fracture process zone requires that $\Delta a << r$ and
2. J must increase rapidly with crack extension so that the region of non-proportionality is small and $dJ/da >> J/r$.

Furthermore, the slope dJ_R/da in Eq. (6.71) can be derived using the power-law expression, found in the ASTM E1820 Standard Test Method and elsewhere [1, 14] for performing nonlinear least-squares analysis. Thus,

$$J_R = C_1 (a - a_o)^{C_2} \tag{6.72}$$

where C_1 and C_2 are curve fitting constants.

Now, take dJ_R/da and manipulate the algebra to get [1]

$$\frac{dJ_R}{da} = C_2 C_1 (a - a_o)^{C_2-1} = C_2 C_1^{1/C_2} J_R^{(C_2-1)/C_2} \tag{6.73}$$

Inserting Eq. (6.73) into (6.71) yields the materials resistance modulus as

$$T_R = \left(\frac{E}{\sigma_{ys}^2} C_2 C_1^{1/C_2} \right) J_R^{(C_2-1)/C_2} \tag{6.74}$$

Solving for J_R gives

$$J_R = \left(\frac{\sigma_{ys}^2 T_R}{E C_2 C_1^{1/C_2}} \right)^{C_2/(C_2-1)} \tag{6.75}$$

Similarly, assume a cracked material under mode I loading condition which develops a small plastic zone so that a plastic J-integral is $J_p \ll J_e$. Under this condition, take the derivative dJ_I/da from Eq. (6.67c) and manipulate the resultant expression to get

$$\frac{dJ_I}{da} = \frac{\pi \sigma^2}{E'} = \frac{J_I}{a} \tag{6.76}$$

Now, substitute Eq.(6.76) into (6.70) to get the applied tearing modulus for plane conditions as

$$T_J = \left(\frac{E}{a \sigma_{ys}^2} \right) J_I \tag{6.77}$$

$$J_I = \left(\frac{a \sigma_{ys}^2}{E} \right) T_J \tag{6.78}$$

Denote that $T_R = f(J_R)$ or $J_R = f(T_R)$ is a nonlinear function that gives an exponential curve, while $T_J = f(J_I)$ or $J_J = f(T_J)$ is a linear function that provides a straight line. Mathematically, T_R, as defined by Eq. (6.74), can be anticipated to have an exponential decay as the applied load is increased.

Additionally, the instability criterion can be defined in terms of these modulus as indicated below:

$$T_J = T_R \tag{6.79}$$

Thus, the J-integrals become nonlinearly related to each other as defined by

$$J_I = \left(a C_2 C_1^{1/C_2} \right) J_R^{(C_2-1)/C_2} \tag{6.80}$$

$$J_R = \left(\frac{J_I}{a C_2 C_1^{1/C_2}} \right)^{C_2/(C_2-1)} \tag{6.81}$$

The above theoretical background can be made clear through an example.

Example 6.3. *A hypothetical solid plate containing a single-edge crack is subjected to a remote stress of unknown value. Assume that this material has power-law behavior described by $J_R = 300\Delta a^{0.485}$ in kJ/m^2 and Δa in meters.* **(a)** *Develop a $J_R = f(T_R)$ diagram and identify the instability point and* **(b)** *find the instability stress for the stable crack growth limit. Given data: $E = 200,000\,MPa$, $\sigma_o = \sigma_{ys} = 400\,MPa$, $a_o = 5\,mm$, and $v = 0.3$.*

Solution. *From the given J_R curve fitting equation, $J_R = 300\Delta a^{0.485}$, one gets $C_1 = 300$ and $C_2 = 0.485$. Carrying out calculations using Eqs. (6.75) and (6.78) gives*

$$J_R = \left(\frac{\sigma_o^2 T_R}{EC_2 C_1^{1/C_2}} \right)^{C_2/(C_2-1)} = 60.259\,(T_R)^{-0.94175} \tag{a}$$

$$J_I = \left(\frac{a\sigma_o^2}{E} \right) T_J = 4.0 T_J \tag{b}$$

From Eq. (6.67c), the J-integral equation becomes

$$J_I = \frac{K_I^2}{E'} = \frac{\pi a \left(1 - v^2\right)\sigma^2}{E}$$

$$J_I = 7.1471 \times 10^{-11}\sigma^2 \tag{c}$$

The figures below show the J-T_R, J-T_J, and J_I-σ plots as per Eqs. (a)–(c).

The most obvious feature shown in the figure below is its intersection point which immediately defines the crack instability point, which represents the stable crack growth limit for a given load. The instability point in this figure is located at $(J, T_x) = (4.04, 16.17)$ This figure depicts also he instability point at $(J, \sigma) = (4.76, 16.17)$. For convenience, the former instability-point coordinates are calculated by letting $J_R = J_I$ so that

$$T_J = T_R = 4.0425$$

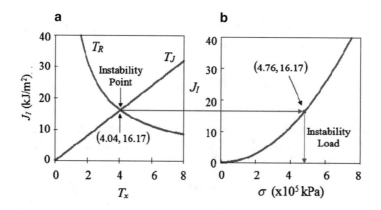

Then, the J-integral and the applied or instability load are, respectively,

$$J_I = J_R = 4.0 T_J = 16.17 \, \text{kJ/m}^2$$

$$\sigma = \sqrt{\frac{EJ_I}{\pi a \, (1 - v^2)}} = 4.76 \times 10^5 \, \text{kPa} = 470 \, \text{MPa}$$

Now that J_R = has a known value, the crack extension Δa at instability can be calculated using the given power-law equation and, subsequently, determine the crack length at the instability load. Thus,

$$J_R = 300 \Delta a^{0.485}$$

$$\Delta a = (J_R / 300)^{1/0.485} = 2.42 \times 10^{-3} \, \text{m} = 2.42 \, \text{mm}$$

$$a = \Delta a + a_o = 2.42 + 5 = 7.42 \, \text{mm} \; @ \; \sigma = 470 \, \text{MPa}$$

Finally, fracture should occur when $\sigma > 470 \, MPa$.

In light of the above J-integral approach, there are other J-theories applied to linear-elastic materials [29]. These are the J-T theory where T is known as T-stress and it is used for quantifying the crack tip constraint effect [13], the J-Q theory which is based on numerical analysis that includes small-scale yielding (SSY) and large-scale yielding (LSY) conditions [16, 17], and the J-A_2 theory which is an asymptotic three-term solution for higher-order crack tip field under elastic-plastic conditions in power-law hardening materials [26–28]. Each of the J-theories has its own theoretical background describing its usefulness and applicability in the field of fracture mechanics.

6.11   Problems

6.1. Use Dugdale's model for a fully developed plane stress yielding confined to a narrow plastic zone. Yielding is localized to a narrow size roughly equal to the sheet thickness (B). This is a fully elastic case in which the plastic strain may be defined as $\epsilon = \delta_t / B$, where δ_t is the crack tip opening displacement. If the J-integral is defined by $dJ = \sigma d\delta$, then show that

$$\delta_t = \frac{\pi \alpha^2 a \sigma^2}{E \sigma_{ys}}$$

6.2. The crack tip opening displacement (δ_t) for perfectly plastic solution to the Dugdale's model was derived by Rice in 1966 [14] as defined by Eq. (5.35). Show that the path-independent J-integral is defined by

$$J = \frac{(\kappa + 1)(1 + v) \pi a \sigma^2}{E}$$

6.3. A bending test specimen made out of carbon steel showed a load-displacement behavior If the area under the curve $P = f(\mu)$ is 10 J at the onset of crack

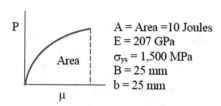

growth, determine **(a)** the fracture toughness in terms of J_{IC} as per ASTM E1820 standard, **(b)** K_{IC} and its validity as per ASTM E399 testing method, and **(c)** δ_{tc} using Eq. (5.31) with $\lambda = \sqrt{3}$, **(d)** the fracture strain ϵ_f, and **(e)** the plastic zone size. Explain the meaning of the results. [Solutions: (a) $J_{IC} = 32\,\text{kJ/m}^2$, (b) $K_{IC} = 81.39\,\text{mpA}\sqrt{m}$, (c) $\delta_{tc} = 0.016\,\text{mm}$, (d) $\epsilon_f = 0.064\,\%$, (e) $r = 0.16\,\text{mm}$.]

6.4. If $J_I = \sigma_{ys}\delta_t$ is used to determine the fracture toughness, will δ_t be a path-independent entity? Explain.

6.5. Assume that crack growth occurs when $J_I \leq J_{IC}$. If a well-developed plastic flow occurs, will this inequality be valid? Explain.

6.6. A double cantilever beam (DCB) is slowly loaded in tension up 10 MN as schematically shown in Fig. 6.5. Assume that there is no rotation at the end of the beam and that the beam is made of an isotropic steel having the following properties: $K_{IC} = 47\,\text{MPa}\sqrt{m}$, $E = 207\,\text{GPa}$, and $\nu = 0.3$. Will fracture occur? Dimensions: $a = 20\,\text{mm}$, $B = 20\,\text{mm}$, and $h = 10\,\text{mm}$.

References

1. T.L. Anderson, *Fracture Mechanics: Fundamentals and Applications*, 3rd edn. (Taylor & Francis, LLC. CRC Press, Boston, 2005) pp. 58, 405–408
2. ASTM 351–94, Annual Book of ASTM Standards, vol. 03.01
3. D. Broek, *Elementary Engineering Fracture Mechanics*, Chap. 4, 4th edn. (Kluwer Academic, Boston, 1986), p. 102
4. J.D. Eshelby, Calculation of energy release rate, in *Prospects of Fracture Mechanics*, ed. by G.C. Sih, H.C. Von Elst, D. Broek (Noordhoff, Growingen, 1974), pp. 69–84
5. D.K. Felbeck, A.G. Atkins, *Strength and Fracture of Engineering Solids*, 2nd edn. (Prentice-Hall, Upper Saddle River, NJ, 1996), pp. 430–443
6. A.A. Griffith, The phenomena of rupture and flow in solids. Philos. Trans. R. Soc. Lond. Ser. A **221**, 163–198 (1921)
7. K. Hellan, *Introduction to Fracture Mechanics* (McGraw-Hill, New York, 1984)
8. R.W. Hertzberg, *Deformation and Fracture Mechanics of Engineering Materials*, 3rd edn. (Wiley, New York, 1989)
9. J.E.W. Hutchinson, P.C. Paris, in *Elastic-Plastic Fracture*, ASTM STP, vol. 668 (1979) 37

10. C.E. Inglis, Stresses in a plate due to the presence of cracks and sharp corners. Trans. Inst. Nav. Arch. **55**, 219–241 (1913)

11. G.R. Irwin, ASTM Bulletin (January 1960), 29

12. J.D. Landes, J.A. Begley, The effect of specimen geometry on JIC. ASTM STP **514**, 24–39 (1972)

13. S.G. Larsson, A.J. Carlsson, Influence of non-singular stress terms and specimen geometry on small-scale yielding at crack tips in elastic–plastic material. J. Mech. Phys. Solids **21**, 263–278 (1973)

14. M.L. Lu, C.B. Lee, F.C. Chang, Fracture toughness of acrylonitrile-butadiene-styrene by j-integral methods. Polym. Eng. Sci. **35**(18), 1433–1439 (1995)

15. E.M. Morozov, G.P. Nikishkov, *Finite Element Method in Fracture Mechanics* (Nauka, Moscow, 1980). Reference cited in [20].

16. N.P. O'Dowd, C.F. Shih, family of crack-tip fields characterized by a triaxiality parameter - i: structure of fields. J. Mech. Phys. Solids **39**, 989–1015 (1991)

17. N.P. O'Dowd, C.F. Shih, Family of crack-tip fields characterized by a triaxiality parameter - ii: fracture applications. J. Mech Phys Solids **40**, 939–963 (1992)

18. M.A. Meyers, K.K. Chawla, *Mechanical Metallurgy: Principles and Applications* (Prentice-Hall, Englewood Cliffs, 1984), pp. 358–359

19. P.C. Paris, H. Tada, A. Zahoor, H. Ernst, Instability of the tearing model of elastic-plastic crack growth. ASTM STP **668**, 5.36 (1979)

20. V.Z. Parton, E.M. Morozov, *Mechanics of Elastic-Plastic Fracture*, 2nd edn. (Hemisphere Publishing Corporation, New York, 1989)

21. S. Rahman, G. Chen, Constraint effects on probabilistic analysis of cracks in ductile solids. Fatigue Fract. Eng. Mater. Struct. **23**, 879–890 (2000)

22. J.R. Rice, A path independent integral and the approximate analysis of strain concentration by notches and cracks. J. Appl. Mech. **35**(2), 379–386 (1968)

23. S. Surch, A.K. Vasudevan, On the relationship between crack initiation toughness and crack growth. Math. Sci. Eng. **79**, 183–190 (1986)

24. H.M. Westergaard, Bearing pressures and cracks. J. Appl. Mech. **G1**, A49–A53 (1939)

25. J.G. Williams, *Fracture Mechanics of Polymers*, Chap. 2 (Ellis Horwood Limited, Halsted Press: A Division of Wiley, New York, 1984)

26. S. Yang, Higher order asymptotic crack-tip fields in a power-law hardening material. Ph.D. Dissertation, University of South Carolina, Columbia, SC (1993)

27. S. Yang, Y.J. Chao, M.A. Sutton, Higher-order asymptotic fields in a power-law hardening material. Eng. Fract. Mech. **45**, 1–20 (1993)

28. S. Yang, Y.J. Chao, M.A. Sutton, On the fracture of solids characterized by one or two parameters: theory and practice. J. Mech. Phys. Solids **42**, 629–647 (1994)

29. X.-K. Zhu, J.A. Joyce, Review of fracture toughness (G, K, J, CTOD, CTOA) testing. Eng. Fract. Mech. **85**, 1–46 (2012)

Elastic-Plastic Fracture Mechanics

7

7.1 Introduction

This chapter describes the stress and strain fields at a crack tip for materials that will obey the Ramberg–Osgood [21] nonlinear stress–strain relation. These materials are considered to be strain hardenable, specially under tension loading. A two-dimensional field equations are characterized for ductile or low-strength materials, which can be strain hardened at a large-scale yielding. In fact, hardening in a polycrystalline material is due to plastic deformation, in which dislocation motion is the primary phenomenon for this irreversible process. However, dislocation-imperfection interactions impede the mobility of dislocations leading to strain hardening. For instance, the dislocation distribution within a plastic zone is normally a complex and mixed phenomenon.

The mechanism of fracture is related to plastic deformation at the crack tip where high stresses and strains are developed. Therefore, the use of a dislocation model for determining stresses and strains would be an ideal mathematical approach for predicting crack instability. Instead, for a cracked elastic-plastic material subjected to an external load, the onset of plastic flow occurs at the crack tip, and the flow criterion that predicts the onset of crack instability is usually the J-integral, which is limited to a stationary crack in a strain hardening material.

According to Rice [22], the J-integral is a particular version of the rate of change in potential energy, and it is mathematically defined as a path-independent line integral. The usefulness of the J-integral in the field of elastic-plastic fracture mechanics is fundamentally significant for determining fracture toughness at the onset of crack growth. Some details on how to interpret the J_R-curve are given in Sect. 3.7.2.

© Springer International Publishing Switzerland 2017
N. Perez, *Fracture Mechanics*, DOI 10.1007/978-3-319-24999-5_7

7.2 J-Controlled Crack Growth

Hutchinson [12] and Rice and Rosengren [24] in separate publications in 1968 showed that the J-integral characterizes the stress and strain fields at the crack tip in nonlinear-elastic materials. Their work is referred to as the HRR theory, which is as an extension of linear fracture mechanics that account for a large-scale yielding (*LSY*) phenomenon and related microscopic fracture mechanisms, such as void formation and void coalescence within the plastic zone. Subsequently, the pertinent mathematical relationships have become known as the HRR field equations applicable near the crack tip within the J-dominated region as schematically illustrated in Fig. 7.1a [6, 13].

Since the J-integral is path independent, a circular path of radius r is convenient for deriving the field equations. This implies that the nonlinear fracture mechanics approach can be used for analyzing rate-independent materials under monotonic loading [22]. Particularly, the J-integral and the Ramberg–Osgood semiempirical uniaxial stress–strain relationship are widely used for characterizing crack growth at a large-scale yielding [21]. Therefore, nonlinear fracture mechanics has a semiempirical foundation.

Furthermore, the crack growth model shown in Fig. 7.1 indicates that the J-dominated region, where microscopic separation occurs, is within a relatively large plastic zone. One particular characteristic of a cracked ductile and strain hardenable material is the occurrence of crack blunting (Fig. 7.1b) before appreciable crack growth takes place under monotonic loading. Despite that the J-integral does not model elastic unloading, the J-controlled crack growth requires that the region of elastic unloading and distinct non-proportional loading be contained with the J-dominated region of the deformation theory [13]. Thus, the amount of crack growth must be related to the plastic zone size r, the J-integral, and the crack tip opening displacement δ_t. For crack extension,

$$\Delta a < r \tag{a}$$

$$J/(dJ/da) < r \tag{b}$$

$$r > \delta_t \tag{c}$$

In fact, the inequality $\Delta a < r$ assures that crack advance and unloading take place within the J-dominance zone.

Fig. 7.1 Schematic process zone within the J-dominant region [6, 13]. (**a**) J-dominance and (**b**) crack-tip blunting

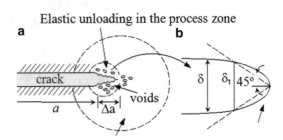

7.3 Plastic Field Equations

The analysis given below is for a study on strain hardening plasticity within the HRR regime. These authors defined the J-Integral as the crack driving force for characterizing plastic solids containing cracks. For instance, Hutchinson's analysis [12] is also found elsewhere [5, 10, 17].

Figure 7.2 schematically shows the stages of a ductile fracture process for load control systems.

It should be pointed out that Begley et al. [2, 3] proposed the use of the J-integral [22] as the crack driving force for characterizing the onset of crack growth (crack extension) at a large-scale yielding. Crack growth may be stable if the concentration of the HRR field on the crack tip region, during deformation, is continuous, irreversible, and maintained, to an extent, a plastic level. This means that the J-integral and the crack tip opening displacement characterize plastic fracture at the crack tip region, where the *HRR* fields are located. Hence, fracture toughness is defined at the onset of stable crack growth for a J-controlled situation [14, 31].

Assuming deformation plasticity as per the multiaxial stress J_2-deformation theory of plasticity and isotropic hardening, the invariant J_2, the nonlinear multiaxial strains (ϵ_{ij}), the first deviatory stress (S_{ij}), and the von Mises effective stress (σ_e) for incompressible materials under tension are, respectively [7],

$$J_2 = \frac{1}{2}S_{ij}S_{ij} \tag{7.1}$$

$$\epsilon_{ij} = \frac{3\alpha'\epsilon_o S_{ij}}{2\sigma_e}\left(\frac{\sigma_e}{\sigma_o}\right)^n \tag{7.2}$$

$$S_{ij} = \sigma_{ij} - \frac{1}{3}\sigma_{kk}\delta_{ij} \tag{7.3}$$

$$\sigma_e = \sqrt{\frac{3}{2}S_{ij}S_{ij}} \tag{7.4}$$

Fig. 7.2 Schematic crack growth behavior of ductile materials [17]

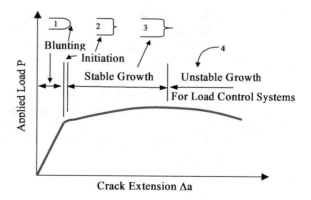

where $\alpha' = $ Constant

$n = $ Strain hardening exponent

$\epsilon_o = $ Yield strain or reference strain $= \sigma_o/E$

$\sigma_o = $ Yield strength or reference stress (MPa)

These are common entities used in deformation theory of plasticity, which is basically a nonlinear-elasticity theory [24]. In fact, plasticity is a term used to describe a solid material undergoing an irreversible process of shape change due to an applied external stress greater than a threshold stress referred to as the yield strength. In general, plastic deformation implies the above, and it is characterized by a particular physical mechanism, such as formation of dislocations due to active slip and possibly twinning systems in crystalline materials (long-range atomic order) and crazing in amorphous materials (lack of long-range order). From an engineering point of view, some failure criteria are used to characterize solid materials subjected external mechanical loads.

Of particular interest is the von Mises failure criterion which utilizes the effective stress for the above purpose, and it can be defined in polar and Cartesian coordinates as

$$\sigma_e = \sqrt{\sigma_r^2 + \sigma_\theta^2 - \sigma_r\sigma_\theta + 3\tau_{r\theta}^2} \quad \text{for plane stress} \tag{7.5}$$

$$\sigma_e = \sqrt{\frac{3}{4}(\sigma_r - \sigma_\theta)^2 + 3\tau_{r\theta}^2} \quad \text{for plane strain} \tag{7.6}$$

$$\sigma_e = \frac{3}{2\sqrt{2}}(\sigma_x - \sigma_y) \quad \text{for plane strain} \tag{7.7}$$

$$\sigma_e = \frac{1}{\sqrt{2}}\sqrt{(\sigma_1 - \sigma_2)^2 + (\sigma_2 - \sigma_3)^2 + (\sigma_3 - \sigma_1)^2} \tag{7.8}$$

$$+3\left(\tau_{12}^2 + \tau_{13}^2 + \tau_{23}^2\right)$$

$$\sigma_e = \frac{\sqrt{3}}{2}(\sigma_1 - \sigma_2); \; \sigma_1 > \sigma_2 \quad \text{for plane strain} \tag{7.9}$$

With regard to the strain hardening exponent n, low-strength (ductile) materials have greater hardening exponents than high-strength ones. As a result, large tensile stresses at the crack front in ductile materials cause nucleation of voids and void coalescence as the source for crack nucleation. Consequently, the effects of yielding and strain hardening at the crack front is a major concern in fracture mechanics.

The phenomena of yielding and strain hardening are schematically shown in Fig. 7.3 for quasi-static tension loading. Common experimental results for characterizing materials at a macroscale are presented as stress–strain curves of crack-free solids. Denote that the strain hardening exponent increases with increasing ductility [12].

Fig. 7.3 Schematic
stress–strain curves showing
the effect of strain hardening
during deformation

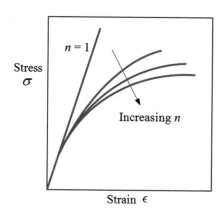

Further, the hydrostatic stress and the Kronecker delta are, respectively,

$$\sigma_{kk} = \sigma_x + \sigma_y + \sigma_z \tag{7.10}$$

$$\delta_{ij} = \begin{bmatrix} 1 & 0 & 0 \\ 0 & 1 & 0 \\ 0 & 0 & 1 \end{bmatrix} \tag{7.11}$$

For an elastic-plastic behavior and a circular plastic zone with radius r, the Airy
stress function defined by Eq. (1.60) can be used here for deriving the plastic stresses
at the crack front [10]. Thus,

$$\phi = r^{\lambda+1} f(\theta) \tag{7.12}$$

The polar stresses are defined by Eq. (1.58), but they are included in this chapter for
convenience. Hence,

$$\sigma_r = \frac{1}{r} \frac{\partial \phi}{\partial r} + \frac{1}{r^2} \frac{\partial^2 \phi}{\partial \theta^2}$$

$$\sigma_\theta = \frac{\partial^2 \phi}{\partial r^2} \tag{7.13}$$

$$\tau_{r\theta} = \frac{1}{r^2} \frac{\partial \phi}{\partial \theta} - \frac{1}{r} \frac{\partial^2 \phi}{\partial r \partial \theta} = -\frac{\partial}{\partial r} \left(\frac{1}{r} \frac{\partial \phi}{\partial \theta} \right)$$

The boundary conditions can be set as

$$\sigma_\theta = \tau_{r\theta} = 0 \qquad \text{for } \theta = \pm \alpha \tag{a}$$

$$f = f(\theta) = \frac{df}{d\theta} = 0 \qquad \text{for } \theta = \pm \alpha \tag{b}$$

The strain compatibility equation that must be satisfied is of the form

$$\frac{\partial^2 \epsilon_\theta}{\partial r^2} + \frac{2}{r}\frac{\partial \epsilon_\theta}{\partial r} - \frac{1}{r}\frac{\partial^2 \gamma_{r\theta}}{\partial r \partial \theta} - \frac{1}{r^2}\frac{\partial \gamma_{r\theta}}{\partial \theta} + \frac{1}{r^2}\frac{\partial^2 \epsilon_r}{\partial \theta^2} - \frac{1}{r}\frac{\partial \epsilon_r}{\partial r} = 0 \tag{7.14}$$

Inserting the partial derivatives of the Airy stress function, Eq. (7.12), into (7.13) yields

$$\sigma_r = r^{\lambda-1}\left[(\lambda+1)f + \frac{d^2 f}{d\theta^2}\right] = r^{\lambda-1}\tilde{\sigma}_r$$

$$\sigma_\theta = r^{\lambda-1}\lambda(\lambda+1)f = r^{\lambda-1}\tilde{\sigma}_\theta \tag{7.15}$$

$$\tau_{r\theta} = r^{\lambda-1}\lambda f = r^{\lambda-1}\tilde{\tau}_{r\theta}$$

Thus, the equivalent stress as defined by Eqs. (7.5) and (7.6), respectively, becomes

$$\sigma_e = r^{\lambda-1}\sqrt{\tilde{\sigma}_r^2 + \tilde{\sigma}_\theta^2 - \tilde{\sigma}_r\tilde{\sigma}_\theta + 3\tilde{\tau}_{r\theta}^2} \quad \text{for plane stress} \tag{7.16}$$

$$\sigma_e = r^{\lambda-1}\sqrt{\frac{3}{4}(\tilde{\sigma}_r - \tilde{\sigma}_\theta)^2 + 3\tilde{\tau}_{r\theta}^2} \quad \text{for plane strain} \tag{7.17}$$

One can observe that these stresses depend on the plastic zone size (r), which, in turn, affects the J-integral defined by Eq. (6.53). The J-integral along with $y = r\sin\theta$, Eq. (4.6), $dy = r\cos\theta d\theta$, and $ds = rd\theta$ can be written as [24]

$$\frac{J}{r} = \int_{-\pi}^{\pi}\left[W(r,\theta)\cos\theta - \vec{T}(r,\theta)\frac{\partial \vec{\mu}(r,\theta)}{\partial x}\right]d\theta \tag{7.18}$$

The remaining analytical procedure for deriving the trigonometric functions for the field equations is lengthy and complicated. Therefore, only relevant results are included as reported in the literature. Important papers on the subject are cited accordingly.

The essential mathematical approach for deriving the *HRR* field equations is a well-developed method, and yet, a sophisticated technique for assessing the plastic J-integral [12, 24]. Essentially, a rigorous computational skill is required to explicitly characterize the crack behavior in the elastic-plastic regime, which is a complex nonlinearity manifested through the strain hardening phenomenon at the crack tip region, where the J-controlled crack growth is characterized.

An elastic-plastic or a ductile crack behavior, as opposed to brittle crack behavior along with a very small plastic size (r), exhibits a significant crack tip blunting before the onset of crack growth (crack extension), complicating the applicability of the aforementioned mathematical technique [17]. Also, the stress–strain state of a practical structural component having a finite thickness must be taken into account due to the constraints either plane stress or plane-strain conditions imposed on any mathematical treatment, which must include single or mixed mode of loading and crack orientation with respect to the load direction. This problem is alleviated using Hutchinson's work [13] on plastic stress field equations.

Plastic field equations: According to Hutchinson [13], the Airy stress function in polar coordinates (r, θ) for a nonlinear or strain hardening material containing a crack is of the form

$$\phi = r^s \tilde{\phi}_1 + r^t \tilde{\phi}_2 + \ldots \ldots \simeq r^s \tilde{\phi}_1 \tag{7.19}$$

Here, s and t are exponents dependent on the Ramberg–Osgood strain hardening exponent n, and $\tilde{\phi}_1$ and $\tilde{\phi}_2$ are dimensionless functions of θ being independent of geometry and boundary conditions [21]. Subsequently, Hutchinson [13] considered the dominant singular term of the asymptotic expansion at the crack tip by assuming that $s < t$ and introducing K_p as the amplitude of the stress function ϕ. In fact, K_p can be defined as the plastic stress intensity factor [20, 26]. Hence, Eq. (7.19) becomes

$$\phi = K_p r^s \tilde{\phi}_1 \tag{7.20}$$

where

$$s = \frac{2n + 1}{n + 1} \tag{7.21}$$

$$K_p = \left(\frac{J}{\alpha' \sigma_o \epsilon_o I_n} \right)^{1/(n+1)} \tag{7.22}$$

Thus, the stress tensor components σ_{ij} and the equivalent stress σ_e are

$$\sigma_{ij} = [\sigma_r, \sigma_\theta, \tau_{r\theta}] = K_p r^{s-2} \tilde{\sigma}(\theta, s) \tag{7.23}$$

$$\sigma_e = K_p r^{s-2} \tilde{\sigma}_e(\theta, s) \tag{7.24}$$

Inserting Eqs. (7.21) and (7.22) into (7.23) and (7.24) gives the *HRR* stress field equations (*HRR* singularities). For mode I loading, the asymptotic crack tip field equations are [13]

$$\sigma_{ij} = [\sigma_r, \sigma_\theta, \tau_{r\theta}] = \sigma_o \left(\frac{J}{\alpha' \sigma_o \epsilon_o I_n r} \right)^{1/(n+1)} . \tilde{\sigma}_{ij}(\theta, n) \tag{7.25a}$$

$$\epsilon_{ij} = [\epsilon_r, \epsilon_\theta, \gamma_{r\theta}] = \alpha' \epsilon_o \left(\frac{J}{\alpha' \sigma_o \epsilon_o I_n r} \right)^{n/(n+1)} . \tilde{\epsilon}_{ij}(\theta, n) \tag{7.25b}$$

$$\mu_i = [\mu_r, \mu_\theta, \mu_z] = \alpha' \epsilon_o r \left(\frac{J}{\alpha' \sigma_o \epsilon_o I_n r} \right)^{n/(n+1)} . \tilde{\mu}_i(\theta, n) \tag{7.25c}$$

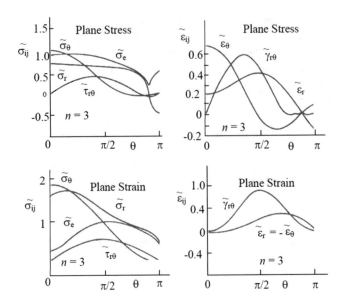

Fig. 7.4 Distribution of plastic stresses and plastic strains in a strain hardening material [12]

Table 7.1 Values of I_n at
$\sigma_e = 1$ [7, 12]

n		3	5	9	13
Plane stress		3.86	3.41	3.03	2.87
$I_n = 4.55 - 0.272\,86n + 1.111\,2 \times 10^{-2}n^2$					
Plane strain		5.51	5.01	4.60	4.40
$I_n = 6.2511 - 0.293\,81n + 1.172\,4 \times 10^{-2}n^2$					

where $\tilde{\sigma}_{ij} = [\tilde{\sigma}_r, \tilde{\sigma}_\theta, \tilde{\tau}_{r\theta}]$ $\sigma_o = \sigma_{ys}$
       $\tilde{\epsilon}_{ij} = [\tilde{\epsilon}_r, \tilde{\epsilon}_\theta, \tilde{\gamma}_{r\theta}]$ $\epsilon_o = \epsilon_{ys}$
       $\tilde{\mu}_i = [\tilde{\mu}_r, \tilde{\mu}_\theta, \tilde{\mu}_z]$

Mathematically, for a power-law hardening material, the crack tip field equations are proportional to the plastic zone size as indicated below [1, 24]

$$\sigma_{ij} \propto r^{-1/(n+1)} \tag{7.26a}$$

$$\epsilon_{ij} \propto r^{-n/(n+1)} \tag{7.26b}$$

$$\mu_i \propto r^{1/(n+1)} \tag{7.26c}$$

Figure 7.4 shows the numerical distribution of dimensionless stress $\tilde{\sigma}_{ij}(n, \theta) = [\tilde{\sigma}_r, \tilde{\sigma}_\theta, \tilde{\tau}_{r\theta}]$ and strain $\tilde{\epsilon}_{ij}(n, \theta) = [\tilde{\epsilon}_r, \tilde{\epsilon}_\theta, \tilde{\gamma}_{r\theta}]$ functions [12], and Table 7.1 gives some numerical results for the constant $I_n = f(n)$.

Table 7.1 gives the numerical results for the constant $I_n = f(n)$. Furthermore, the interpretation of the strain hardening exponent (n) is based on the type of mechanical behavior of the solid material being tested and the size of the plastic zone in terms of the crack length, such as $r \leq 0.02a$ for the J-controlled condition [8]. For fully linear elasticity, the strain hardening is $n = 1$ and the field equations have a singularity in the order of $\sigma_{ij} \propto r^{-1/2}$, $\epsilon_{ij} \propto r^{-1/2}$, and $\mu_{ij} \propto r^{-1/2}$ which is exactly the order as determined in Chap. 4 . Conversely, $n = \infty$ for fully nonlinear plasticity so that $\sigma_{ij} \geq \sigma_o$.

Furthermore, the interpretation of the strain hardening exponent (n) is based on the type of mechanical behavior of the solid material being tested and the size of the plastic zone in terms of the crack length, such as $r \leq 0.02a$ for the J-controlled condition [8].

For fully linear elasticity, the strain hardening is $n = 1$, and the field equations have a singularity in the order of $\sigma_{ij} \propto r^{-1/2}$, $\epsilon_{ij} \propto r^{-1/2}$, and $\mu_i \propto r^{1/2}$ which is exactly the order as determined in Chap. 4. Conversely, $n = \infty$ for fully nonlinear plasticity yields $\sigma_{ij} \geq \sigma_o$.

For convenience, divide Eq. (7.25c) by (7.25b) to get the plastic zone size as

$$r = \frac{\mu_i \, \tilde{\mu}_i \, (\theta, n)}{\epsilon_{ij} \, \tilde{\epsilon}_{ij} \, (\theta, n)} \tag{7.28}$$

Substituting Eq. (7.28) into (7.25a) yields the J-integral as

$$J = \alpha' \sigma_o \mu_i I_n \left(\frac{\epsilon_o}{\epsilon_{ij}} \right) \left(\frac{\sigma_{ij}}{\sigma_o} \right)^{n+1} \tilde{f}_{ij} \tag{7.29}$$

where

$$\tilde{f}_{ij} = \frac{\tilde{\epsilon}_{ij} \, (\theta, n)}{\tilde{\mu}_i \, (\theta, n)} \left[\frac{1}{\tilde{\sigma}_i \, (\theta, n)} \right]^{n+1} \tag{7.30}$$

It is clear from Eq. (7.29) that the plastic J-integral can be regarded as a measure of the intensity of the field parameters at the crack tip and accounts for a large-scale yielding tied to the small-strain theory. Therefore, J characterizes the crack tip field parameters.

Most efforts to assess the J-dominance, modeled in Fig. 7.1 as a circular region, have been focused on plane-strain condition under mode I loading [13]. Both K_I and J_I characterize local field parameters, but K_I is strictly used for a small-scale yielding (SSY) and J_I for small-scale and large-scale yielding cases.

The complete assessment of the fracture process includes fracture mechanisms because fracture toughness is a measure of fracture ductility of a material having a particular crack configuration within its geometrical shape. The concept of fracture ductility, as reported by Liu [19], depends on hydrostatic tensile stresses during deformation, during which the tearing modulus is a measure of the increase in fracture ductility.

For instance, combining Eqs. (7.25a) and (7.25b) for tension loading yields the widely used Ramberg–Osgood relationship for plastic flow [21]. Thus,

$$\epsilon = \alpha' \epsilon_o \left(\frac{\sigma}{\sigma_o} \right)^n \frac{\tilde{\epsilon}(\theta, n)}{[\tilde{\sigma}(\theta, n)]^n} \quad \text{for } \sigma \geq \sigma_o \text{ (Ramberg–Osgood)} \tag{7.31}$$

If $\tilde{\epsilon}(\theta, n) / [\tilde{\sigma}(\theta, n)]^n = 1$ (vanishes), then Eq. (7.31) becomes the Ramberg–Osgood equation, which is compared with the Hollomon equation for comparison purposes. Hence,

$$\epsilon = \alpha' \epsilon_o \left(\frac{\sigma}{\sigma_o} \right)^n \quad \text{for } \sigma \geq \sigma_o \text{ (Ramberg–Osgood)} \tag{7.32a}$$

$$\epsilon = \left(\frac{\sigma}{k_y} \right)^{1/m} \quad \text{for } \sigma \geq \sigma_o \text{ (Hollomon)} \tag{7.32b}$$

Rearranging Eqs. (7.32a) and (7.32b) gives the monotonic plastic stress equations as

$$\sigma = \sigma_o \left(\frac{\epsilon_{ij}}{\alpha' \epsilon_o} \right)^{1/n} \quad \text{for } \sigma \geq \sigma_o \text{ (Ramberg–Osgood)} \tag{7.33}$$

$$\sigma = k_y \epsilon^m \quad \text{for } \sigma \geq \sigma_o \text{ (Hollomon)} \tag{7.34}$$

Denote that both exponents in Eqs. (7.31) and (7.32) or (7.33) and (7.34) have the same meaning, but differ in magnitude since $1 < n < \infty$ and $0 < m < 1$. Typically, $3 < n < 6$ for high hardening and $n \leq 20$ for low hardening [13]. The k_y term in Eqs. (7.32) and (7.34) is called the strength coefficient in the literature. Both Ramberg–Osgood [21] and Hollomon [11] equations are used to predict the plastic uniaxial stress–strain relationship schematically depicted in Fig. 7.3. These power-law relationships indicate that the contributions to the strains that depend linearly on the stress are simply negligible.

Moreover, for pure tension loading, the equivalent stress, Eq. (7.5) or (7.6), is simply defined by

$$\sigma_e = \psi \sigma \tag{7.35}$$

where $\psi = 1$ for plane stress
$\psi = \sqrt{3}/2$ for plane strain

Example 7.1. *This example requires use of the J-integral concept in the HRR field equation for a power-law strain hardening ASTM A533B steel being subjected to a monotonic stress at infinity. Consider a large plate having an infinitely long single-edge crack. Calculate the plastic zone size r for compact tension C(T) specimen made of ASTM A533B steel at a strain of 0.1. Consider plane-strain condition and use the data reported by Kumar [17]. The steel obeys the Ramberg–Osgood stress–strain relation [21]. Assume that $J = J_e + J_p = 22.22 \text{ MJ/m}^2$. Data:*

Solution. *From Hooke's,*

$\sigma_o = 414\,MPa$	$n = 9.71$	$a_o = 117\,mm$
$E = 207\,GPa$	$\alpha' = 1.12$	$a_c = 121\,mm$
$T = 93°C$	$w = 203\,mm$	$\mu = 4.7\,mm$
$\sigma =?$	$B = 0.5w$	$J_{IC} = 1.00\,MJ/m^2$

$$\epsilon_o = \frac{\sigma_o}{E} = \frac{414\,MPa}{207,000\,MPa} = 0.002$$

Plot the stress–strain curve using Eq. (7.32). This is to verify that the given strain corresponds to an applied stress greater than the yield strength. Hence,

$$\sigma = \sigma_o \left(\frac{\epsilon}{\alpha'\epsilon_o}\right)^{1/n} = 776.05\epsilon^{1/9.71}$$

$$\sigma = 612.21\,MPa \; @ \; \epsilon = 0.1$$

Therefore, $\sigma > \sigma_o$. The stress–strain plot is

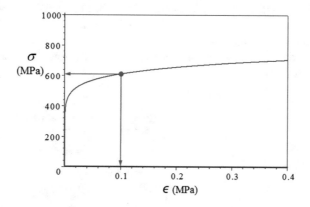

Therefore, the requirement that $\sigma > \sigma_{ys}$ has been met. Now, using Eq. (7.25a) and Table 7.1 yields the constant I_n and r.

$$I_n = 6.2511 - 0.293\,81n + 1.172\,4 \times 10^{-2}n^2 = 4.5036$$

$$r = \frac{J}{\alpha'\sigma_o\epsilon_o I_n}\left(\frac{\sigma_o}{\sigma}\right)^{n+1}$$

$$r = \frac{(22.22\,MPa\,m)}{(1.12)\,(414\,MPa)\,(0.002)\,(4.5036)}\left(\frac{414\,MPa}{612.21\,MPa}\right)^{10.71}$$

$$r = 80.60\,mm$$

This is a large plastic zone. From Eq. (7.25c), the displacement becomes

$$\mu = \alpha'_o \epsilon_o r \left(\frac{J}{\alpha' r \sigma_o \epsilon_o I_n} \right)^{1/(n+1)}$$

$$\mu = 0.27 \, \text{mm}$$

Therefore, 612.21MPa applied stress induces a sufficiently large plastic zone and a small crack mouth displacement.

7.3.1 Strain Hardening Effects

The general approach for modeling a strain hardening material during plastic deformation is to represent its stress–strain behavior through a power-law strain hardening equation, such as the Ramberg–Osgood [21] defined by Eq. (7.33), for crack-free specimens and its crack growth through the J-integral when the specimen contains a specific crack configuration. The most suitable approach is to determine the total J-integral near the crack tip by incorporating the elastic and plastic contributions.

Characterization of the J-integral, the crack opening displacement ($\delta = COD$), and the load-line displacement (μ) has numerically been established as an engineering approach (an approximation scheme) for elastic [25, 29, 32] and fully plastic [3, 15, 16, 22] fracture analyses.

Figure 7.5 depicts the schematic sequence of plastic zone size formation and related J-integral [17].

Fig. 7.5 Schematic elastic-plastic fracture behavior [17]

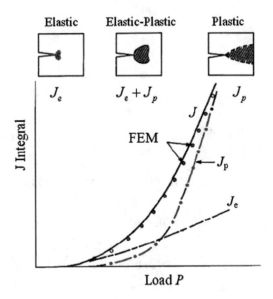

The fully plastic analysis is strictly based on the Ramberg–Osgood power-law equation (Eq. (7.33) [21]). However, the Hollomon stress–strain power law [Eq. (7.34)] can also be used for the same purpose [11].

The elastic-plastic J-integral analysis can be defined in a general form as the sum of the elastic and plastic parts [30]

$$J = J_e\,(a_e) + J_p\,(a, n) \tag{7.36}$$

$$\delta = \delta_e\,(a_e) + \delta_p\,(a, n) \tag{7.37}$$

$$\mu = \mu_e\,(a_e) + \mu_p\,(a, n) \tag{7.38}$$

where a_e = Irwin's effective crack length [17]

a = Original crack length

In addition, Irwin's effective crack length accounts for strain hardening effects mathematically defined as [1, 17]

$$a_e = a + \varpi r \tag{7.39a}$$

with

$$\varpi = \frac{1}{1 + (P/P_o)^2} \tag{7.39b}$$

$$r = \left(\frac{1}{\pi\beta}\right)\left(\frac{n-1}{n+1}\right)\left(\frac{K_I}{\sigma_o}\right)^2 \tag{7.39c}$$

where $\beta = 2$ for plane stress

$\beta = 6$ for plane strain

7.3.2   Near-Field J-Integral Approximation

Consider the single-edge notched (SEN) specimen shown in Fig. 7.6 containing Γ_1 and Γ_2 as the near-field and far-field contours for predicting the J-integral. The specimen edges constitute the contour Γ_2. Use of half of the entire contour due to symmetry is a practical assumption.

Kang and Kobayashi [15] developed a J-estimation procedure for two dimensional states of stress and strains. Consider the near-field J-Integral and a strain hardening material that obeys the Ramberg–Osgood [21] relation defined by Eq. (7.30) in order to solve the integral for the plastic strain energy density, which is needed in the J-integral equation. Thus,

$$W_p = \int \sigma\, d\epsilon = \int \sigma_o\left(\frac{\epsilon}{\alpha'\epsilon_o}\right)^{1/n} d\epsilon \tag{7.40}$$

$$W_p = \left(\frac{n}{n+1}\right)\frac{\sigma_o}{(\alpha'\epsilon_o)^{1/n}}\epsilon^{(n+1)/n} \tag{7.41}$$

Fig. 7.6 SEN specimen
having two contours:
near-field Γ_1, 1-2-3-4-5-6,
and far-field Γ_2, a-b-c-d-e-f

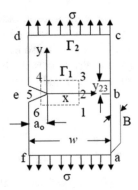

Substituting Eq. (7.30) back into (7.41) along with Hooke's law yields the most practical expression for the plastic strain energy density

$$W_p = \frac{\alpha'\sigma_o^2}{E}\left(\frac{n}{n+1}\right)\left(\frac{\sigma}{\sigma_o}\right)^{n+1} \tag{7.42}$$

The evaluation of the plastic J-integral (J_p) is restricted to the square segment illustrated in Fig. 7.6 as Γ_1 or Γ_2. For the contour Γ_1, the J-integral is applied on each segment of the upper half of the contour. Hence,

$$J_p = \int_{23}\left(Wdy - T_i\frac{\partial\mu}{\partial x}dx\right) + \int_{45}\left(Wdy - T_i\frac{\partial\mu}{\partial x}dx\right) \tag{7.43}$$
$$+ \int_{34}\left(Wdy - T_i\frac{\partial\mu}{\partial x}dx\right) + \int_{52}\left(Wdy - T_i\frac{\partial\mu}{\partial x}dx\right)$$

where $T_{23} = T_{45} = 0$, $ds = dy$ and $y_{23} = y_{45}$
  $dy_{34} = dy_{52} = 0$ and $d\mu_{34} = d\mu_{52} = 0$
Thus,

$$J_p = \int_{23} W_p dy + \int_{45} W_p dy = 2\int_0^{y_{23}} W_p dy = 2W_p y_{23} \tag{7.44}$$

Substituting Eq. (7.42) into (7.44) yields

$$J_p = \frac{2\alpha'\sigma_o^2}{E}\left(\frac{n}{n+1}\right)\left(\frac{\sigma}{\sigma_o}\right)^{n+1} y_{23} \tag{7.45}$$

For pure tension, the fundamental elastic J-integral is defined by Eq. (6.61) as

$$J_e = \frac{K_I^2}{E'} = \frac{\pi a\sigma^2}{E'} \tag{7.46}$$

Finally, the total J-integral, $J_I = J_e + J_p$, becomes

$$J = \frac{\pi a \sigma^2}{E'} + \frac{2\alpha' \sigma_o^2}{E} \left(\frac{n}{n+1} \right) \left(\frac{\sigma}{\sigma_o} \right)^{n+1} y_{23} \tag{7.47}$$

Similarly, the contour Γ_2 for the far-field condition yields

$$J_I = \frac{\pi a \sigma^2}{E'} + \frac{2\alpha' \sigma_o^2}{E} \left(\frac{n}{n+1} \right) \left(\frac{\sigma}{\sigma_o} \right)^{n+1} y_{bc} \tag{7.48}$$

The only difference between Eqs. (7.47) and (7.48) is the height of the upper vertical segments; that is, $y_{bc} > y_{23}$.

In summary, the J-integral components are generalized as $J_e = f(a, \sigma)$ and $J_p = f(n, \sigma)$ functions. This provides information based on strain hardening effects.

Example 7.2. **(a)** *Calculate the total J-integral (J_I) for a* steel *plate under plane stress conditions (Fig. 7.6). Determine* **(b)** *if the elastic J-integral contributes significantly to the total value of J_I.* **(c)** *What does the plastic strain energy density W measure?* **(d)** *Plot the plastic stress–strain and J = f(σ). Use the following data to carry out all calculations: a = 1.40 mm, w = 19 mm, B = 0.8 mm, L = 10 cm, σ_o = 64 MPa, E = 207,000 MPa, α' = 0.35, n = 5, P = 1.01 kN (load) and $y_{23} = w/3$.*

Solution. **(a)** *The following calculations are self-contained. Thus,*

$$\epsilon_o = \frac{\sigma_o}{E} = \frac{64}{207000} = 3.0918 \times 10^{-4}$$

$$A = wB = 15.20 \times 10^{-6} \, \text{m}^2$$

$$\sigma = \frac{P}{A} = \frac{1.01 \times 10^{-3} \, \text{MN}}{15.20 \times 10^{-6} \, \text{m}^2} = 66.45 \, \text{MPa} > \sigma_{ys}$$

$$y_{23} = \frac{w}{3} = \frac{19 \, \text{mm}}{3} = 6.33 \, \text{mm}$$

From Eq. (1.21),

$$W_e = \frac{\sigma_o^2}{2E} = \frac{(64 \, \text{MPa})^2}{2 \, (207,000 \, \text{MPa})} = 9.89 \times 10^{-3} \, \text{MPa} = 9.89 \, \text{kPa}$$

$$W_e = 9.89 \, \text{kJ/m}^3$$

From Eq. (7.42),

$$W_p = \frac{\alpha' \sigma_o^2}{E} \left(\frac{n}{n+1} \right) \left(\frac{\sigma}{\sigma_o} \right)^{n+1}$$

$$W_p = \frac{(0.35)\,(64\,\text{MPa})^2}{(207,000\,\text{MPa})} \left(\frac{5}{6} \right) \left(\frac{66.45}{64} \right)^6 (10^3) = 7.23\,\text{kPa}$$

$$W_p = 7.23\,\text{kJ/m}^3$$

and

$$W = W_e + W_p = 9.89\,\text{kJ/m}^3 + 7.23\,\text{kJ/m}^3$$

$$W = 17.12\,\text{kJ/m}^3$$

From Eq. (7.44),

$$J_p = 2W_p y_{23} = (2) \left(7.23\,\text{kJ/m}^3 \right) \left(\frac{19}{3} \times 10^{-3}\,\text{m} \right)$$

$$J_p = 91.58\,\text{J/m}^2$$

For a single-edge crack with $\alpha \simeq 1$,

$$J_e = \frac{K_I^2}{E} = \frac{\pi a \alpha^2 \sigma^2}{E} = \frac{\pi \left(1.40 \times 10^{-3}\,\text{m} \right) (1)^2 (66.45\,\text{MPa})^2}{207,000\,\text{MPa}}$$

$$J_e = 93.82\,\text{J/m}^2$$

The total J-integral is

$$J_I = J_e + J_p = 93.82\,\text{J/m}^2 + 91.58\,\text{J/m}^2 = 185.40\,\text{J/m}^2$$

(b) *These results indicate that $J_e = 0.51J$ and $J_p = 0.49J$. Therefore, J_e and J_p contribute about the same amount to the total J-integral J. More accurate results can be obtained if the geometry factor $\alpha = f(a/w)$ is calculated using the equation in Table 3.2.*

(c) *The meaning of the plastic strain energy density (W_p) is that it measures the toughness of a material as the area under the stress–strain curve. This area represents the strain energy per unit volume (17.12 kJ/m³) the material absorbs during straining irreversibly. Obviously, the plastic strain energy density depends on the applied load, and it represents the total amount of energy stored in the material.*

(d) *The required stress–strain curve can be determined using Eq. (7.33). Thus,*

$$\sigma = \sigma_o \left(\frac{\epsilon}{\alpha' \epsilon_o} \right)^{1/n} = (322.08 \text{ MPa}) \sqrt[5]{\epsilon}$$

The stress–strain plot is shown below

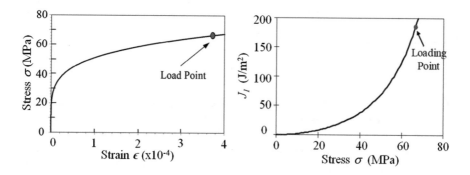

The J-integral plot along with the load point is based on the following equation:

$$J_I = \frac{\pi a \sigma^2}{E} + \frac{2\alpha' \sigma_o^2}{E} \left(\frac{n}{n+1} \right) \left(\frac{\sigma}{\sigma_o} \right)^{n+1} . y_{23}$$

$$J_I = 2.1247 \times 10^{-2} \sigma^2 + 1.0638 \times 10^{-9} \sigma^6 \quad in \ J/m^2$$

The J-integral vs. applied stress plot is given above. It is clear that the trend of the $J_I = f(\sigma)$ nonlinear as expected, and it resembles the trend shown in Fig. 7.5. This result proves that the J-integral theory given above is correct or acceptable.

7.3.3 Far-Field J-Integral Approximation

Consider the single-edge notched (SEN) specimen with Γ_2 contour as large as the specimen shown in Fig. 7.6 for characterizing the far-field J-integral using only half of the entire contour due to symmetry. The present stress field requires that $\sigma_y = \sigma$ and $\sigma_x = \tau_{xy} = 0$ for the assumed contour a-b–c–d–e–f with segments ab-bc-cd-de-ef and fa. Thus, the far-field J-integral evaluation can be treated as a reasonable approximation for the near field. For an axial tension loading, Kang and Kobayashi [15] using Moiré interferometry evaluated the J-integral as the sum of the vertical and horizontal parts of half the contour due to symmetry. Thus, the J-integral for the traction free vertical edges of the segments bc and de becomes

$$J_v = \int_{bc} W dy + \int_{de} W dy \tag{7.49}$$

$$J_v \simeq \sum (W_i \Delta y_i)_{bc} + \sum (W_i \Delta y_i)_{de} \tag{7.50}$$

and that for the horizontal part along with $\sigma_x = \tau_{xy} = 0$ and $\mu_x = 0$ on the crack line is

$$J_h = \int_{cd} T_y \frac{\partial \mu}{\partial x} dx = \int_{cd} \sigma_y \frac{\partial \mu}{\partial x} dx \qquad (7.51)$$

$$J_h \simeq \sum \left[\left(\sigma_y \frac{\Delta \mu_y}{\Delta x_i} \right) \Delta x_i \right]_{cd} \qquad (7.52)$$

The total J-integral due to symmetry becomes [15]

$$J_I = 2 \left(J_v + J_h \right) \qquad (7.53)$$

Once more, the J-integral represents the strain energy release rate, which is also referred to as mechanical work per unit surface area during the onset of crack growth. With respect to the concept that the J-integral contour can have an arbitrary shape, thus the rectangular contour shown in Fig. 7.6 agrees with such a concept.

Nevertheless, the above approximation is a suitable approach for defining the J-integral in simplistic mathematical manner by using the approximated integral method.

It should be mentioned that the concept of a path-independent integral determines the intensity of a singularity of a field quantity of an arbitrary field shape. Apparently, Cherepanov [4] and Rice [22–23] were the pioneers who introduced this concept into fracture mechanics as the J-integral. Subsequently, the field of fracture mechanics has evolved since then in a fast mode and besides complicated mathematical treatments available in the literature for finding solutions to fracture mechanics problem, simplified mathematical approaches seem to gain momentum for the same purpose. Therefore, the above integral approximation demonstrates the use of path-independent J-integral in a simple manner.

7.4   Engineering Approach

Kumar et al. [17] developed another approximation scheme for characterizing simple specimen configurations. Assume that a material obeys the Ramberg–Osgood [21] power law [Eq. (7.30)] and that the plastic field is controlled by the J-integral for limited crack growth. It is also assumed that unloading does not occur behind the crack tip during the irreversible plastic deformation [14] and that the onset of unstable crack growth (crack propagation) occurs at the maximum load carrying capability of the component. The measurement of the J-integral is carried out by determining the displacement from the deformation properties of the Ramberg–Osgood material [27].

The simplified methodology for predicting fracture of structural components is based on Eqs. (7.36) through (7.38) along with modified (7.30), and $\sigma_{ys} = \sigma_o$ and $\epsilon_{ys} = \epsilon_o$. The general field equations for elastic-plastic fracture analysis are defined as [1, 17]

$$J = \frac{P^2}{E'}f_1 + \alpha'\sigma_o\epsilon_o\lambda_1 h_1 \left(\frac{P}{P_o}\right)^{n+1} \tag{7.54a}$$

$$\delta = \frac{P}{E'}f_2 + \alpha' a\epsilon_o h_2 \left(\frac{P}{P_o}\right)^{n} \tag{7.54b}$$

$$\mu = \frac{P}{E'}f_3 + \alpha' a\epsilon_o h_3 \left(\frac{P}{P_o}\right)^{n} \tag{7.54c}$$

where P = Load per unit thickness (MN/m)
P_o = Limit load per unit thickness (MN/m)
h_1, h_2, h_3 = Constant tabulated in Ref. [1, 17]
f_1, f_2, f_3 = Constant tabulated in Ref. [32]

The constants λ_1 and P_o are defined in Table 7.2. The above field equations are strongly dependent on increasing load raised to the power n. Therefore, the plastic part of these equations are dominant in strain hardenable materials [17].

Figure 7.7 shows a crack driving force diagram for an axially cracked and internally pressurized cylinder made of ASTM A533B steel [18].

This diagram represents a complete history of deformation and crack growth. Thus, the onset of crack growth can be predicted when $J = J_R$ at the intercept of J_R-P_{fixed}. This intercept is the instability point where

$$J_{IC} \simeq 10\,\text{in.kips/in}^2 \simeq 21\,\text{kJ/m}^2$$

$$P \simeq 5.5\,\text{kips} = 24.47\,\text{kN}$$

$$a_c \simeq 2.6\,\text{in.} = 6.60\,\text{cm}$$

$$a_o \simeq 2.25\,\text{in.} = 5.72\,\text{cm}$$

$$\Delta a = 0.88\,\text{cm}$$

Furthermore, careful attention to the data in Fig. 7.7 indicates stable crack growth is $\Delta a = 0.88$ mm for the monotonic load-controlled system. This amount of crack growth is considered very small prior to crack propagation. If the load is $P > P_c$, then the crack becomes unstable and grows very rapidly.

Additionally, He and Hutchinson [8] derived principles associated with upper and lower bounds on the J-integral for a finite crack in an infinite plane (central crack case) and edge crack in a semi-infinite plane. These principles are based on the complementary potential energy and potential energy theories. The resultant J-integral equation [8] is also given in Refs. [9, 27, 28]. Thus,

$$J_p = \frac{3\pi a\epsilon_e\sigma_e\sqrt{n}}{4}\left(\frac{\sigma}{\sigma_e}\right)^2 \tag{7.55}$$

Let Eq. (7.30) be an equivalent expression so that

Table 7.2 Crack configurations and related geometric correction factors for J_p, δ_p and μ_p [1, 17]

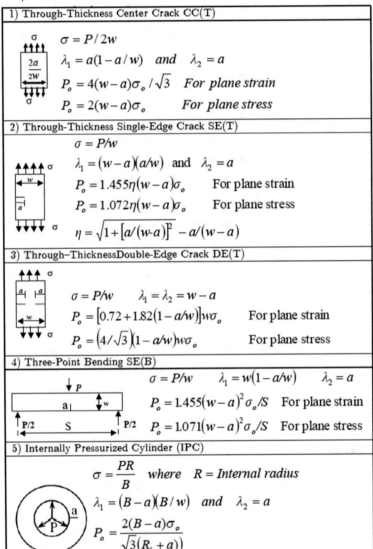

1) Through-Thickness Center Crack CC(T)

$\sigma = P/2w$

$\lambda_1 = a(1 - a/w)$ and $\lambda_2 = a$

$P_o = 4(w - a)\sigma_o /\sqrt{3}$ For plane strain

$P_o = 2(w - a)\sigma_o$ For plane stress

2) Through-Thickness Single-Edge Crack SE(T)

$\sigma = P/w$

$\lambda_1 = (w - a)(a/w)$ and $\lambda_2 = a$

$P_o = 1.455\eta(w - a)\sigma_o$ For plane strain

$P_o = 1.072\eta(w - a)\sigma_o$ For plane stress

$\eta = \sqrt{1 + [a/(w-a)]^2} - a/(w - a)$

3) Through–Thickness Double-Edge Crack DE(T)

$\sigma = P/w$ $\lambda_1 = \lambda_2 = w - a$

$P_o = [0.72 + 1.82(1 - a/w)]w\sigma_o$ For plane strain

$P_o = (4/\sqrt{3})(1 - a/w)w\sigma_o$ For plane stress

4) Three-Point Bending SE(B)

$\sigma = P/w$ $\lambda_1 = w(1 - a/w)$ $\lambda_2 = a$

$P_o = 1.455(w - a)^2\sigma_o/S$ For plane strain

$P_o = 1.071(w - a)^2\sigma_o/S$ For plane stress

5) Internally Pressurized Cylinder (IPC)

$\sigma = \dfrac{PR}{B}$ where $R = $ *Internal radius*

$\lambda_1 = (B - a)(B/w)$ and $\lambda_2 = a$

$P_o = \dfrac{2(B - a)\sigma_o}{\sqrt{3}(R_i + a)}$

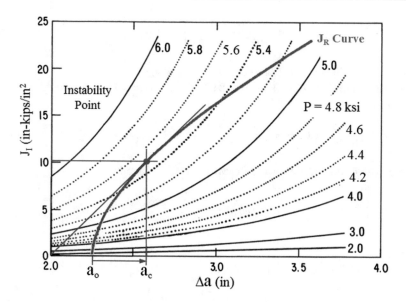

Fig. 7.7 Experimental J-integral diagram for an axially cracked and internally pressurized cylinder made of ASTM A533B steel [18]

$$\epsilon_e = \alpha' \epsilon_o \left(\frac{\sigma_e}{\sigma_o} \right)^n \qquad (7.56)$$

Substituting Eqs. (7.34) and (7.56) into (7.55) yields the upper bound J-integral for plane conditions

$$J_p = \frac{3\pi a \sqrt{n}}{4\psi^{1-n}} \alpha' \epsilon_o \sigma \left(\frac{\sigma}{\sigma_o} \right)^n \quad \text{(upper bound)} \qquad (7.57)$$

He and Hutchinson [8, 9] also derived an upper bound J-integral as

$$J_p = \pi a h \sqrt{n} \epsilon_e \sigma_e \qquad (7.58)$$

where $h = 1$ for a central crack
$h = 1.26$ for an edge crack
Similarly, substitute Eqs. (7.34) and (7.56) into (7.58) to get

$$J_p = \pi a h \psi \alpha' \sqrt{n} \epsilon_o \sigma \left(\frac{\psi \sigma}{\sigma_o} \right)^n \quad \text{(upper bound)} \qquad (7.59)$$

The constant h_1: The constant h_1 in Eq. (7.54a) can be derived for the specimen configurations illustrated in Table 7.2. This can be done very easily by equating Eqs. (7.54a) and (7.57) and using the proper definitions of

$$h_1 = \frac{3\pi a \sqrt{n}}{4\psi^{1-n}\lambda_1} \left(\frac{\sigma}{\sigma_o}\right)^{n+1} \left(\frac{P_o}{P}\right)^{n+1} \tag{7.60}$$

where ψ is defined by Eq. (7.34) and $h_1 = h_1\,(a/w, n)$.

For the central cracked plate given in Table 7.2, $\lambda_1 = a\,(w - a)$ and this constant becomes

$$h_1 = \pi \sqrt{n} \left(1 - \frac{a}{w}\right)^n \quad \text{CC(T) specimen} \tag{7.61}$$

Rearranging Eq. (7.60) gives a different constant defined by

$$H\,(a/w, n) = H_1 = \frac{3\pi a \sqrt{n}}{4\psi^{1-n}\lambda_1} \tag{7.62}$$

Then,

$$H_1 = h_1 \left(\frac{\sigma_o}{\sigma}\right)^{n+1} \left(\frac{P}{P_o}\right)^{n+1} \tag{7.63}$$

This expression, Eq. (7.63), can be derived by defining the J-integral in a general form for mode I as

$$J = \sigma \epsilon a H \left(\frac{a}{w}, n\right) \tag{7.64}$$

where the strain ϵ for linear and nonlinear materials is defined by Hooke's law and Ramberg–Osgood equation [21], respectively,

$$\epsilon = \frac{\sigma}{E} \quad \text{for } \sigma < \sigma_o \ \& \ n = 1 \tag{7.65}$$

$$\epsilon = \alpha' \epsilon_o \left(\frac{\sigma}{\sigma_o}\right)^n \quad \text{for } \sigma > \sigma_o \ \& \ n > 1 \tag{7.66}$$

Substituting these strain ϵ equations into Eq. (7.64) yields the elastic and plastic J-integrals as

$$J_e = \epsilon_o \sigma_o a \left(\frac{\sigma}{\sigma_o}\right)^2 H \left(\frac{a}{w}, n\right) \quad \text{for } n = 1 \tag{7.67}$$

$$J_p = \alpha' \epsilon_o \sigma_o a \left(\frac{\sigma}{\sigma_o}\right)^{n+1} H \left(\frac{a}{w}, n\right) \quad \text{for } n > 1 \tag{7.68}$$

Inserting Eq. (7.63) into (7.68) yields the plastic J-integral as

$$J_p = \alpha' \epsilon_o \sigma_o a \left(\frac{P}{P_o} \right)^{n+1} h_1 \left(\frac{a}{w}, n \right) \quad \text{for } n > 1 \tag{7.69}$$

Equating Eqs. (7.68) and (7.69) gives

$$\left(\frac{\sigma}{\sigma_o} \right)^{n+1} H \left(\frac{a}{w}, n \right) = \left(\frac{P}{P_o} \right)^{n+1} h_1 \left(\frac{a}{w}, n \right) \tag{7.70}$$

and solving for $H(a/w, n)$ yields Eq. (7.63). For linear (elastic) materials, $H(a/w) = \pi f(a/w)^2 = \pi \alpha^2$, $n = 1$, and $f(a/w)$ are the usual geometry correction factor listed in Table 3.2. Denote also that if $n = 1$ in Eq. (7.68), then $\alpha' = H(a/w, n) = 1$ and the resultant expression becomes Eq. (7.67).

Example 7.3. *Use the Hollomon equation to derive the J-integral similar to Eq. (7.69).*

Solution. *From Eq. (7.31),*

$$\epsilon = \left(\frac{\sigma}{k_y} \right)^{1/m} \tag{a}$$

$$\epsilon_o = \left(\frac{\sigma_o}{k_y} \right)^{1/m} \tag{b}$$

$$\frac{\epsilon}{\epsilon_o} = \left(\frac{\sigma}{\sigma_o} \right)^{1/m} \tag{c}$$

$$\epsilon = \epsilon_o \left(\frac{\sigma}{\sigma_o} \right)^{1/m} \tag{d}$$

Substitute the last ϵ equation into Eq. (7.64), multiply the resultant expression by σ_o/σ_o, and replace the strain hardening exponent n for m to get the plastic J-integral as

$$J_p = \sigma_o \epsilon_o a \left(\frac{\sigma}{\sigma_o} \right)^{(m+1)/m} H \left(\frac{a}{w}, m \right) \tag{e}$$

$$J_p = \sigma_o \epsilon_o a \left(\frac{P}{P_o} \right)^{(m+1)/m} h_1 \left(\frac{a}{w}, m \right) \tag{f}$$

where

$$(\sigma/\sigma_o)^{(m+1)/m} H(a/w, m) = (P/P_o)^{(m+1)/m} h_1(a/w, m) \tag{g}$$

was used in Eq. (e). Apparently, results for $h_1\left(\frac{a}{w}, m\right)$ using the Hollomon equation are not available in the literature at this moment [11]. The Hollomon strain hardening exponent m was assumed to be a constant; however, it may be strain dependent, $n = f(\epsilon)$.

Example 7.4. *Calculate* **(a)** *the total J-integral* $(J = Je + Jp)$ *for a 2024 Al-alloy plate having a through the thickness central crack under plane conditions. Assume that the material obeys the Ramberg–Osgood relation. Use the following data:* $a = 1.40\ mm$, $w = 19\ mm$, $B = 8\ mm$, $\sigma_o = 64\ MPa$, $E = 72,300\ MPa$, $\alpha' = 0.35$, $n = 5$, $v = 0.3$,*and* $\sigma = 80\ MPa$. **(b)** *Plot* J_I *vs.* σ *for plane-strain conditions. Explain.*

Solution. **(a)** *The solution requires the following parameters:*

$$\frac{a}{w} = 0.0737 \quad and \quad \left(1 - \frac{a}{w}\right) = 0.9263$$

$$\epsilon_o = \frac{\sigma_o}{E} = 0.0009$$

From Eq. (3.29) and Table 3.1, the applied stress intensity factor is

$$K_I = \alpha\sigma\sqrt{\pi a} \tag{3.29}$$

$$\alpha = \left[1 + 0.5\left(\frac{a}{w}\right)^2 + 20.46\left(\frac{a}{w}\right)^4 + 81.72\left(\frac{a}{w}\right)^6\right]^{1/2} \simeq 1$$

$$K_I = (80\ \text{MPa})\sqrt{\pi\,(1.40 \times 10^{-3}\ m)} = 5.31\ \text{MPa}\sqrt{\text{m}}$$

and from Eq. (6.61) for pure tension loading, the elastic J-integral becomes

$$J_e = \frac{K_I^2}{E'} = \frac{\left(1 - v^2\right)K_I^2}{E} = 3.55 \times 10^{-4}\ \text{MPa m} = 0.36\ \text{kJ/m}^2 \tag{6.61}$$

Let us determine J_p *using Eq. (7.54a) and Table 7.2*

$$J_p = \alpha'\sigma_o\epsilon_o\lambda_1 h_1\left(\frac{P}{P_o}\right)^{n+1} \tag{7.51}$$

Using Table 7.2 for plane-strain condition yields

$$\lambda_1 = a\left(1 - \frac{a}{w}\right) = (1.40 \times 10^{-3}\ m)(1 - 0.0737) = 1.30 \times 10^{-3}\ m$$

$$P = 2w\sigma = 2\left(19 \times 10^{-3}\ m\right)(80\ \text{MPa}) = 3.04\ \text{MN/m}$$

$$P_o = \frac{4\,(w - a)\,\sigma_o}{\sqrt{3}} = \frac{4\left(19 \times 10^{-3}\ m - 1.4 \times 10^{-3}\ m\right)(64\ \text{MPa})}{\sqrt{3}}$$

$$P_o = 2.60\ \text{MN/m}$$

From Eq. (7.61), the constant h_1 along with $\psi = \sqrt{3}/2$ and $n = 5$ is

$$h_1 = h_1 = \pi \sqrt{n} \left(1 - \frac{a}{w}\right)^n \tag{7.61}$$

$$h_1 = \pi \sqrt{5} \, (1 - 0.0737)^5$$

$$h_1 = 4.7906$$

and Eq. (7.63) gives

$$H_1 = h_1 \left(\frac{\sigma_o}{\sigma} \frac{P}{P_o}\right)^{n+1} = (4.7906) \left(\frac{64}{80} \frac{3.04}{2.60}\right)^{5+1}$$

$$H_1 = 3.2087 < h_1$$

Thus,

$$J_p = \alpha' \sigma_o \epsilon_o \lambda_1 h_1 \left(\frac{P}{P_o}\right)^{n+1}$$

$$J_p = (0.35) \, (64 \, \text{MPa}) \, (0.0009) \, (1.30 \times 10^{-3} \, \text{m}) \, (4.7906) \left(\frac{3.04}{2.60}\right)^{5+1}$$

$$J_p = 3.2079 \times 10^{-4} \, \text{MPa m} \simeq 0.32 \, \text{kJ/m}^2$$

The plastic J-integral is now calculated using Eqs. (7.57) and (7.59) for comparison purposes. Thus,

$$J_p = \frac{3\pi a \sqrt{n}}{4 \psi^{1-n}} \alpha' \epsilon_o \sigma \left(\frac{\sigma}{\sigma_o}\right)^n \quad \text{(Upper bound)} \tag{7.57}$$

$$J_p = \frac{3\pi \left(1.40 \times 10^{-3} \, \text{m}\right) \sqrt{5}}{4 \left(\sqrt{3}/2\right)^{1-5}} (0.35) \, (0.0009) \, (80 \, \text{MPa}) \left(\frac{80 \, \text{MPa}}{64 \, \text{MPa}}\right)^5$$

$$J_p = 3.1908 \times 10^{-4} \, \text{MPa m} \simeq 0.32 \, \text{kJ/m}^2$$

and the plastic J-integral is

$$J_p = \pi a h \psi \alpha' \sqrt{n} \epsilon_o \sigma \left(\frac{\psi \sigma}{\sigma_o}\right)^n \quad \text{(Upper bound)} \tag{7.59}$$

$$J_p = \pi \left(1.4 \times 10^{-3} \, \text{m}\right) (1) \left(\frac{\sqrt{3}}{2}\right) (0.35) \sqrt{5} \, (0.0009) \, (80 \, \text{MPa}) \left(\frac{\sqrt{3} \times 80 \, \text{MPa}}{2 \times 64 \, \text{MPa}}\right)^5$$

$$J_p = 3.1908 \times 10^{-4} \, \text{MPa m} \simeq 0.32 \, \text{kJ/m}^2$$

Denote that Eqs. (7.54a), (7.57) and (7.59) yield similar results up to two decimal places. The former gives approximately 0.5 % higher value than the latter equation. The total J-integral is

$$J_I = J_e + J_p = 0.36 \text{ kJ/m}^2 + 0.32 \text{ kJ/m}^2 = 0.68 \text{ kJ/m}^2$$

(b) *The second part of the example requires the determination of a function $J_I = f(\sigma)$ with the stress σ in MPa. From part (a), the elastic J-integral becomes*

$$J_e = \frac{\pi \left(1 - v^2\right) a\sigma}{E}$$

$$J_e = \frac{\pi \left(1 - 0.3^2\right) \left(1.40 * 10^{-3}\right) \sigma^2}{72,300} = 5.5358 \times 10^{-8} \sigma^2$$

and the plastic J-integral gives

$$J_p = \frac{3\pi a \sqrt{n}}{4\psi^{1-n}} \alpha' \epsilon_o \sigma \left(\frac{\sigma}{\sigma_o}\right)^n$$

$$J_p = \frac{3\pi \left(1.40 \times 10^{-3}\right) \sqrt{5}}{4 \left(\sqrt{3}/2\right)^{-4}} (0.35)(0.0009)(80 \text{ MPa}) \left(\frac{\sigma}{64}\right)^5$$

$$J_p = 9.7375 \times 10^{-14} \sigma^5$$

$$J_I = 5.5358 \times 10^{-8} \sigma^2 + 9.7375 \times 10^{-14} \sigma^5 \quad \text{in kJ/m}^2$$

The required plot for plane-strain condition is shown below.

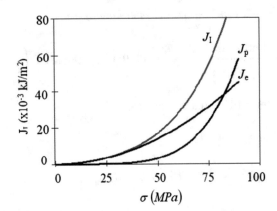

This plot indicates that the elastic contribution to J is significant up to approximately 10×10^{-3} kJ/m^2 at 40 MPa. Beyond this stress value, the plastic contribution becomes apparent as shown above at $\sigma = 80$ MPa.

Furthermore, the resultant plot in this example resembles Fig. 7.5. Recall from Chap. 3 that the critical value of the total J-integral is, in essence, the strain energy release rate per unit surface area. The critical J-integral, say, for mode I, is J_{IC} at onset of crack growth. In this example, J_{IC} is the energy for a large-scale yielding (LSY).

The above approach for characterizing the J-integral of an elastic-plastic material excludes a more realistic evaluation based on uncertainty of random loads encountered in engineering structures. Hence, the application of conventional J-integral engineering approaches for determining fracture toughness must include statistics to assure structural integrity of linear-elastic materials. This implies that a deterministic fracture mechanics model must include all possible loads, crack orientation and crack geometry, and environmental effects. Subsequently, a large safety factor required by fracture mechanics design application is important in selecting the proper material for a specific structural component.

7.4.1 Practical Aspects

In light of the above, it appears obvious that the J-integral offers the most realistic engineering approach for characterizing structural damage due to the presence of cracks. Nondestructive evaluation (NDE) of structural components may reveal cracks using visual inspection, acoustic emission, and the like. Once a crack is detected, the load and the J-integral are estimated and compared their values against a design code in order to determine the degree of damage in a structural component.

For mode I loading, if $J_I < J_{IC}$, then the critical load $P = P_c$ must be determined so that $J_I = J_{IC}$ for the onset of crack growth. The disadvantage of the J-integral approach is that it does not predict the amount of crack extension prior to crack propagation or fracture. Therefore, the practical application of fracture mechanics in industry for assessing structural integrity depends on the design application guidelines. For instance, the basic engineering stress component for structural integrity is the ultimate strength of the undamaged structure for specified design applications. Hence, the design life associated with a limit damage is determined prior to crack detection, followed by crack repair or component replacement.

Moreover, assessing the safety and reliability of a structure, a failure assessment diagram (FAD) can be useful as a failure curve relating brittle fracture and general yielding of structures containing flaws or cracks. Development of a FAD curve is included in Chap. 10.

7.5 Problems

7.1. Determine (**a**) the J-integral (J) and (**b**) dJ/da for a hypothetical steel plate containing a central crack of 114-mm long loaded at 276 MPa and exposed to room temperature air. What does $T_J < T_R$ mean? Assume plane-strain conditions and that the stainless steel obeys the Ramberg–Osgood relation with curve fitting parameters such as $\alpha' = 1.69$ and $n = 5.42$. Explain the results based on the strain hardening effect. Data: $\sigma_o = 207\,\text{MPa}$, $E = 206,850\,\text{MPa}$, $\nu = 0.3$, and $J_{IC} = 130\,\text{kJ/m}^2$. Dimensions: $w = 4a$, $L = 2w$, and $B = a$. [Solution: a) $J = 248.29\,\text{kJ/m}^2$ and $dJ/da = 4.05\,\text{MJ/m}^3$.]

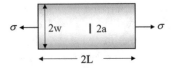

7.2. A single-edge cracked plate made out of ASTM A533B steel is loaded in tension at 93 °C. Plot the given uniaxial stress–strain data and perform a regression analysis based on the Ramberg–Osgood equation. Determine the elastic-plastic J-integral. Will the crack grow in a stable manner? Why? or Why not? Assume plane-strain conditions and necessary assumptions. Let $a = 100\,\text{mm}$, $\nu = 0.30$, $E = 207\,\text{GPa}$, $L = 3w$, $w = 400\,\text{mm}$, $J_{IC} = 1.20\,\text{m MPa}$, and $B = 150\,\text{mm}$.

$\epsilon\ (\times 10^{-3})$	0	1.00	2.24	2.30	5.00	7.50	10	20	40
$\sigma\ (MPa)$	0	381	414	415	450	469	483	519	557

7.3. Repeat Problem 7.2 using the Hollomon equation, $\sigma = k_y \epsilon^m$, for the plastic region. Curve fitting should be performed using this equation for obtaining k_y and m. Assume plane strain conditions and make the necessary assumptions. Compare the results from Problem 7.2. What can you conclude from these results? Assume a contour as shown in Fig. 7.6 with $y_{23} = 70\,\text{mm}$. [Solution: $J_I = 1.86\,\text{m MPa}$.]

7.4. A steel plate having a single-edge crack is loaded in tension at room temperature. Calculate **(a)** the load-line displacement (μ) and **(b)** the crack opening displacement (δ) that corresponds to a point on the resistance curve (not included). Assume plane-strain conditions and use the following data: $J_{IC} = 1.20\,\text{m MPa}$, $E = 206,850\,\text{MPa}$, and $\sigma = 500\,\text{MPa}$. Variables: $n = 0.3$, $h_1 = 0.523$, $h_2 = 1.93$, $h_3 = 3.42$, $B = 150\,\text{mm}$, $w = 400\,\text{mm}$, $L = 1.20\,\text{m}$, and $a = 0.1\,\text{m}$.

7.5. Determine the strain hardening exponent for a steel with $\sigma_o = 400\,\text{MPa}$, $E = 207\,\text{GPa}$. Assume that it obeys the Hollomon equation, $\sigma = k_y \epsilon^m$, with $k_y = 700\,\text{MPa}$. Consider the maximum plastic stress in your calculations. [Solution: $n = 0.0895$.]

7.6. **(a)** Derive an expression for the J-integral ratio, $J_p/J_e = f(\sigma/\sigma_o)$, using Rice model. **(b)** Plot the resultant expression for a remote stress ratio range $0 \le \sigma/\sigma_0 \le 1$. Explain the resultant plot.

7.7. (a) Calculate the total J-integral (J_I) for a 2024 *Al-alloy* plate containing a single-edge crack under plane stress conditions. **(b)** Plot the $\sigma = f(\epsilon)$ and $J = (\sigma)$. Use the following data to carry out all calculations: $a = 1.40$ mm, $w = 19$ mm, $B = 0.8$ mm, $L = 10$ cm, $\sigma_o = 64$ MPa, $E = 72,300$ MPa, $\alpha' = 0.35$, $n = 5$, and $F = 1.01$ kN. [Solution: $J_I = 539.46$ J/m^2]

7.8. The plastic J-integral (J_p) for some configurations can be defined by $J = \eta W/(Bb)$, where W = absorbed energy, Bb = cross-sectional area, $b = (w - a)$ = ligament, and η = constant. This integral can then be separated into elastic and plastic components. For pure tension,

$$J = J_e + J_p = \frac{K_I^2}{E'} + \frac{\eta_p W_p}{Bb}$$

Consider a strain hardenable material and a specimen with unit thickness B. If the plastic load is defined by

$$P = C\mu_p$$

where C is the compliance and n is the strain hardening exponent, then the load displacement and *J-P* profiles are schematically shown below:

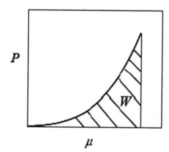
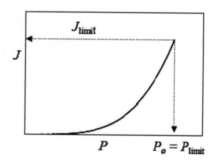

Derive an expression for η_p.

References

1. T.L. Anderson, *Fracture Mechanics: Fundamentals and Applications*, Chaps. 3 and 9, 3rd edn. (Taylor & Francis Group, LLC/CRC Press, Boca Raton, 2005)
2. J.A. Begley, J.D. Landes, American Society for Testing Materials (ASTM), STP 514 (1972)
3. J.A. Begley et al., ASTM STP 560, (1974), p. 155
4. C.P. Cherepanov, Crack propagation in continuous media. Appl. Math. Mech. **31**, 476–488 (1967)

5. M.S. Dadhah, A.S. Kobayashi, *Further Studies of the HRR Fields of a Moving Crack: An Experimental Analysis*, Office of the Chief of Naval Research, Contract N00014-85-0187, Technical Report No. UWA/DME/TR-88/61 (1988)
6. J.W. Dally, W.F. Riley, *Experimental Stress Analysis*, 3rd edn. (McGraw-Hill, Inc., New York, 1991)
7. J. Galkiewicz, M. Graba, *Algorithm for determination of* $\tilde{\sigma}_{ij}(n, \theta)$, $\tilde{\xi}_{ij}(n, \theta)$, $\tilde{\mu}_i(n, \theta)$, $d_n(n)$ *and* $I_n(n)$ *functions in Hutchinson-Rice-Rosengren solution and its 3D generalization*. J. Theor. Appl. Mech. **44**(1), 19–30 (2006)
8. M.Y. He, J.W. Hutchinson, The penny-shaped crack and the plane strain crack in an infinite body of power-law material. Trans. ASME: J. Appl. Mech. **48**, 830–840 (1981)
9. M.Y. He, J.W. Hutchinson, *Bounds for Fully Plastic Crack Problems for Infinite Bodies*, ASTM STP 803, vol. I (1983), pp. 277–290
10. K. Hellan, *Introduction to Fracture Mechanics* (McGraw-Hill Book Company, New York, 1984)
11. J.H. Hollomon, Tensile deformation. Trans. Am. Inst. Min. Metall. Pet. Eng. **162**, 268–290 (1945)
12. J.W. Hutchinson, Singular behavior at the end of a tensile crack in a hardening material. J. Mech. Phys. Solids **16**, 13–31 (1968)
13. J.W. Hutchinson, *Fundamentals of the phenomenological theory of nonlinear fracture mechanics*. J. Appl. Mech. **50**, 1042–1051 (1983)
14. J.W. Hutchinson, P.C. Paris, American Society for Testing Materials (ASTM), STP 668 (1979), p. 37
15. B.S.J. Kang, A.S. Kobayashi, *J-Estimation Procedure based on Moiré Interferometry Data*, Report UWA/DME/TR-87/58, Office of Naval Research (ONR) (August 1987)
16. V. Kumar, C.F. Shih, *Fully Plastic Crack Solutions, Estimates Scheme and Stability Analyses for Compact Specimens*, American Society for Testing Materials (ASTM), STP 700 (1980), p. 406
17. V. Kumar, M.D. German, C.F. Shih, *An Engineering Approach for Elastic-Plastic Fracture Analysis*, General Electric Company, Report Project 1237-1 (July 1981)
18. V. Kumar, M.D. German, C.F. Shih, Elastic-plastic and fully plastic analysis of crack initiation, stable growth, and instability in flawed cylinders, in *Elastic-Plastic Fracture: Second Symposium*, ASTM STP 803, vol. I (1983), pp. 306–353
19. H.W. Liu, *On the fundamentals basis of fracture mechanics*. Eng. Fract. Mech. **17**(5), 425–438 (1983)
20. F.A. McClintock, *Plasticity aspects of fracture,* in Fracture: An Advanced Treatise, ed. by H. Liebowitz, Academic Press, New York, Vol. III (1971) pp. 47–225
21. W. Ramberg, W.R. Osgood, *Description of Stress-Strain Curves by Three Parameters*, Technical Note No. 902, National Advisory Committee For Aeronautics, Washington, DC (1943)
22. J.R. Rice, A path independent integral and the approximate analysis of strain concentrations by notches and cracks. J. Appl. Mech. **35**, 379–386 (1968)
23. J.R. Rice, Mathematical analysis in the mechanics of fracture, in *Fracture*, vol. II, ed. by H. Liebowitz (Academic, New York, 1968), p. 191
24. J.R. Rice, G.F. Rosengren, Plane strain deformation near a crack tip in a power-law hardening material. J. Mech. Phys. Solids **16** (1968)
25. D.P. Rooke, J.C. Cartwright, *Compendium of Stress Intensity Factors* (Her Majesty's Stationary Office, London, 1976)
26. M.H. Sadd, *Elasticity: Theory, Applications and Numerics*, 2nd edn. (Elsevier Inc., New York, 2009), p. 161
27. A. Saxena, *Nonlinear Fracture Mechanics for Engineers* (CRC Press LLC, New York, 1998), p. 126
28. C.F. Shih, *Tables of Hutchinson-Rice-Rosengren Singular Field Quantities*, Division of Engineering, Brown University, Province, RI (June 1983)
29. G.C. Sih, *Handbook of Stress Intensity Factors* (Institute of Fracture and Solid Mechanics, Bethlehem, PA, USA, 1973)

30. C.F. Shih, J.W. Hutchinson, Fully plastic solutions and large-scale yielding estimate for plane stress crack problems. Trans. Am. Soc. Mech. Eng.: J. Eng. Mater. Technol. Ser. H **98**(4), 289–295 (1976)
31. C.F. Shih et al., ASTM STP 668 (1979), p. 65
32. H. Tada et al., *The Stress Analysis of Cracks Handbook* (Del Research Corporation, Hellertown, PA, 1973)

Mixed-Mode Fracture Mechanics

8

8.1   Introduction

Practical structures are not only subjected to tension but also experience shear and torsion loading leading to a mixed-mode interaction. Correspondingly, the stress state ahead of a crack is frequently based on mixed-mode I/II type of interactions, which designate the amplitude of the crack tip stresses because of skew-symmetric loading. Problems of this type are encountered in multiphase materials such as welded structures, adhesive joints, composite materials, plain and reinforced concrete structures, bridges, aircrafts, and so forth. A mixed-mode interaction can also arise when crack branching occurs, that is, when a crack changes direction in which the classical energy balance of Griffith can no longer be carried out in a simple manner since cracking is not collinear as it has been assumed in previous chapters.

In addition, cracks may develop in the skin of aircraft fuselages and can be subjected to mixed-mode loading. In general, crack initiation and growth must be correlated with the governing stress intensity factors in a complex state of stress. This means that the crack tip fields are inherently three dimensional with varying distribution through the thickness of the solid component. Therefore, the field equations must be determined for a better understanding of mixed-mode fracture mechanics [1–7]. Nonetheless, cracks loaded in tension and shear which may exhibit crack branching or kinking must be characterized using traditional singular and high-order terms in the stress field. The mixed-mode analysis of branched cracks requires the determination of stress intensity factors for the original and branched crack parts in terms of the stress field surrounding the crack tip [1–28]. Therefore, an optimal kink angle and an additional T-stress component can be determined based on far-field boundary conditions in homogeneous or heterogeneous materials.

© Springer International Publishing Switzerland 2017
N. Perez, *Fracture Mechanics*, DOI 10.1007/978-3-319-24999-5_8

8.2　Elastic State of Stresses

Among many structural parts containing skewed cracks subjected to asymmetry loading inducing inherent mixed-mode fracture, Fig. 8.1 shows two ideal specimen configurations containing symmetric cracks with respect to specimen dimensions and directions of the remote applied stress loading. In order to develop mathematical procedures for characterizing the mixed-mode crack behavior and for determining the magnitude of the local stress intensity factors, it is assumed that the specimens are subjected to quasi-static and steady loadings that induce crack extension, and subsequently, crack propagation occurs when the local crack tip principal stresses reach a critical state. These specimens are symmetric representing ideal models of real situations encountered in structural components subjected to inherent internal stresses that weaken the structural integrity and eventually provoke crack propagation.

Using the linear superposition of stresses in rectangular or polar coordinates, the elastic state of stress at the crack tip is obtained very easily. For a mixed-mode I/II shown in Fig. 8.1a, the total stress in rectangular coordinates is the sum of the stresses of each loading mode component derived in Chap. 4. For a brittle, isotropic, homogeneous, and semi-infinite plate, the stresses in Cartesian coordinates are

$$
\sigma_x = \frac{K_I}{\sqrt{2\pi r}} \cos\frac{\theta}{2} \left[1 - \sin\frac{\theta}{2}\sin\frac{3\theta}{2}\right] - \frac{K_{II}}{\sqrt{2\pi r}} \sin\frac{\theta}{2} \left[2 + \cos\frac{\theta}{2}\cos\frac{3\theta}{2}\right]
$$

$$
\sigma_y = \frac{K_I}{\sqrt{2\pi r}} \cos\frac{\theta}{2} \left[1 + \sin\frac{\theta}{2}\sin\frac{3\theta}{2}\right] + \frac{K_{II}}{\sqrt{2\pi r}} \sin\frac{\theta}{2}\cos\frac{\theta}{2}\cos\frac{3\theta}{2}
$$

$$
\tau_{xy} = \frac{K_I}{\sqrt{2\pi r}} \cos\frac{\theta}{2}\sin\frac{\theta}{2}\cos\frac{3\theta}{2} + \frac{K_{II}}{\sqrt{2\pi r}} \cos\frac{\theta}{2} \left[1 - \sin\frac{\theta}{2}\sin\frac{3\theta}{2}\right] \qquad (8.1)
$$

Considering a liner elastic two-dimensional analysis in Cartesian coordinates, the near-tip displacements can be determined by integrating the strains, but they

Fig. 8.1 Mixed-mode interactions. (**a**) Plate and (**b**) solid cylinder

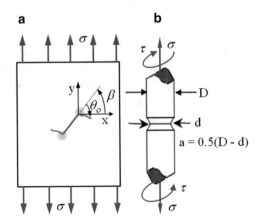

can also be defined by the sum of Eqs. (4.18) and (4.37). Hence, the displacements along the x, y, and z coordinates in an ideally elastic solid materials under uniform and quasi-static loading are

$$\mu_x = \frac{2K_I}{E} \sqrt{\frac{r}{2\pi}} \cos \frac{\theta}{2} \left[(1 - v) + (1 + v) \sin^2 \frac{\theta}{2} \right]$$

$$+ \frac{2(1 + v)K_{II}}{E} \sqrt{\frac{r}{2\pi}} \sin \frac{\theta}{2} \left[\frac{2}{1 + v} + \cos^2 \frac{\theta}{2} \right]$$

$$\mu_y = \frac{2K_I}{E} \sqrt{\frac{r}{2\pi}} \sin \frac{\theta}{2} [2 - (1 + v)] \cos^2 \frac{\theta}{2} \qquad (8.2)$$

$$+ \frac{2K_{II}}{E} \sqrt{\frac{r}{2\pi}} \cos \frac{\theta}{2} \left[(v - 1) + (1 + v) \sin^2 \frac{\theta}{2} \right]$$

$$\mu_z = - \frac{2vBK_I}{E} \sqrt{\frac{r}{2\pi}} \cos \frac{\theta}{2} + \frac{2vBK_{II}}{E} \sqrt{\frac{r}{2\pi}} \sin \frac{\theta}{2}$$

In general, the transformation of rectangular into polar coordinates requires use of the transformation relations between the stress components of the two coordinate systems. In light of this, the transformation of Eq. (8.2) requires that the derivatives with respect to x and y be carried out in terms of r and θ. In order to avoid a lengthy mathematical manipulation, simply adopt the polar stress components from Chap. 4, and add them up for the current mixed-mode I/II problem.

For polar coordinates, adding Eqs. (4.58) and (4.59) yields

$$\sigma_r = \frac{K_I}{\sqrt{2\pi r}} \left(\frac{5}{4} \cos \frac{\theta}{2} - \frac{1}{4} \cos \frac{3\theta}{2} \right) + \frac{K_{II}}{\sqrt{2\pi r}} \left(-\frac{5}{4} \sin \frac{\theta}{2} + \frac{3}{4} \sin \frac{3\theta}{2} \right)$$

$$\sigma_\theta = \frac{K_I}{\sqrt{2\pi r}} \left(\frac{3}{4} \cos \frac{\theta}{2} + \frac{1}{4} \cos \frac{3\theta}{2} \right) + \frac{K_{II}}{\sqrt{2\pi r}} \left(-\frac{3}{4} \sin \frac{\theta}{2} - \frac{3}{4} \sin \frac{3\theta}{2} \right) \qquad (8.3)$$

$$\tau_{r\theta} = \frac{K_I}{\sqrt{2\pi r}} \left(\frac{1}{4} \sin \frac{\theta}{2} + \frac{1}{4} \sin \frac{3\theta}{2} \right) + \frac{K_{II}}{\sqrt{2\pi r}} \left(\frac{1}{4} \cos \frac{\theta}{2} + \frac{3}{4} \cos \frac{3\theta}{2} \right)$$

Rearranging Eq. (8.3) yields [1]

$$\sigma_r = \frac{2}{\sqrt{2\pi r}} \left[K_I (3 - \cos \theta) \cos \frac{\theta}{2} + K_{II} (3 \cos \theta - 1) \right] \qquad (8.4a)$$

$$\sigma_\theta = \frac{2}{\sqrt{2\pi r}} \left[K_I (1 + \cos \theta) \cos \frac{\theta}{2} + 3K_{II} \sin \theta \cos \frac{\theta}{2} \right] \qquad (8.4b)$$

$$\tau_{r\theta} = \frac{2}{\sqrt{2\pi r}} \left[K_I \sin \theta \cos \frac{\theta}{2} + K_{II} (3 \cos \theta - 1) \cos \frac{\theta}{2} \right] \qquad (8.4c)$$

Now the displacements are obtained from Eqs. (4.63) and (4.64)

$$
\mu_r = \frac{K_I}{4G} \sqrt{\frac{r}{2\pi}} \left[(2\kappa - 1) \cos\frac{\theta}{2} - \cos\frac{3\theta}{2} \right]
$$

$$
+ \frac{K_{II}}{4G} \sqrt{\frac{r}{2\pi}} \left[-(2\kappa - 1) \sin\frac{\theta}{2} + 3\sin\frac{3\theta}{2} \right]
$$

$$
\mu_\theta = \frac{K_I}{4G} \sqrt{\frac{r}{2\pi}} \left[-(2\kappa - 1) \sin\frac{\theta}{2} + \sin\frac{3\theta}{2} \right] \tag{8.5}
$$

$$
+ \frac{K_{II}}{4G} \sqrt{\frac{r}{2\pi}} \left[-(2\kappa - 1) \cos\frac{\theta}{2} + 3\cos\frac{3\theta}{2} \right]
$$

$$
\mu_z = -\frac{2\upsilon B K_I}{E} \sqrt{\frac{r}{2\pi}} \cos\frac{\theta}{2} + \frac{2\upsilon B K_{II}}{E} \sqrt{\frac{r}{2\pi}} \sin\frac{\theta}{2}
$$

Here, plane stress and plane-strain conditions are controlled by the plane factor κ defined by

$$
\kappa = \frac{3 - \upsilon}{1 + \upsilon} \quad \text{for plane stress} \tag{8.6}
$$

$$
\kappa = 3 - 4\upsilon \quad \text{for plane strain} \tag{8.7}
$$

where υ is the Poisson's ration. Recall that plane stress is for thin bodies with $\sigma_z = 0$ and plane strain is for thick bodies with $\sigma_z > 0$.

From Chap. 4, the stresses and displacements for the specimen configuration shown in Fig. 8.1b are

$$
\sigma_r = \frac{2K_I}{\sqrt{2\pi r}} (3 - \cos\theta) \cos\frac{\theta}{2} + \frac{K_{III}}{2} \sqrt{\frac{2}{\pi r}} \sin\frac{\theta}{2}
$$

$$
\sigma_\theta = \frac{2K_I}{\sqrt{2\pi r}} (1 + \cos\theta) \cos\frac{\theta}{2} + \frac{K_{III}}{2} \sqrt{\frac{2}{\pi r}} \cos\frac{\theta}{2} \tag{8.8}
$$

$$
\mu_z = -\frac{2\upsilon B K_I}{E} \sqrt{\frac{r}{2\pi}} \cos\frac{\theta}{2} + \frac{K_{III}}{G} \sqrt{\frac{2r}{\pi}} \sin\frac{\theta}{2}
$$

For mode III only in Cartesian coordinates,

$$
\mu_z = \frac{K_{III}}{G} \sqrt{\frac{2r}{\pi}} \sin\frac{n\theta}{2}
$$

$$
\tau_{rz} = \frac{K_{III}}{2} \sqrt{\frac{2}{\pi r}} \sin\frac{n\theta}{2} \tag{8.9}
$$

$$
\tau_{\theta z} = \frac{K_{III}}{2} \sqrt{\frac{2}{\pi r}} \cos\frac{n\theta}{2}
$$

Therefore, the stress state ahead of the crack tip is three dimensional, and the concept of plane stress or plane strain does not apply in this case.

8.3 Strain Energy Methodology

8.3.1 Mode-Mode I and II

This section includes the strain energy release rate associated with mixed-mode stress intensity factor interactions during elastic deformation of a solid containing an inclined crack. This is the available energy for crack growth, but it must overcome the material resistance, such as surface energy, energy dissipation, and plastic work.

The main objective in this section is to develop fracture criteria based on the strain energy release rate for mixed I and II interaction.

In principle, assume that the basic three modes interact on an elastic component so that the elastic J-integral and the elastic strain energy release rate become equal. The J-integral is used as a critical property for determining the onset of stable crack growth. Thus, Eq. (6.61) for the elastic mixed-mode interaction is applicable for stable crack motion in its tangent plane

$$J_i = G_i = G_I + G_{II} + G_{III} \tag{8.10}$$

$$G_i = \frac{K_I^2}{E'} + \frac{K_{II}^2}{E'} + \frac{(1+v)\,K_{III}^2}{E} \tag{8.11}$$

where $E' = E$ for plane stress
$E' = E/\left(1 - v^2\right)$ for plane strain

Recall that for pure mode I loading at fracture (crack propagation), the plane-strain fracture toughness expression is

$$G_{IC} = G_I = \frac{K_{IC}^2}{E'} \tag{8.12}$$

Substituting Eq. (8.12) into (8.11) gives the mixed-mode fracture criterion based on the strain energy release rate G-criterion

$$K_{IC}^2 = K_I^2 + K_{II}^2 + \frac{E'\,(1+v)\,K_{III}^2}{E} \quad \text{(G-criterion)} \tag{8.13}$$

This equation indicates that any combination of the stress intensity factors may give the value for fracture toughness K_{IC}.

Now, consider the mixed-mode I/II configuration shown in Fig. 8.1a, where the stress tensor components are defined by

$$\sigma_x = \sigma \sin^2 \beta \tag{8.14a}$$

$$\sigma_y = \sigma \cos^2 \beta \tag{8.14b}$$

$$\tau_{xy} = \sigma \sin \beta \cos \beta \tag{8.14c}$$

and the applied stress intensity factors are

$$K_I = \sigma_y \sqrt{\pi a} = \sigma \sqrt{\pi a} \cos^2 \beta \tag{8.15}$$

$$K_{II} = \tau_{xy} \sqrt{\pi a} = \sigma \sqrt{\pi a} \sin \beta \cos \beta \tag{8.16}$$

The G-criterion as a circle model: In this case, the G-criterion, Eq. (8.13), becomes the equation of a circle with radius K_{IC}

$$K_I^2 + K_{II}^2 = K_{IC}^2 \quad \text{(circle)} \tag{8.17}$$

This fracture criterion can be named because of the nature of Eq. (8.17) defining the mixed-mode I/II interaction.

Inserting Eqs. (8.15) and (8.16) into (8.17) yields the fracture stress as a function of plane-strain fracture toughness and inclined crack angle

$$\sigma_f = \frac{K_{IC}}{\sqrt{\pi a}} \frac{1}{\cos^2 \beta} = \frac{K_{IC}}{\sqrt{\pi a}} \sec^2 \beta \quad \text{(circle)} \tag{8.18}$$

This is a fundamental equation for predicting the plane-strain fracture toughness where $\cos^2 \beta$ is a correction factor.

Figure 8.2 shows how the fracture stress, Eq. (8.18), varies with increasing incline angle at a fixed crack length and various levels of the plane-strain fracture toughness. Denote how σ_f under the mixed-mode interaction I/II strongly depends on $\beta > 40°$ and also on the K_{IC} value.

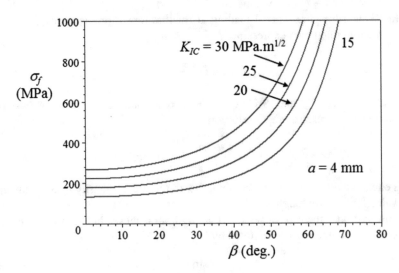

Fig. 8.2 Effect of inclined angle β on fracture stress for a fixed crack length and varying fracture toughness

Assume that a solid has a small or large degree of ductility so that the prediction on crack growth behavior due to mixed-mode loading conditions must include crack instability dominated by tensile stress and shear stress ahead of the crack tip. According to the above mathematical analysis, mode I loading seems to dominate crack growth after some crack extension has occurred. This is indirectly predicted by Eq. (8.13) which is based on the Griffith strain energy release rate (G-criterion) approach for a small amount of crack extension along the original crack plane at an inclined angle β.

In reality, the main goal in this section is to show that crack growth in a self-similar manner is no longer the predictable crack behavior; instead crack growth takes place at a fracture angle θ_o. This is accomplished next.

Fracture angle in a circle model: Consider the incline crack shown in Fig. 8.1a and the tangential stress σ_θ. For the mixed-mode defined by Eq. (8.17), solve Eqs. (4.58b) and (4.59b) for K_I and K_{II}, respectively, and substitute them into (8.17) and then solve for σ_θ. The resultant expression is of the form

$$\sigma_\theta = \frac{K_{IC}}{4\sqrt{\pi r}} \left[\left(3\cos\frac{\theta}{2} + \cos\frac{3\theta}{2} \right)^{-2} + \left(\sin\frac{\theta}{2} + \sin\frac{3\theta}{2} \right)^{-2} \right]^{-1/2} \tag{8.19}$$

For pure mode I, σ_θ, $\partial\sigma_\theta/\partial\theta = 0$ and $\theta = \theta_o$ at fracture, respectively, are

$$\sigma_\theta = \frac{K_{IC}}{4\sqrt{\pi r}} \left[\left(3\cos\frac{\theta}{2} + \cos\frac{3\theta}{2} \right)^{-2} \right]^{-1/2} = \frac{K_{IC}}{4\sqrt{\pi r}} \left(3\cos\frac{\theta}{2} + \cos\frac{3\theta}{2} \right)$$

$$\left.\frac{\partial\sigma_\theta}{\partial\theta}\right|_{\theta=\theta_o} = -\frac{3}{2}\sin\frac{\theta}{2} - \frac{3}{2}\sin\frac{3\theta}{2} = 0$$

$$\theta_o = 0 \text{ (crack in the x-direction)}$$

Therefore, crack propagation occurs in a self-similar manner since $\theta_o = 0$ as per G-criterion.

For pure mode II, $\partial\sigma_\theta/\partial\theta = 0$ and $\theta = \theta_o$ at fracture gives

$$\sigma_\theta = \frac{K_{IC}}{4\sqrt{\pi r}} \left[\left(\sin\frac{\theta}{2} + \sin\frac{3\theta}{2} \right)^{-2} \right]^{-1/2} = \frac{K_{IC}}{4\sqrt{\pi r}} \left(\sin\frac{\theta}{2} + \sin\frac{3\theta}{2} \right)$$

$$\left.\frac{\partial\sigma_\theta}{\partial\theta}\right|_{\theta=\theta_o} = \cos\frac{\theta_o}{2} + 3\cos\frac{3\theta_o}{2} = \cos\frac{\theta_o}{2} + 3\cos\frac{\theta_o}{2}\left(1 - 4\sin^2\frac{\theta_o}{2} \right) = 0$$

$$\sin^2\frac{\theta_o}{2} = 1/2$$

$$\theta_o = \pi/2 = 90° \text{ (crack in the y-axis)}$$

Again, crack propagation occurs in a self-similar manner since $\theta_o = 90°$ as per G-criterion.

Real inclined cracks or mixed-mode cracks, on the other hand, have the tendency to become mode I cracks so that $0 < |\theta_o| < 90°$.

The G-criterion as a ellipse model: Dividing K_I and K_{II} in Eq. (8.17) by their respective plane-strain fracture toughness and letting $K_{IIC} = \sqrt{3/2}K_{IC}$ as a practical definition [4] give the equation of an ellipse under mixed-mode I/II interaction:

$$\left(\frac{K_I}{K_{IC}}\right)^2 + \left(\frac{K_{II}}{K_{IIC}}\right)^2 = 1 \qquad \text{(ellipse)} \tag{8.20}$$

$$K_I^2 + \frac{2}{3}K_{II}^2 = K_{IC}^2 \quad \text{(ellipse)} \tag{8.21}$$

In light of the above, Eq. (8.21) can be defined in a general form by introducing an arbitrary constant $C_e \neq 1$ so that

$$K_I^2 + C_e K_{II}^2 = K_{IC}^2 \quad \text{(ellipse)} \tag{8.21a}$$

Figure 8.3 shows the trend of Eqs. (8.17) and (8.21) for a quarter of an ellipse. The elliptical fracture criterion is based on the assumption that crack propagation takes place in a self-similar manner; the mixed-mode crack tends to become a mode I crack when the principal stress σ_1 dominates the stress state [21, 22].

Substituting Eqs. (8.15) and (6.16) into (8.21) gives

$$\sigma_f = \frac{K_{IC}}{\sqrt{\pi a}}\left[\frac{1}{3}(2 + \cos\beta)\right]^{-1/2} \qquad \text{(ellipse)} \tag{8.22}$$

where σ is the critical or fracture stress when $K_I = K_{IC}$.

The advantage of these criteria is that they can be combined with material properties, the tensile and shear strengths, and that there is no restriction on specimen size.

Most safety evaluations and life predictions of cracked components are based on mode I loading, but many engineering structural components are subjected to mixed-mode I/II interaction. Consequently, crack growth may not occur along the crack plane, but at a fracture angle θ_o.

Fig. 8.3 Locus of the *circular* and *elliptic* fracture criteria

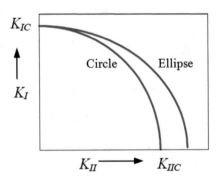

Fracture angle in a ellipse model: Similarly, combining Eqs. (4.58b), (4.59b), and (8.21) yields a slight different σ_θ equation for an elliptical equation describing a fracture criterion. Hence,

$$\sigma_\theta = \frac{K_{IC}}{4\sqrt{2\pi r}} \left[\left(3\cos\frac{\theta}{2} + \cos\frac{3\theta}{2} \right)^2 + \frac{2}{3} \left(\sin\frac{\theta}{2} + \sin\frac{3\theta}{2} \right)^{-2} \right]^{-1/2} \tag{8.23}$$

Letting $\partial\sigma_\theta/\partial\theta = 0$ at fracture for individual modes gives the same results as the G-criterion as a circle model above.

Example 8.1. *A large steel plate has a 8-mm long and inclined central crack at 20°. If the applied stress is 200 MPa, determine whether or not the plate will fracture. The fracture toughness of the steel is $K_{IC} = 30\,MPa\sqrt{m}$.*

Solution. *From Eqs. (8.15) and (8.16),*

$$K_I = \sigma\sqrt{\pi a}\cos^2\beta = (200\,\text{MPa})\left(\sqrt{4\pi \times 10^{-3}\,\text{m}}\right)\cos^2(\pi/9)$$

$$K_I = 19.80\,\text{MPa}\sqrt{\text{m}}$$

$$K_{II} = \sigma\sqrt{\pi a}\sin\beta\cos\beta = (200\,\text{MPa})\left(\sqrt{4\pi \times 10^{-3}\,\text{m}}\right)\sin(\pi/9)\cos(\pi/9)$$

$$K_{II} = 7.21\,\text{MPa}\sqrt{\text{m}}$$

Thus,

$$\sqrt{K_I^2 + K_{II}^2} = 21.07\,\text{MPa}\sqrt{\text{m}} < K_{IC}$$

The fracture stress should be calculated using Eq. (8.18)

$$\sigma_f = \frac{K_{IC}}{\sqrt{\pi a}\cos^2\beta} = \frac{30\,\text{MPa}\sqrt{\text{m}}}{\left(\sqrt{4\pi \times 10^{-3}\,\text{m}}\right)\cos^2(\pi/9)} = 300.07\,\text{MPa}$$

Therefore, the plate will not fracture because $K_{IC} > 21.07\,MPa\sqrt{m}$ and $\sigma < \sigma_f$ for a fixed crack length.

Example 8.2. *A brittle steel plate containing a 6-mm long and inclined central crack (Fig. 8.1a)is loaded in tension. Use both the circular and the elliptical fracture criteria to determine the fracture stress when $\beta = \pi/4$ and $\pi/3$. Given data: $K_{IC} = 40\,MPa\sqrt{m}$ and $E = 206,850\,MPa$.*

Solution. *It is expected that* $\sigma_f^{(c)}$ *(circle)* $\neq$ $\sigma_f^{(e)}$ *(ellipse). Thus, Eqs. (8.18) and (8.22) yield, respectively,*

$$\sigma_f^{(c)} = \frac{K_{IC}}{\sqrt{\pi a} \cos^2 \beta} = \frac{40 \, \text{MPa} \sqrt{\text{m}}}{\sqrt{\pi \, (3 \times 10^{-3} \, \text{m})}. \cos^2 (\pi/4)}$$

$$\sigma_f^{(c)} = 824.05 \, \text{MPa}$$

$$\sigma_f^{(e)} = \sqrt{\frac{3}{\pi a \, (\cos^2 \beta + 2)}} \frac{K_{IC}}{\cos \beta}$$

$$\sigma_f^{(e)} = \sqrt{\frac{3}{\pi \, (3 \times 10^{-3} \, \text{m}) \, [\cos^2 (\pi/4) + 2]}} \frac{(40 \, \text{MPa} \sqrt{\text{m}})}{\cos (\pi/4)}$$

$$\sigma_f^{(e)} = 638.31 \, \text{MPa} > \sigma_f^{(c)}$$

Therefore, the fracture stress values are different as expected.

8.3.2   Mode-Mode I and III

In practice, structures are not only subjected to tension loading (mode I) but to shear (mode II) and torsion (mode III) loadings. When a structural component is subjected to mixed-mode I/III loading, the crack simultaneously experiences a mixed tension and torsion interactions, and as a result, the component is exposed to a mixed-mode cracking failure mode.

In principle, the direction of crack growth and the orientation of the crack tip in a mixed-mode I/III interaction in a cylindrical specimen are expected to be orthogonal due to the nature of the specimen configuration. In such a case, the crack grows in mode I, but the principal stress field is expected to rotate, inducing crack splitting at a critical angle of rotation σR of the principal stress field [29].

In any event, it is expected that crack growth or fracture of elastic and elastic-plastic solids under mixed-mode I/III loading starts along the radial direction. This implies that crack propagation occurs along the crack plane where the maximum tensile and shear stresses must remain perpendicular and parallel to the crack plane, respectively.

The dependency of fracture toughness on the stress-based mixed-mode I/III plastic mixity parameter (M_{p13}) and displacement-based I/III mixity parameter (β_{p13}) of a elastic-plastic solid material is fundamentally inherent in mixed-mode experiments using, for example, symmetric and asymmetric bend specimens. The β_{p13} parameter arises due to the out-of-plane displacement in torsional loading (mode III). This topic can be found elsewhere [30]. Despite that mode I tensile loading is the most common in fracture mechanics, mode II being the shear sling loading and mode III being the anti-plane loading have to be considered to an extent in real engineering situations. In addition, mode I and III and mode II and III interactions have the apparent tendency to be overstresses by the mode I after some crack growth. This means that mode I is the apparent dominant mode in a mixed-mode situation.

Consider the crack configuration for a solid cylinder shown in Fig. 8.1b. For a mixed-mode I/III, the cylinder is subjected to a biaxial loading ahead of the crack tip. Thus, Eq. (8.13) reduces to

$$K_{IC}^2 = K_I^2 + \frac{E'\,(1+v)\,K_{III}^2}{E} \tag{8.24}$$

$$K_{IC}^2 = K_I^2 + \frac{K_{III}^2}{1-v} \qquad \text{for plane strain} \tag{8.25}$$

$$K_C^2 = K_I^2 + (1+v)\,K_{III}^2 \qquad \text{for plane stress} \tag{8.26}$$

Furthermore, denote that K_{IC} is changed to K_C for plane stress. These expressions indicate that crack growth occurs in a self-similar manner or crack motion manifests itself along its tangent plane. This is in accord with the basis for deriving Eq. (8.13). Therefore, the crack fracture angle is $\theta_o = 0$.

8.4 Principle Stress Criterion

The principal stress theory is a fundamental theory of failure for controlling the maximum principal stress reaching the tensile yield strength of brittle materials. In the literature, this theory is also known as normal stress theory.

In principle, this theory has been implemented in mixed-mode fracture mechanics as a fracture criterion because in many practical situations structural components experience tensile and shear loading simultaneously.

The mixed-mode fracture criterion, σ_θ-criterion, is equivalent to the mixed-mode strain energy release rate criterion [3, 5, 6]. The principal stress theory postulates that crack growth takes place in a direction perpendicular to the maximum principal stress. Hence, the fracture criterion requires maximum principal tension stress for opening the crack along its plane.

$$\frac{\partial \sigma_\theta}{\partial \theta} = \tau_{r\theta}\bigg|_{\theta=\theta_o} = 0 \tag{8.27a}$$

$$\frac{\partial^2 \sigma_\theta}{\partial \theta^2} < 0 \quad \text{for } \sigma_\theta > 0 \tag{8.27b}$$

Consider the tension loading that produces a mixed-mode I/II interaction as shown in Fig. 8.1a and that the stress σ_θ in Eq. (8.4) is a principal stress if $\partial \sigma_\theta / \partial \theta = \tau_{r\theta} = 0$. Setting $\tau_{r\theta} = 0$ in Eq. (8.4) makes σ_θ a principal stress, and using Eqs. (8.15) and (8.16) yields

$$K_I \sin \theta_o + K_{II} (3 \cos \theta_o - 1) = 0 \tag{8.28a}$$

$$\sin \theta_o + (3 \cos \theta_o - 1) \tan \beta = 0 \quad \text{for } \beta \neq 0 \tag{8.28b}$$

Using $\sin \theta_o = 2 \sin (\theta_o/2) \cos (\theta_o/2)$ and $(3 \cos \theta_o - 1) = 2[\cos^2 (\theta_o/2) - 2 \sin (\theta_o/2)]$, letting $x = K_I/K_{II}$ in Eq. (8.28b), and solving for θ_o yields [3, 7]

$$2K_{II} \tan^2 (\theta_o/2) - K_I \tan (\theta_o/2) - K_{II} = 0 \tag{8.29a}$$

so that

$$\tan (\theta_o/2) = \frac{1}{4} \frac{K_I}{K_{II}} + \frac{1}{4} \sqrt{\left(\frac{K_I}{K_{II}} \right)^2 + 8} \tag{2.29b}$$

$$\tan (\theta_o/2) = \frac{x}{4} + \sqrt{(x/4)^2 + 1/2} \quad \text{for } x = K_I/K_{II} = \cot \beta \tag{2.29c}$$

$$\theta_o = 2 \arctan \left[\frac{1}{4} \cot \beta + \sqrt{\left(\frac{1}{4} \cot \beta \right)^2 + 1/2} \right] \tag{2.30}$$

For pure mode II loading, Eqs. (8.28a) and (2.29b) give the fracture angle as

$$\cos \theta_o = \frac{1}{3} \quad \rightarrow \quad \theta_o = 70.53° \tag{8.30a}$$

$$\tan (\theta_o/2) = \sqrt{2}/2 \quad \rightarrow \quad \theta_o = 70.53° \tag{8.30b}$$

The maximum principal stress is defined, after manipulating the trigonometric functions in Eq. (8.4), by

$$\sigma_1 = \sigma_\theta (\theta = \theta_o) = \frac{4}{\sqrt{2\pi r}} \cos^3 \frac{\theta_o}{2} \left[K_I \cos \frac{\theta_o}{2} + 3K_{II} \sin \frac{\theta_o}{2} \right] \tag{8.31}$$

For pure mode I loading at fracture with $K_I = K_{IC}$, Eq. (8.31) becomes

$$\sigma_1 = \frac{4K_{IC}}{\sqrt{2\pi r}} \tag{8.32}$$

Equating Eqs. (8.31) and (8.32) yields the fracture criterion as a trigonometric function

$$K_{IC} = K_I \cos^4 \frac{\theta_o}{2} + 3K_{II} \sin \frac{\theta_o}{2} \cos^3 \frac{\theta_o}{2} \quad (\sigma_\theta\text{-criterion}) \tag{8.33}$$

Fig. 8.4 Distribution of b_{ij} as functions of fracture angle θ_o

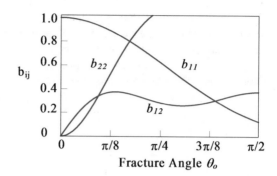

Furthermore, the fracture toughness ratio can be determined by substituting the fracture angle $\theta_o = -70.53$ into Eq. (8.33) for pure mode II at fracture. Thus, $K_{II} = K_{IIC}$ and

$$K_{IIC} = \sqrt{\frac{3}{4}}K_{IC} \simeq \frac{13}{15}K_{IC} \simeq 1.15K_{IC} \quad (\sigma_\theta\text{-criterion}) \tag{8.34}$$

For convenience, squaring Eq. (8.33), manipulating, and simplifying the trigonometric identities yield

$$K_{IC}^2 = b_{11}K_I^2 - 2b_{12}K_IK_{II} + b_{22}K_{II}^2 \quad (\sigma_\theta\text{-criterion}) \tag{8.35}$$

where the b_{ij} are defined by

$$b_{11} = \frac{1}{8}(1 + \cos\theta_o)^3 \tag{8.36a}$$

$$b_{12} = \frac{3}{8}\sin\theta_o(1 + \cos\theta_o)^2 \tag{8.36b}$$

$$b_{22} = \frac{9}{8}\sin^2\theta_o(1 + \cos\theta_o) \tag{8.36c}$$

The distribution of each constant b_{ij} is shown in Fig. 8.4.

8.5 Strain Energy Density Factor

Sih [2] proposed this criterion using the strain energy density factor (S) for a two-dimensional stress field. This criterion states that the initial crack growth takes place in the direction along which the strain energy-density factor reaches a minimum stationary value so that

$$\frac{\partial S}{\partial \theta} = 0 \quad \text{for } \theta = \theta_o \tag{8.37}$$

$$\frac{\partial^2 S}{\partial \theta^2} > 0 \quad \text{for } -\pi < \theta_o < \pi \tag{8.38}$$

This theory can predict crack propagation in an arbitrary direction, and it states that crack propagation occurs when $S = S_c$ at $\theta = \theta_o$, where S_c is the critical strain energy density factor and θ_o is the fracture angle.

The development of S-criterion is achieved by inserting the previously derived stresses and displacements into the strain energy density equations (with unit thickness). This criterion is related to the work done by the an external load, which is the stored strain energy density in solids. Thus, Eq. (6.58) in rectangular and polar coordinates becomes

$$\frac{dW}{dV} = \frac{1}{2E} \left(\sigma_x^2 + \sigma_y^2 + \sigma_z^2 \right) \tag{8.39}$$

$$-\frac{v}{E} \left(\sigma_x \sigma_y + \sigma_y \sigma_z + \sigma_z \sigma_x \right)$$

$$+\frac{1}{E} (1 + v) \left(\tau_{xy}^2 + \tau_{yz}^2 + \tau_{zx}^2 \right)$$

$$\frac{dW}{dA} = \frac{1}{2} \left[\sigma_r \frac{\partial \mu_r}{\partial r} + \sigma_\theta \left(\frac{\mu_r}{r} + \frac{1}{r} \frac{\partial \mu_\theta}{\partial \theta} \right) \right] \tag{8.40}$$

$$+\frac{1}{2} \tau_{r\theta} \left(\frac{1}{r} \frac{\partial \mu_r}{\partial \theta} + \frac{1}{r} \frac{\partial \mu_\theta}{\partial r} - \frac{\mu_\theta}{r} \right)$$

Here, V is the volume and $dA = rd\theta dr$ is the inner area (region) of the plastic zone model shown in Fig. 8.5 [2]. This region should not have its sides coincide with the free crack surface which corresponds to the trivial case when $S = 0$.

The strain energy density factor is derived by combining Eqs. (8.5) and (8.40). Thus,

$$S = r \frac{dW}{dA} = a_{11} K_I^2 + 2a_{12} K_I K_{II} + a_{22} K_{II}^2 \tag{8.41}$$

where a_{ij} = constants
  $G = E / [2 (1 + v)]$ = Shear Modulus
  E = Modulus elasticity in tension
  v = Poisson's ratio
The constants a_{ij} are

$$a_{11} = \frac{1}{16G} [(1 + \cos \theta) (\kappa - \cos \theta)]$$

$$a_{12} = \frac{1}{16G} \sin \theta [2 \cos \theta - (\kappa - 1)] \tag{8.42}$$

$$a_{22} = \frac{1}{16G} [(\kappa + 1) (1 - \cos \theta) + (1 + \cos \theta) (3 \cos \theta - 1)]$$

Figure 8.6 shows the distribution of the constants a_{ij} as a function of θ for Poisson's ratio of $v = 0.3$ and modulus of elasticity of $E = 70\,GPa$ under plane-strain condition.

Fig. 8.5 Stress field of an element at the circular plastic zone boundary in polar coordinates [2]

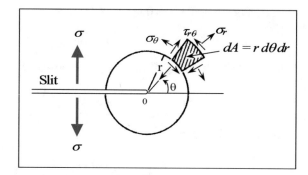

Fig. 8.6 Distribution of a_{ij} for a material having $v = 0.3$ under plane-strain condition

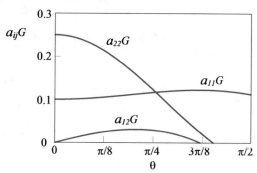

For pure mode II, $K_I = 0$ and Eq. (8.41) yields

$$S_{II} = a_{22}K_{II}^2 \qquad (8.43)$$

Letting $\partial S/\partial\theta = 0$ at $\theta = -\theta_o$ yields

$$\cos\theta_o = \frac{\kappa - 1}{6} \qquad (8.44)$$

$$\theta_o = -\arccos\frac{\kappa - 1}{6} \qquad (8.45)$$

and Eq. (8.43) becomes

$$S_{IIC} = \frac{(v + 1)\left(-\kappa^2 + 14\kappa - 1\right)}{96E}K_{IIC}^2 \qquad (8.46)$$

For instance, if Poisson's ratio is $v = 0.3$, then the fracture angle under plane conditions become

$$\theta_o = -\arccos\frac{1 - v}{3} = -76.51° \qquad \text{for plane stress} \qquad (8.47)$$

$$\theta_o = -\arccos\frac{1 - 2v}{3} = -82.34° \qquad \text{for plane strain} \qquad (8.48)$$

On the other hand, if $K_{II} = 0$ in Eq. (8.41), then

$$S_I = a_{11}K_I^2 \qquad (8.49)$$

Letting $\partial S/\partial \theta = 0$ in Eq. (8.49) at $\theta = -\theta_o$ gives

$$\cos \theta_o = \frac{\kappa - 1}{2} \qquad (8.50)$$

Thus, Eq. (8.49) becomes

$$S_{IC} = \frac{(\nu + 1)(\kappa + 1)^2}{32E}K_{IC}^2 \qquad (8.51)$$

Letting $\partial S/\partial \theta = 0$ at $\theta = -\theta_o$ in Eq. (8.41) for $K_I > 0$ and $K_{II} > 0$ gives the equation for the fracture angle

$$
\begin{aligned}
0 = {} & [(1 - \kappa + 2\cos\theta_o)\sin\theta_o]K_I^2 \\
& + 2\left[(2\cos\theta_o - \kappa + 1)\cos\theta_o - 2\sin^2(\theta_o)\right]K_I K_{II} \\
& + [(\kappa - 1 - 6\cos\theta_o)\sin\theta_o]K_{II}^2
\end{aligned}
\qquad (8.52)
$$

Simplifying Eq. (8.52) gives

$$
\begin{aligned}
0 = {} & [(1 - \kappa + 2\cos\theta_o)\sin\theta_o] \\
& + 2\left[(2\cos\theta_o - \kappa + 1)\cos\theta_o - 2\sin^2(\theta_o)\right]\left(\frac{K_{II}}{K_I}\right) \\
& + [(\kappa - 1 - 6\cos\theta_o)\sin\theta_o]\left(\frac{K_{II}}{K_I}\right)^2
\end{aligned}
\qquad (8.53)
$$

This equation can be used to determine θ_o knowing K_{II}/K_I values. The S-criterion for a mixed-mode I/II can be determined by inserting Eq. (8.51) into (8.41). Thus,

$$K_{IC}^2 = \frac{32E}{(\nu + 1)(\kappa + 1)^2}\left[a_{11}K_I^2 + 2a_{12}K_I K_{II} + a_{22}K_{II}^2\right]_{\theta=\theta_o} \qquad (8.54)$$

Applying Eq. (8.54) at fracture under mode II loading yields

$$K_{IIC}^2 = \frac{(\nu + 1)(\kappa + 1)^2}{32Ea_{22}}K_{IC}^2 \qquad (8.55)$$

If

$$a_{22} = \frac{(\nu + 1)\left(-\kappa^2 + 14\kappa - 1\right)}{96E} \qquad (8.56)$$

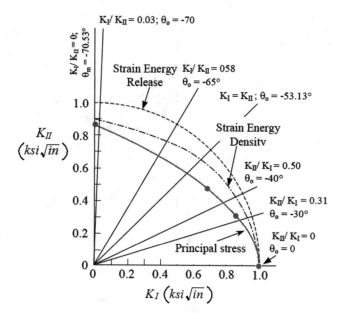

Fig. 8.7 Some mixed-mode I/II fracture criteria [7]

Then,

$$K_{IIC} = \sqrt{\frac{3(\kappa + 1)^2}{14\kappa - \kappa^2 - 1}} K_{IC} \tag{8.57}$$

For plane-strain conditions along with $v = 1/3$ and $\kappa = 3 - 4v$, Eq. (8.57) yields exactly the same result as predicted by the principal stress theory given by Eq. (8.34).

The numerical results of Eq. (8.54) along with other criteria for negative values of θ_o (tension) and positive applied stress are shown in Fig. 8.7 [7].

On the other hand, Fig. 8.8 shows the theoretical relationship between the fracture and inclined angles for tension loading [2].

Note that these numerical results indicate that both θ_o and β theoretically depend on the Poisson's ratio (v) level. Hence, the higher v, the higher θ_o at fixed value of β.

It is usually assumed that crack propagation occurs in a self-similar manner, but in reality this takes place at a fracture angle (θ_o) as shown numerically in Fig. 8.8 and experimentally illustrated in Table 7.1.

In addition, notice the remarkable agreement on the fracture angle θ_o between experimental and theoretical results for Plexiglas given in Table 8.1 [2]. This illustrates the usefulness of these criteria for predicting the fracture angle in elastic solids.

Example 8.3. *A large 2024 Al-alloy plate containing a central crack inclined at an angle β is subjected to a combined mode I/II loading. The plate fractures at a tensile*

Fig. 8.8 Theoretical results for an inclined central crack uniformly loaded in tension mode [2]

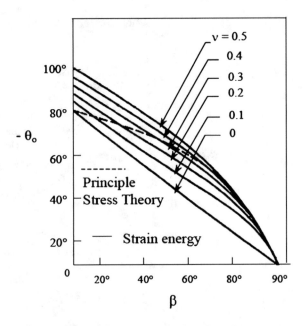

Table 8.1 Experimental and theoretical values of the fracture angle for plexiglas plates with a crack length $a = 25.4$ mm each and w, L $\gg$ a (dimensions: 228.60 mm $\times$ 4570.20 mm $\times$ 4.76 mm) [2]

β	+30.0°	+40.0°	+50.0°	+60.0°	+70.0°	+80.0°
θ_o (Exp.)	−62.4°	−55.6°	−51.6°	−43.1°	−30.7°	−17.3°
σ_θ-criterion	−60.2°	−55.7°	−50.2°	−43.2°	−33.2°	−19.3°
S-criterion	−63.5°	−56.7°	−49.5°	−41.5°	−31.8°	−18.5°
G-criterion	−60.0°	−50.0°	−40.0°	−30.0°	−20.0°	−10.0°

stress of $\sigma_y = 138$ MPa and a shear stress of $\tau_{xy} = 103$ MPa. Use **(a)** the maximum principal stress and **(b)** the strain energy density factor criteria to calculate the fracture angle θ_o, the incline angle β, and the plane-strain fracture toughness K_{IC} and K_{IIC} and the critical strain energy density factor. Use the following data: crack length $2a = 76$ mm, $v = 1/3$, and $E = 72,300$ MPa.

Solution. **(a)** *Maximum Principle Stress Criterion: If* $a = 38$ mm, $\sigma_y = 138$ MPa, $\tau_{xy} = 103$ MPa, the stress intensity factors are

$$K_I = \sigma_y \sqrt{\pi a} = (138 \text{ MPa}) \sqrt{\pi \, (38 \times 10^{-3} \text{ m})} = 47.68 \text{ MPa} \sqrt{m}$$

$$K_{II} = \tau_{xy} \sqrt{\pi a} = (103 \text{ MPa}) \sqrt{\pi \, (38 \times 10^{-3} \text{ m})} = 35.59 \text{ MPa} \sqrt{m}$$

$$x = \frac{K_I}{K_{II}} = \frac{4}{3}$$

From Eq. (8.30),

$$\theta_o = 2\arctan\left[\frac{x}{4} + \sqrt{\left(\frac{x}{4}\right)^2 + 1/2}\right] = 2\arctan\left[\frac{1}{3} + \sqrt{\left(\frac{1}{3}\right)^2 + 1/2}\right]$$

$$\theta_o = 1.6795\,\text{rad} = 96.23°$$

The crack inclined angle β *is*

$$\frac{K_I}{K_{II}} = \frac{\sigma\sqrt{\pi a}\cos^2\beta}{\sigma\sqrt{\pi a}\sin\beta\cos\beta} = \cot\beta$$

$$\beta = \text{arccot}\left(\frac{K_I}{K_{II}}\right) = \text{arccot}\left(\frac{4}{3}\right) = 0.6435\,\text{rad} = 36.87°$$

Inserting the value of θ_o, K_I *and* K_{II} *into Eq. (8.33) yields*

$$K_{IC} = K_I\cos^4\frac{\theta_o}{2} + 3K_{II}\sin\frac{\theta_o}{2}\cos^3\frac{\theta_o}{2} \quad (\sigma_\theta\text{-criterion})$$

$$K_{IC} = \cos^4\left(\frac{1.6795}{2}\right)\left(47.68\,\text{MPa}\sqrt{m}\right)$$

$$+3\sin\left(\frac{1.6795}{2}\right)\cos^3\left(\frac{1.6795}{2}\right)\left(35.59\,\text{MPa}\sqrt{m}\right)$$

$$K_{IC} = 33.13\,\text{MPa}\sqrt{m}$$

From Eq. (8.34),

$$K_{IIC} = \sqrt{\frac{3}{4}}K_{IC} = \sqrt{\frac{3}{4}}\left(33.13\,\text{MPa}\sqrt{m}\right) = 28.69\,\text{MPa}\sqrt{m}$$

(b) *Strain Energy Density Factor Criterion: For plane-strain condition,*

$$\kappa = 3 - 4\nu = 3 - 4/3 = 5/3$$

Using an iteration procedure on Eq. (8.53) yields the fracture angle

$$0 = [(1 + \cos\theta_o)\sin\theta_o - (\sin\theta_o)(\kappa - \cos\theta_o)]$$

$$+2[(\cos\theta_o(2\cos\theta_o - \kappa + 1))(2\cos\theta_o - \kappa + 1 - 2\theta_o\sin\theta_o)]\left(\frac{3}{4}\right)$$

$$+ [(\kappa + 1)\sin\theta_o - (\sin\theta_o)(3\cos\theta_o - 1) - 3(1 + \cos\theta_o)\sin\theta_o]\left(\frac{3}{4}\right)^2$$

$$\theta_o = 0.5130\,\text{rad} = 29.39°$$

From Eqs. (8.42) and (8.54),

$$a_{11} = \frac{1+v}{8E} [(1 + \cos\theta)(\kappa - \cos\theta)] = 1.1485 \times 10^{-5} \, \text{MPa}^{-1}$$

$$a_{12} = \frac{1+v}{8E} \sin\theta [2\cos\theta - (\kappa - 1)] = 4.0744 \times 10^{-6} \, \text{MPa}^{-1}$$

$$a_{22} = \frac{1+v}{8E} [(\kappa + 1)(1 - \cos\theta) + (1 + \cos\theta)(3\cos\theta - 1)]$$

$$a_{22} = 2.595 \times 10^{-5} \, \text{MPa}^{-1}$$

and

$$K_{IC} = \sqrt{\frac{32E}{(v+1)(\kappa+1)^2} \left[a_{11}K_I^2 + 2a_{12}K_IK_{II} + a_{22}K_{II}^2 \right]}$$

$$K_{IC} = 72.85 \, \text{MPa}\sqrt{\text{m}}$$

The fracture toughness for mode II, Eq. (8.57), and the critical strain energy density factor, Eq. (8.41), are respectively,

$$K_{IIC} = \sqrt{\frac{3(\kappa+1)^2}{14\kappa - \kappa^2 - 1}} K_{IC} = 76.09 \, \text{MPa}\sqrt{\text{m}}$$

and

$$a_{11} = \frac{1+v}{8E} [(1 + \cos\theta)(\kappa - \cos\theta)] = 1.1485 \times 10^{-5} \, \text{MPa}^{-1}$$

$$a_{12} = \frac{1+v}{8E} \sin\theta [2\cos\theta - (\kappa - 1)] = 4.0744 \times 10^{-6} \, \text{MPa}^{-1}$$

$$a_{22} = \frac{1+v}{8E} [(\kappa + 1)(1 - \cos\theta) + (1 + \cos\theta)(3\cos\theta - 1)]$$

$$a_{22} = 2.595 \times 10^{-5} \, \text{MPa}^{-1}$$

$$S_c = a_{11}K_{IC}^2 + 2a_{12}K_{IC}K_{IIC} + a_{22}K_{IIC}^2$$

$$S_c = \left(1.1485 \times 10^{-5} \, \text{MPa}^{-1}\right) \left(72.85 \, \text{MPa}\sqrt{\text{m}}\right)^2$$

$$\quad - 2\left(4.0744 \times 10^{-6} \, \text{MPa}^{-1}\right) \left(72.85 \, \text{MPa}\sqrt{\text{m}}\right) \left(76.09 \, \text{MPa}\sqrt{\text{m}}\right)$$

$$\quad + \left(2.595 \times 10^{-5} \, \text{MPa}^{-1}\right) \left(76.09 \, \text{MPa}\sqrt{\text{m}}\right)^2$$

$$S_c = 0.17 \, \text{kPa}\,\text{m} = 0.17 \, \text{kJ/m}^2 \quad \textit{(Critical)}$$

Therefore, crack propagation takes place when $S_c = 0.17 \ kJ/m^2$ *at* $\theta_o = 29.39°$*. The critical strain energy density factor* S_C *surface profile is given below. This 3D plot represents the critical state of the current mixed-mode interaction.*

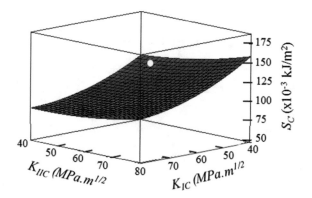

The mixed-mode fracture criterion provides a 3-D surface plot showing that the critical strain energy density factor S_c *increases with increasing fracture toughness values. Nonetheless, this plot represents the fracture resistance of the material.*

In addition, the figure below exhibits the trend of the fracture toughness ratio with increasing Poisson's ratio v. *Note that there is a theoretical maximum limit for the Poisson's ratio of approximately 0.7. At* $v > 0.7$, $K_{IIC}/K_{IC} \to \infty$.

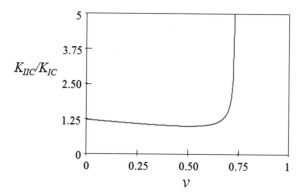

Example 8.4. *Suppose that a thin-wall cylindrical pressure vessel made of AISI 4340 steel with* $K_{IC} = 50 \ MPa\sqrt{m}$, $E = 207 \ GPa$, $\sigma_{ys} = 1,793 \ MPa$, $v = 1/3$, *and* $S_c = 1080 \ N/m$ *contains an internal semielliptical surface crack (2a = 4 mm deep and 2c = 12 mm long) inclined at 30° with respect to the circumferential (hoop) stress* σ_θ. *This does not coincide with the principle stresses, and consequently, it will*

grow at a fracture angle θ_o as shown in the figure below. The cylindrical pressure vessel containing such an inclined crack can be found elsewhere [31]. The vessel thickness and inside diameter are $B = 6$ mm and $d_i = 500$ mm, respectively. Calculate the fracture internal pressure P_c.

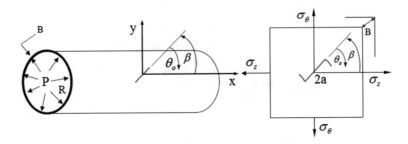

Solution. *The hypothetical pressure vessel above is just a closed structure to store liquids or gases under pressure $P > P_{atm}$, and it is treated as a thin-walled structure. In addition, the direction of crack extension in mixed-mode fracture is assumed to occur at a fracture angle θ_o. Thus, a self-similar crack grown along the crack plane no longer prevails in the analysis required in this mixed-mode I/II example problem. The hoop stress σ_θ in mixed-mode I/II fields compares with the stress in an unconstrained mode I field for crack growth along the crack plane. From Chap. 3, the circumferential (hoop stress) and the longitudinal stresses are*

$$\sigma_\theta = \frac{Pd}{2B} \quad \& \quad \sigma_z = \frac{\sigma_\theta}{2} \tag{a}$$

From Eqs. (8.15) and (8.16),

$$K_I = \sigma_\theta \sqrt{\pi a} \cos^2 \beta$$
$$K_{II} = \sigma_\theta \sqrt{\pi a} \sin \beta \cos \beta$$

A more exact treatment of the problem using the hoop stress gives

$$K_I = \frac{Pd}{2B} \sqrt{\pi a} \cos^2 \beta$$
$$K_{II} = \frac{Pd}{2B} \sqrt{3\pi a} \sin \beta \cos \beta$$

where the stress intensity factor ratio becomes

$$\frac{K_I}{K_{II}} = \cot (\beta = \pi/6) = \sqrt{3}$$

$$x = \frac{K_{II}}{K_I} = \tan (\beta = \pi/6) = \frac{\sqrt{3}}{3}$$

Now consider the strain energy density factor in light of the above

$$S = a_{11}K_I^2 + 2a_{12}K_IK_{II} + a_{22}K_{II}^2$$

$$S = \pi a \left(\frac{Pd}{2B}\right)^2 f(\theta, \beta)$$

where the function $f(\theta, \beta)$ is defined by

$$f(\beta) = a_{11}\cos^4\beta + 2a_{12}\sin\beta\cos^3\beta + a_{22}\sin^2\beta\cos^2\beta$$

$$f(\beta) = 0.5625a_{11} + 0.64952a_{12} + 0.1875a_{22}$$

since $\beta = 30° = \pi/6$, $\sin\beta = 1/2$, $\sin^2\beta = 1/4$, $\cos\beta = \sqrt{3}/2$ and $\cos^2\beta = 3/4$. Before determining the coefficients a_{ij}, let's find the value of the fracture angle by letting $dS/d\theta = 0$ and $\theta = \theta_o$ at fracture. Hence, iterate the following equation to find θ_o

$$0 = [(1 - \kappa + 2\cos\theta_o)\sin\theta_o]$$

$$+2\left[(2\cos\theta_o - \kappa + 1)\cos\theta_o - 2\sin^2(\theta_o)\right]\left(\frac{K_{II}}{K_I}\right)$$

$$+ [(\kappa - 1 - 6\cos\theta_o)\sin\theta_o]\left(\frac{K_{II}}{K_I}\right)^2$$

$$0 = \frac{4}{3}\sin\theta - 8\sin2\theta - \frac{4}{3}(\cos\theta)\sqrt{3} + 4\sqrt{3}\cos2\theta$$

$$\theta_o = 28.91°$$

Now that θ_o is known, the constants a_{ij} follow

$$a_{11} = \frac{1+\nu}{8E}[(1 + \cos\theta)(\kappa - \cos\theta)]$$

$$a_{12} = \frac{1+\nu}{8E}\sin\theta[2\cos\theta - (\kappa - 1)]$$

$$a_{22} = \frac{1+\nu}{8E}[(\kappa + 1)(1 - \cos\theta) + (1 + \cos\theta)(3\cos\theta - 1)]$$

and

$$a_{11} = 1.1948 \times 10^{-6}\,\text{MPa}^{-1}$$

$$a_{12} = 4.1878 \times 10^{-7}\,\text{MPa}^{-1}$$

$$a_{22} = 2.7230 \times 10^{-6}\,\text{MPa}^{-1}$$

Thus,

$$f(\beta) = a_{11} \cos^4 \beta + a_{12} \sin \beta \cos^3 \beta + a_{22} \sin^2 \beta \cos^2 \beta$$
$$f(\beta) = 1.3186 \times 10^{-6} \, \text{MPa}^{-1}$$

The fracture pressure is now calculated based on the critical strain energy density factor. Hence,

$$S_c = \pi a \left(\frac{P_c d}{2B}\right)^2 f(\theta, \beta)$$

$$P_c = \frac{2B}{d} \sqrt{\frac{S_c}{\pi a f(\theta, \beta)}}$$

so that

$$P_c = \frac{(2)(6\,\text{mm})}{500\,\text{mm}} \sqrt{\frac{1080 \times 10^{-6} \, \text{MPa m}}{\pi \, (2 \times 10^{-3}\,\text{m}) \left(1.3186 \times 10^{-6} \, \text{MPa}^{-1}\right)}}$$

$$P_c = 8.67 \, \text{MPa}$$

Calculate the stress intensity factor for mode I based on the critical pressure P_c in order to determine if fracture occurs in mode I loading. Thus,

$$K_I = \frac{P_c d}{8B} \sqrt{\pi a} \cos^2 \beta$$

$$K_I = \frac{(8.67\,\text{MPa})(500\,\text{mm})}{2\,(6\,\text{mm})} \sqrt{\pi\,(2 \times 10^{-3}\,\text{m})} \cos^2 (\pi/6)$$

$$K_I = 21.48 \, \text{MPa}\sqrt{\text{m}} < K_{IC} = 50 \, \text{MPa}\sqrt{\text{m}}$$

Therefore, fracture does not occur in this mode because $K_I < K_{IC}$.

Will fracture occur in mode II? Let us find out by estimating the corresponding fracture toughness and by calculating the applied stress intensity factor for mode II. Thus, from Eq. (8.57),

$$K_{IIC} = \sqrt{\frac{3(\kappa+1)^2}{14\kappa - \kappa^2 - 1}} K_{IC} = 22.44 \, \text{MPa}\sqrt{\text{m}}$$

and

$$K_{II} = \frac{P_c d}{8B} \sqrt{\pi a} \sin(\beta) \cos(\beta)$$

$$K_{II} = \frac{(8.67\,\text{MPa})(500\,\text{mm})}{2\,(6\,\text{mm})} \sqrt{\pi\,(2 \times 10^{-3}\,\text{m})} \sin(\pi/6) \cos(\pi/6)$$

$$K_{II} = 12.40 \, \text{MPa}\sqrt{\text{m}}$$

$$K_{II} = \left(\sqrt{3}/3\right) K_I = \left(\sqrt{3}/3\right) (21.\,48\,\text{MPa}\sqrt{\text{m}}) = 12.40\,\text{MPa}\sqrt{\text{m}}$$

$$K_{II} = 12.40\,\text{MPa}\sqrt{\text{m}} < K_{IIC} \ \textit{No fracture occurs in this mode}$$

Therefore, fracture does not occur in this mode II. In light of the above calculations, the pressure vessel hypothetically can withstand a higher pressure before fracture or burst occurs.

8.6 Crack Branching

It is recognized that the tensile cracks in solids can significantly branch out due to mechanical, microstructural, or environmental effects. Changes in crack path are normally induced by (1) multiaxial far-field stresses, (2) interaction of the crack tip with microstructural defects, and (3) sudden changes in load and embrittlement effects of an aggressive environment [8–12].

Figure 8.9 illustrates some examples of severely branched cracks in ductile materials during different conditions, which can be found in the cited references.

The effects of crack branching can be rationalized based on the stress intensity factors for the small-branched cracks shown in Fig. 8.10. However, numerous solutions for kinked and forked cracks have been proposed, but there have been considerable disagreements.

The stress intensity factors K_1 and K_2 for kinked and forked cracks are smaller than the nominal K_I and K_{II}. Based on the projected length of the crack, K_1 and K_2 are meaningful if the plastic zone is smaller than the zone of dominance of the K_I and K_{II} singular fields.

The stress intensity factor solutions for kinked and forked elastic cracks under mixed-mode I/II loading are based on the models shown in Fig. 8.10 [13–16].

Thus,

$$K_1 = a_{11}(\alpha) K_I + a_{12}(\alpha) K_{II} \tag{8.58}$$

$$K_2 = a_{21}(\alpha) K_I + a_{22}(\alpha) K_{II}$$

where $a_{ij}(\alpha)$ are defined as

$$a_{11}(\alpha) = \frac{1}{4}\left(3\cos\frac{\alpha}{2} + \cos\frac{3\alpha}{2}\right)$$

$$a_{12}(\alpha) = -\frac{3}{4}\left(\sin\frac{\alpha}{2} + \sin\frac{3\alpha}{2}\right) \tag{8.59}$$

$$a_{21}(\alpha) = \frac{1}{4}\left(\sin\frac{\alpha}{2} + \sin\frac{3\alpha}{2}\right)$$

$$a_{22}(\alpha) = \frac{1}{4}\left(\cos\frac{\alpha}{2} + 3\cos\frac{3\alpha}{2}\right)$$

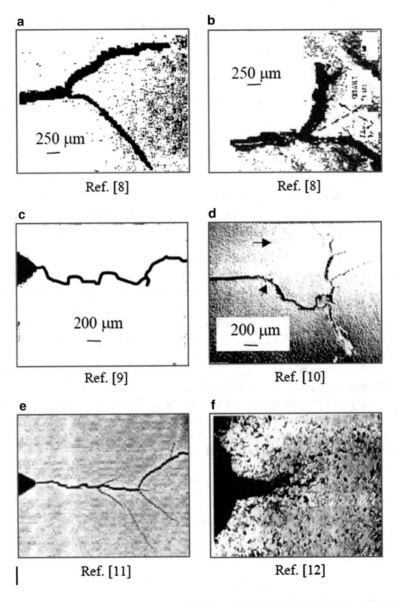

Fig. 8.9 (**a**), (**b**), and (**c**) mode I crack branching in an Al-2.9Cu-2.1Li-0.12Zr alloy, (**d**) kinking of a fatigue crack in 2020-T651 Al-alloy, (**e**) stress corrosion crack branching in 9Ni-4Co-0.45C martensitic steel, and (**f**) creep crack branching in copper

These coefficients are the solutions for an infinitesimal kink or branch crack $b/a \rightarrow$ 0. A simple analysis of these equations implies that if $a = 0$, then the stress intensity factors for crack kinking become equal to the nominal counterparts; that is, $K_1 = K_I$

Fig. 8.10 Nomenclature for mode I/II crack branching [7]. (**a**) Kinked crack and (**b**) forked cracks

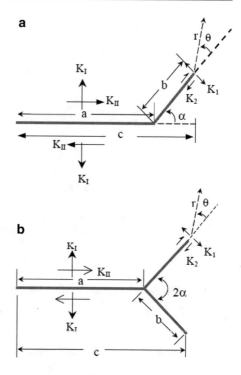

and $K_2 = K_{II}$. The profiles for $a_{ij}(\alpha)$ and normalized stress intensity factors for the branched crack are depicted in Fig. 8.11.

The strain energy release rate for crack extension in a self-similar manner (in the plane of the original crack plane) was derived previously as Eq. (6.34). For an infinitesimally small crack branched out, the strain energy release rate, Eq. (6.34), is now defined in terms of the stress intensity factor for the kinked crack tip

$$G(\alpha)_i = \frac{K_1^2}{E'} + \frac{K_2^2}{E'} + \frac{(1+v)K_3^2}{E} \qquad (8.60)$$

The stress intensity factors K_1 and K_2 derived by Hussain et al. [17] and K_3 by Sih [18] are

$$K_1 = \left(\frac{4}{3 + \cos^2\alpha}\right)\left(\frac{\pi - a}{\pi + \alpha}\right)^{\alpha/2\pi}\left(K_I \cos\alpha + \frac{3}{2}K_{II}\sin\alpha\right)$$

$$K_2 = \left(\frac{4}{3 + \cos^2\alpha}\right)\left(\frac{\pi - a}{\pi + \alpha}\right)^{\alpha/2\pi}\left(-\frac{1}{2}K_I \sin\alpha + K_{II}\cos\alpha\right) \qquad (8.61)$$

$$K_2 = \left(\frac{\pi - a}{\pi + \alpha}\right)^{\alpha/2\pi} K_{III}$$

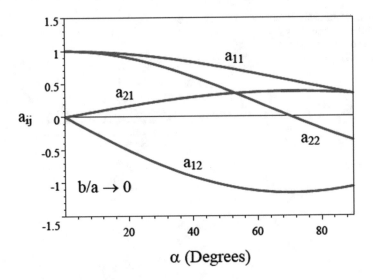

Fig. 8.11 Variation of constants a_{ij} with kink angle α

Substituting Eq. (8.61) into (8.60) yields an expression for the strain energy release rate as a function of kink fracture angle α. Thus,

$$G(\alpha)_i = \frac{4\lambda}{(3+\cos^2\alpha)^2 E'}\left[\begin{array}{c}(1+3\cos^2\alpha)\,K_I^2+(4\sin 2\alpha)\,K_I K_{II}\\+(9-5\cos^2\alpha)\,K_{II}^2\end{array}\right]$$

$$+\frac{\lambda(1+v)\,K_{III}^2}{E} \tag{8.62}$$

where

$$\lambda = \left(\frac{\pi - a}{\pi + \alpha}\right)^{\alpha/\pi} \tag{8.62a}$$

The hypothesis states that crack extension takes place in a direction of maximum strain energy release rate $G(\alpha)_{\max}$ and that crack branching occurs at a fracture angle $\alpha = \alpha_o$. Hence,

$$\frac{\partial G(\alpha)}{\partial \alpha} = 0 \quad \text{for } \alpha = \alpha_o \tag{8.63a}$$

$$\frac{\partial^2 G(\alpha)}{\partial \alpha^2} < 0 \quad \text{for } \alpha = \alpha_o \tag{8.63b}$$

Crack propagation takes place when $G(\alpha)_i = G(\alpha)_c$. For pure mode I at fracture, Eq. (8.62) becomes

$$G(\alpha)_{IC} = \frac{K_{IC}^2}{E'} \tag{8.64}$$

Letting $G(\alpha)_{IC} = G(\alpha)_i$ in Eq. (8.62) yields the fracture criterion

$$K_{IC}^2 = \frac{4\lambda}{(3 + \cos^2 \alpha)^2} \left[(1 + 3\cos^2 \alpha) K_I^2 + (4\sin 2\alpha) K_I K_{II} \right]$$

$$+ \frac{4\lambda}{(3 + \cos^2 \alpha)^2} \left[+ (9 - 5\cos^2 \alpha) K_{II}^2 \right] \qquad (8.65)$$

$$+ \frac{E'\lambda (1 + v) K_{III}^2}{E}$$

Denote that if $\alpha = 0$ in Eqs. (8.62) and (8.65), the former reduces to Eq. (6.34) and the latter becomes Eq. (8.13).

Furthermore, the above mathematical treatment of mixed-mode fracture mechanics problems has been based on the stress field surrounding the crack tip where stress singularities exist as $r \to 0$. Including the influence of the T-stress (T_x) in the mixed-mode I/II process, the generalized stress based on Williams [23] K-field solution for unkinked crack is defined as [24]

$$\sigma = \frac{K_I f_I(\theta) + K_{II} f_{II}(\theta)}{\sqrt{2\pi r}} + T_x \qquad (8.66)$$

Recall from Chap. 4 that T_x is a second-order term in the series expansion of the stress field and it is non-singular, but specimen size and geometry dependent. As pointed out by Becker et al. [24], T_x represents the strength in the crack x-direction. For mode I loading, $T_x < 0$ and $K_{II} = 0$ for stabilizing the crack. Conversely, crack branching may occur when $T_x > 0$ and $K_{II} > 0$ at angle α as indicated in Fig. 8.10. Further theoretical or numerical details on crack kinking can be found elsewhere [25–28].

8.7 Wood

Wood is an important material in our modern world. For instance, wood can be used as a material for construction, tools, decoration, and furniture. It comes in a variety of natural colors ranging from white (sapwood) to black (ebony). In a single sample of wood, coloration can also change. Figure 8.12 illustrates a transverse section of a stem of wood exhibiting the characteristics of the wood fibers along the principle axes [32, 33].

Thus, wood is characterized by its concentric layers known as annual rings or growth rings, which in turn are used for characterizing softwoods and hardwoods. Each of these types of woods are also classified as earlywood and latewood or as sapwood and heartwood, respectively. Nonetheless, wood consists of hard hexagonal like cells. The microscopic, macroscopic, and physical characteristics of wood can be found elsewhere [34].

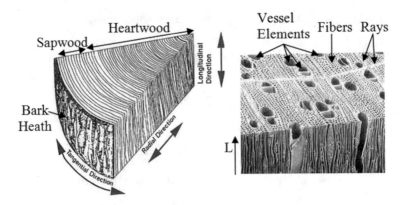

Fig. 8.12 Characteristics of wood texture

The bear tree trunk can be treated as a fiber-reinforced composite or a cylindrical orthotropic material solid that responses to external stresses, temperature, and environment. Thus, the structural integrity can be assessed by determining the fracture behavior of wood. Basically, wood has a heterogeneous internal structure, and it is anisotropic with respect to the cylindrical principle axes, which are denoted as radial (R), tangential (T), and longitudinal (L) directions. Consequently, the mechanical properties and the fracture behavior of wood are strongly affected by the internal structure of the material and the direction of applied forces. For instance, crack growth direction in a wooden component or specimen must be characterized according to the fiber orientations illustrated in Fig. 8.13 [35].

Furthermore, a wooden specimen can be subjected to a mixed-mode loading as shown in Fig. 8.14 for ponderosa pine specimens. The wood photo was taken from Ref. [36].

Thus, K_I and K_{II} are the main parameters in characterizing the mixed-mode fracture, and their magnitudes strongly depend on the crack orientation with respect to the fiber direction. Denote that the specimen in Fig. 8.14 does not exhibit knots which are the bases for lateral tree branches and are obstacles for crack growth. Thus, this material may be classified as a clear wooden specimen which would experience crack growth along the radial (R) direction at a fracture angle θ_o. This implies that cracks in clear wood usually grow along the fibers [37].

The reader interested in further details on fracture mechanics or fracture of wood should consult relevant references [37–45]. Nonetheless, the experimental fracture data from Mall, Murphy, and Schottafer [41] for Eastern red spruce wood and Hunt and Croager [42] for Scots pine wood has been used by Jernkvist [37] to derive fracture mechanics criteria based on the linear-elastic fracture mechanics (LEFM) framework. Further use of the same experimental data gives the mixed-mode fracture behavior shown in Fig. 8.15.

Denote in Fig. 8.15 that an empirical mixed-mode fracture criterion was fitted to the experimental data described above. The mixed-mode fracture criterion is of the form

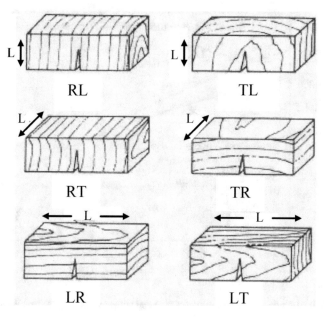

Fig. 8.13 Directions of crack growth in wood [35]

Fig. 8.14 LR specimens made of ponderosa pine wood. The photo was taken from Ref. [36]. (**a**) CCT and (**b**) SE(T)

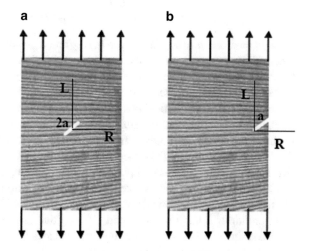

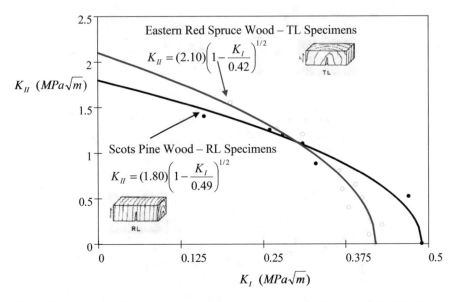

Fig. 8.15 Mixed-mode fracture criterion for pine and spruce woods. Data from Refs. [41, 42]

$$\left(\frac{K_I}{K_{IC}}\right)^m + \left(\frac{K_{II}}{K_{IIC}}\right)^n = 1 \qquad (8.67)$$

Here, $m = 1$ and $n = 2$. The values of K_{IC} and K_{IIC} are given in Fig. 8.15 for both pine and spruce woods. It is clear that this empirical model predicts the mixed-mode fracture behavior of these anisotropic materials [37].

The mixed-mode fracture criteria discussed in previous sections are not suited for these data. The reader should confirm this statement by working out Problem 8.7.

Furthermore, the mixity parameter K_{II}/K_I increases with increasing mode II loading due to an increase in fracture energy [37]. Thus, $K_{IIC}/K_{IC} = 5.0$ for the Eastern red spruce (TL specimens) and $K_{IIC}/K_{IC} \approx 4$ for the Scots pine (RL specimens) represent relatively high fracture toughness ratios, which indicate that different fracture process takes place at the crack tip in mode I and II, regardless of the type of specimens made of different softwoods [37, 39].

In addition, it has been reported [39, 45] that smooth surface cracks occur in softwood due to crack growth through weak planes along fibers (tracheids) creating a low energy consumption fracture process, while rough surface cracks develop in mode II due to a high fracture energy consumption for opening and coalescence of microcracks ahead of the main crack tip.

Figure 8.16 shows the fracture angle predicted by the principal stress fracture criterion, Eq. (8.29). Denote that the fracture angle θ_o for both data sets follows a similar trend and exhibits a strong dependency on the mixity parameter at $0 < K_{II}/K_I < 5$ or $0 < K_{II}/K_I < 80°$ which diminishes for $K_{II}/K_I \geq 8$ [37].

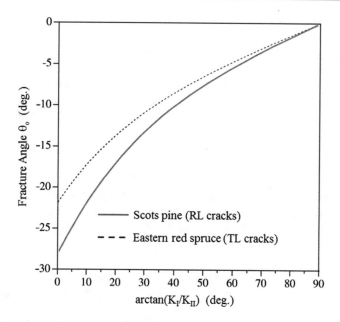

Fig. 8.16 Fracture angle for Scots pine and Eastern red spruce woods [37]

Table 8.2 Stress and displacement equations per mode of loading

	Mode I	Mode II	Mode III
$\sigma_x =$	$\frac{K_I}{\sqrt{2\pi r}} \cos\frac{\theta}{2} \left(1 - \sin\frac{\theta}{2}\sin\frac{3\theta}{2}\right)$	$-\frac{K_{II}}{\sqrt{2\pi r}} \sin\frac{\theta}{2} \left(2 + \cos\frac{\theta}{2}\cos\frac{3\theta}{2}\right)$	$+0$
$\sigma_y =$	$\frac{K_I}{\sqrt{2\pi r}} \cos\frac{\theta}{2} \left(1 + \sin\frac{\theta}{2}\sin\frac{3\theta}{2}\right)$	$\frac{K_{II}}{\sqrt{2\pi r}} \sin\frac{\theta}{2}\cos\frac{\theta}{2}\cos\frac{3\theta}{2}$	$+0$
$\sigma_z =$	$\begin{cases} 0 & \text{Plane stress} \\ v(\sigma_x + \sigma_y) & \text{Plane strain} \end{cases}$	$\begin{cases} 0 & \text{Plane stress} \\ v(\sigma_x + \sigma_y) & \text{Plane strain} \end{cases}$	$+0$
$\tau_{xy} =$	$\frac{K_I}{\sqrt{2\pi r}} \cos\frac{\theta}{2}\sin\frac{\theta}{2}\cos\frac{3\theta}{2}$	$+\frac{K_{II}}{\sqrt{2\pi r}} \cos\frac{\theta}{2} \left(1 - \sin\frac{\theta}{2}\sin\frac{3\theta}{2}\right)$	$+0$
$\tau_{xz} =$	0	$+0$	$-\frac{K_{III}}{\sqrt{2\pi r}} \sin\frac{\theta}{2}$
$\tau_{yz} =$	0	$+0$	$+\frac{K_{III}}{\sqrt{2\pi r}} \cos\frac{\theta}{2}$
$u_x =$	$\frac{K_I}{2G}\sqrt{\frac{r}{2\pi}} \cos\frac{\theta}{2} \left(\kappa - 1 + 2\sin^2\frac{\theta}{2}\right)$	$\frac{K_{II}}{2G}\sqrt{\frac{r}{2\pi}} \sin\frac{\theta}{2} \left(\kappa + 1 + 2\cos^2\frac{\theta}{2}\right)$	$+0$
$u_y =$	$\frac{K_I}{2G}\sqrt{\frac{r}{2\pi}} \sin\frac{\theta}{2} \left(\kappa + 1 - 2\cos^2\frac{\theta}{2}\right)$	$+\frac{K_{II}}{2G}\sqrt{\frac{r}{2\pi}} \cos\frac{\theta}{2} \left(\kappa - 1 - 2\sin^2\frac{\theta}{2}\right)$	$+0$
$u_z =$	$\begin{cases} -\frac{vB}{E}(\sigma_x + \sigma_y) & \text{Plane stress} \\ +0 & \text{Plane strain} \end{cases}$	$\begin{cases} -\frac{vB}{E}(\sigma_x + \sigma_y) & \text{Plane stress} \\ +0 & \text{Plane strain} \end{cases}$	$+\frac{2K_{III}}{G}\sqrt{\frac{r}{2\pi}} \sin\frac{\theta}{2}$

As a final remark, Table 8.2 summarizes the most common stress and displacement equations for the three modes of loading. It is convenient and advantageous to make use of these equations for solving a problem related to a particular infinity loading system.

8.8   Problems

8.1. A large plate (2024-0 Al-alloy) containing a central crack is subjected to a combined mode I/II loading. The internal stresses are $\sigma_{sy} = 138\,\text{MPa}$ and $\tau_{xy} = 103\,\text{MPa}$. Use the maximum principle stress criterion (σ_θ-criterion) to calculate the fracture toughness (K_{IC} and K_{IIC}). Data: crack length $2a = 76\,\text{mm}$, $\nu = 1/3$, and $E = 72,300\,\text{MPa}$. [Solution: $K_{IC} = 72.5\,\text{MPa}\sqrt{\text{m}}$ and $K_{IIC} = 62.77\,\text{MPa}\sqrt{\text{m}}$.]

8.2. Repeat Problem 8.1 using the strain energy density factor criterion (S-criterion).

8.3. Determine **(a)** the applied tensile stress and **(b)** the safety factor for a loaded plate containing a 76-mm central crack inclined at $53.25°$. The strain energy release rate mode I is $9.642\,\text{kJ/m}^2$. Data: $E = 207\,\text{GPa}$, $\sigma_{ys} = 958\,\text{MPa}$, and $\nu = 0.35$. [Solution: $\sigma = 201.38\,\text{MPa}$ and $S_F = 4.46$.]

8.4. Calculate the critical stress (fracture stress) for the problem described in Problem 8.3 according to the σ_θ-criterion. Will crack propagation take place?

8.5. Repeat Problem 8.4 using the S-criterion with $\sigma = 215\,\text{MPa}$. [Solution: $\sigma_c = 204\,\text{MPa}$].

8.6. Show that the stress intensity factor is $K_I = \left(\sqrt{8/11}\right)K_{IC}$ when $K_I = 2K_{III}$ and the Poisson's ration is $\nu = 1/3$.

8.7. Use the experimental mixed-mode fracture data for Solid A and Solid B given below to compare the mixed-mode fracture criteria discussed in this chapter. Determine which criterion is the most suited for predicting the mixed-mode fracture behavior of these solids.

Solid A		Solid B	
K_{II}	K_I	K_{II}	K_I
$(\text{MPa}\sqrt{\text{m}})$	$(\text{MPa}\sqrt{\text{m}})$	$(\text{MPa}\sqrt{\text{m}})$	$(\text{MPa}\sqrt{\text{m}})$
0.49	0	0.42	0
0.43	0.52	0.41	0.10
0.33	1.02	0.43	0.21
0.31	1.10	0.38	0.40
0.28	1.20	0.40	0.50
0.26	1.25	0.37	0.60
0.16	1.40	0.39	0.65
0	1.50	0.37	0.80
		0.31	1.20
		0.27	1.26
		0.20	1.55
		0	1.65

8.8. A large and wide plate has a small through-thickness center crack as shown in the figure below. Use the free-body diagram (FBD) to derive expressions for (a) the normal (σ_N) and shear (τ) stresses and (b) the intensity factors K_I and K_{II} as functions of β.

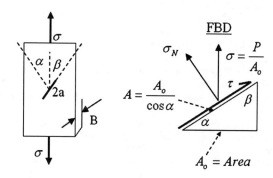

8.9. Two identical cracked plates as shown in Problem 8.8 are to be tested to determine K_{IC} and K_{IIC}. The observed critical tension loads and the incline angles were 120 MPa @ $\alpha = 0$ and 130 MPa @ $\alpha = \pi/4$, respectively. Use the equations given below to determine the fracture toughness for mode I and II.

$$\frac{K_I}{K_{IC}} + \left(\frac{K_{II}}{K_{IIC}}\right)^2 = 1$$

$$\left(\frac{K_I}{K_{IC}}\right)^2 + \left(\frac{K_{II}}{K_{IIC}}\right)^2 = 1$$

References

1. G.C. Sih, H. Liebowitz, *Mathematical Fundamental of Fracture* (Academic, New York, 1968), pp. 67–190
2. J.A. Begley, J.D. Landes, The J-integral as a fracture criterion. ASTM STP **514**, 1–26 (1972)
3. J.A. Begley, J.D. Landes, W.K. Wilson, An estimation model for the application of the J-integral. ASTM STP **560**, 155–169 (1974)
4. K. Hellan, *Introduction to Fracture Mechanics*, chap. 7 (McGraw-Hill Book Company, New York, 1984)
5. F. Erdogan, G.C. Sih, On the crack extension in plates under plane loading and transverse shear. J. Basic Eng. **85**, 519–527 (1963)
6. J.W. Dally, W.F. Riley, *Experimental Stress Analysis*, 3rd edn. (McGraw-Hill, New York, 1991), pp. 115–116
7. D. Broek, *Elementary Engineering Fracture Mechanics*, 4th edn. (Kluwer Academic Publisher, Boston, 1986), p. 379
8. M.Y. He, J.W. Hutchinson, The penny-shaped crack and the plane strain crack in an infinite body of power-law material. Trans. ASME J. Appl. Mech. **48**, 830–840 (1981)

9. M.Y. He, J.W. Hutchinson, Bounds for fully plastic crack problems for infinite bodies. Volume I. ASTM STP **803**, I-277ÚI-290 (1983)
10. K. Hellan, *Introduction to Fracture Mechanics* (McGraw-Hill, New York, 1984), p. 8
11. J.H. Hollomon, Tensile deformation. Trans. Am. Inst. Min. Metall. Pet. Eng. **162**, 268–290 (1945)
12. J.W. Hutchinson, Singular behavior at the end of a tensile crack in a hardening material. J. Mech. Phys. Solids **16**, 13–31 (1968)
13. J.W. Hutchinson, Fundamentals of the phenomenological theory of nonlinear fracture mechanics. J. Appl. Mech. **50**, 1042–1051 (1983)
14. J.W. Hutchinson, P.C. Paris, Stability analysis of J-controlled crack growth. ASTM STP **668**, 37–64 (1979)
15. B.S.J. Kang, A.S. Kobayashi, J-Estimation Procedure Based on Moiré Interferometry Data. Contract N00014-85-K-0187, Technical Report No. UWA/DME/TR-87/58, Office of Naval Research, Arlington, VA (August 1987), pp. 1–19; J. Press. Vessel Technol. ASME (1988) pp. 291–300
16. V. Kumar, C.F. Shih, Fully plastic crack solutions, estimates scheme and stability analyses for compact specimens. ASTM STP **700**, 406–438 (1980)
17. M.A. Hussain, S.L. Pu, J. Underwood, ASTM STP 560 (1973), pp. 2–28
18. V. Kumar, M.D. German, C.F. Shih, Elastic-plastic and fully plastic analysis of crack initiation, stable growth, and instability in flawed cylinders. Volume I. ASTM STP **803**, 306–353 (1983)
19. K. Palaniswamy, Ph.D. thesis, California Institute of Technology (1972)
20. B.C. Hoskin, D.G. Graff, P.J. Foden, Aer. Res. Lab., Melbourne, Report S.M. 305 (1965)
21. W. Ramberg, W.R. Osgood, Description of Stress-Strain Curves by Three Parameters. Technical Note No. 902, National Advisory Committee For Aeronautics, Washington, DC (1943), pp. 1–22
22. J.R. Rice, A path independent integral and the approximate analysis of strain concentrations by notches and cracks. J. Appl. Mech. **35**, 379–386 (1968)
23. M.L. Williams, On the stress distribution at the base of a stationary crack. J. Appl. Mech. **24**, 109–114 (1957)
24. T.L. Becker Jr., R.M. Cannon, R.O. Ritchie, Finite crack kinking and T-stresses in functionally graded materials. Int. J. Solids Struct. **38**, 5545–5563 (2001)
25. B. Cottrell, Notes on the paths and instability of cracks. Int. J. Fract. **2**, 526–533 (1966)
26. M.Y. He, J.W. Hutchinson, Kinking of a crack out of an interface. J. Appl. Mech. **56**, 270–278 (1989)
27. M.Y. He, A. Bartlett, A. Evans, J.W. Hutchinson, Kinking of a crack out of an interface: role of plane stress. J. Am. Ceram. Soc. **74**, 767–771 (1991)
28. K. Palaniswamy, W.G. Knauss, On the problem of crack extension in brittle solids under general loading, in *Mechanics Today*, vol. 4, ed. by S. Namat-Nassar (Pergamon Press, New York, 1978)
29. E. Sommer, Formation of fracture lances in glass. Eng. Fract. Mech. **1**, 539–546 (1969)
30. J. Li, Estimation of the mixity parameter of a plane strain elastic-plastic crack by using the associated J-integral. Eng. Fract. Mech. **61**(3–4), 355–368 (1998)
31. G.C. Sih, *Handbook of Stress Intensity Factors: Stress-Intensity Factor Solutions and Formulas for Reference*, vol. 1 (Institute of Fracture and Solid Mechanics, Lehigh University, Bethlehem, PA, 1973), pp. xxxi–xxxiii
32. J.E. Reeb, *Wood and Moisture Relationships*, EM 8600, Oregon State University Extension Services (June 1995)
33. Society of Wood Science and Technology, Teaching Unit No. 1, swst.org (2005)
34. G. Tsoumis, *Wood as Raw Material* (Pergamon Press, New York, 1968)
35. M. Patton-Mallory, S.M. Cramer, Fracture mechanics:a tool for predicting wood component strength. For. Prod. J. **37**(7/8), 39–47 (1987)
36. *The Anatomy of Wood: Microscopic Structure and Grain of Wood* (2005). www.waynesword.palomar.edu

37. L.O. Jernkvist, Fracture of wood under mixed mode loading I. Derivation of fracture criteria. Eng. Fract. Mech. **68**, 549–563 (2001)
38. L.O. Jernkvist, Fracture of wood under mixed mode loading II. Experimental investigation of *Picea abies*. Eng. Fract. Mech. **68**, 565–576 (2001)
39. D.M. Tan, S.E. Stanzi-Tschegg, E.K. Tschegg, Models of wood fracture in mode I and mode II. Holz als Roh-und Werkstoff **53**(3), 159–164 (1995)
40. G. Valentin, P. Caumes, Crack propagation in mixed mode in wood: a new specimen. Wood Sci. Technol. **23**, 43–53 (1989)
41. S. Mall, J.F. Murphy, J.E. Shottafer, Criterion for mixed mode fracture in wood. J. Eng. Mech. **109**(3), 680–690 (1983)
42. D.G. Hunt, W.P. Croager, Mode II fracture toughness of wood measured by a mixed-mode test method. J. Mater. Sci. Lett. **1**, 77–79 (1982)
43. E.M. Wu, Application of fracture mechanics to anisotropic plates. J. Appl. Mech. **34**(4), 967–974 (1967)
44. J.G. Williams, M.W. Birch, *Mixed Mode Fracture in Anisotropic Media*. ASTM STP 601 (1976), pp. 125–137
45. S. Mindess, A. Bentur, Crack propagation in notched wood specimens with different grain orientations. Wood Sci. Technol. **20**, 145–155 (1986)

Fatigue Crack Growth

9

9.1   Introduction

Fatigue in materials subjected to repeated cyclic loading can be defined as a
progressive failure due to crack initiation (stage I), crack growth (stage II), and
crack propagation (stage III) or instability stage. For instance, crack initiation of
crack-free solids may be characterized by fatigue crack nuclei due to dislocation
motion, which generates slip bands at the surface having slip steps in the order of
$0.1\,\mu m$ in height [11, 80] or slip may occur at matrix-inclusion interfaces. These
steps produce surface intrusions and extrusions as schematically indicated below
for stages I and II. These intrusions caused by reversed slip due to load reversal
are the source for crack initiation, which may consume most of the solid life before
crack growth. This crack initiation may occur along the slip direction due to a local
maximum shear stress. After the consumption of many cycles, the crack may change
in direction when the maximum principal normal stress (in the vicinity of the crack
tip) governs crack growth. In this stage II, some materials show striations and beach
marks as common surface features of fatigue fracture.

In general, fatigue is a form of failure caused by fluctuating or cyclic loads over
a short or prolong period of time. Therefore, fatigue is a time-dependent failure
mechanism related to microstructural features. The fluctuating loading condition
is not a continuous failure process as opposed to cyclic loading. The former is
manifested in bridges, aircraft, and machine components, while the latter requires
a continuous constant or variable stress amplitude until fracture occurs. It is also
important for the reader to know that fatigue failure or fracture can occur at a
maximum stress below the static yield strength of a particular material. Obviously,
temperature effects must be considered in fatigue failure characterization. From an
engineering point of view, predicting fatigue life is major a requirement.

© Springer International Publishing Switzerland 2017
N. Perez, *Fracture Mechanics*, DOI 10.1007/978-3-319-24999-5_9

9.2   Cyclic Stress History

Figure 9.1 shows schematic cyclic stress fluctuating curves with constant stress amplitude (symmetrical) and random loading as function of time. These schematic curves represent cyclic stress histories from which the number of cycles is accounted for fatigue life (N).

The stress history or stress spectrum can be

- Axial due to tension-compression applied stresses
- Flexural due to bending applied stress
- Torsion due to twisting

From Fig. 9.1, the stress range ($\Delta\sigma$) is the algebraic difference between the maximum and minimum forces in a cycle expressed as $\Delta\sigma = \sigma_{max} - \sigma_{min}$. Other common stress parameters extracted from a stress spectrum are the mean stress (σ_m), alternating stress (σ_a), the stress ratio, and the stress amplitude (A_s)

$$\sigma_m = \frac{\sigma_{max} + \sigma_{min}}{2} \tag{9.1}$$

$$\sigma_a = \frac{\Delta\sigma}{2} = \frac{\sigma_{max} - \sigma_{min}}{2} \tag{9.2}$$

$$R = \frac{\sigma_{min}}{\sigma_{max}} = \frac{K_{min}}{K_{max}} \tag{9.3}$$

Fig. 9.1 Schematic stress histories. (**a**) Symmetrical and (**b**) asymmetrical

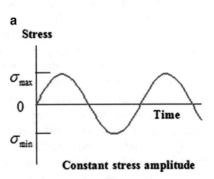

Constant stress amplitude

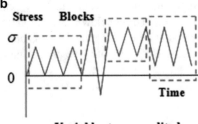

Variable stress amplitude

$$A_s = \frac{\sigma_a}{\sigma_m} = \frac{\sigma_{max} - \sigma_{min}}{2\sigma_m} = \frac{1-R}{1+R} \qquad (9.4)$$

These stress parameters can be varied while conducting fatigue tests for characterizing materials having specific geometries, weldments, or microstructural features. In fact, varying stress ratio is the most common parameter in determining the fatigue behavior of crack-free and cracked specimens.

The mechanics of fatigue and fracture is concerned with the reliability and effectiveness of structural components in engineering applications. Thus, a precise determination of fatigue crack growth rate (da/dN) and fatigue remaining life (N) and a detailed understanding of how materials respond to fluctuating or cyclic loads are important for assuring structural integrity.

In reality, structural components are generally subjected to a wide spectrum of stresses over their lifetime that strongly affects fatigue life. The wide spectrum on a structural component may consists of constant stress amplitude (*CSA*) and variable stress amplitude (*VSA*) blocks as schematically shown in Fig. 9.1b.

For crack-free or notched specimens, the usual characterization of fatigue behavior is through a stress-cycle curve, commonly known as a S-N diagram. Figure 9.2 shows two schematic S-N curves for two hypothetical materials.

For crack-free specimens, the number of cycles to initiate a fatigue crack is known as the fatigue-crack initiation life (N_i), which can have a very large magnitude representing most of the usual life of a component. The remaining fatigue life (N_p) is related to stable fatigue crack growth till the crack reaches a critical length, and consequently, crack propagation occurs very rapidly without any warning. As a result, a component can have a fatigue life defined by the total

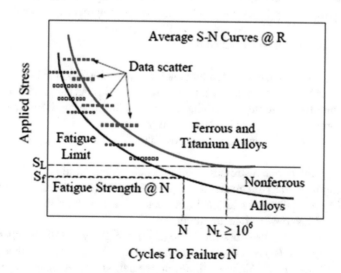

Fig. 9.2 Schematic fatigue S-N curves

number of cycles, $N = N_i + N_p$, consumed during testing or service. Conversely, a preexisting crack reduces fatigue life because $N_i = 0$ and $N = N_p$.

Despite that fatigue represents a cumulative damage in structural components, it is the fluctuating or cyclic local stresses and strains imparted by an external or nominal loading mode that are the primary factors for localized crack initiation and growth. Therefore, fatigue life can be prolonged if the nominal fluctuating or cyclic stress level is reduced or eliminated, if the microstructure is homogeneous, if dimensional changes are not severe enough, or if the environment is not significantly corrosive.

It should be pointed out that the fatigue limit (for infinite life) was formerly called the endurance limit, and it is the stress level below which fatigue failure does not occur. Generally, ferrous alloys such as steels exhibit a stress limit, while nonferrous do not show an asymptote, and the stress continues to decrease with increasing cycles to failure. The latter materials are characterized by determining the fatigue strength at a specific life (N). This stress is meaningless if the specific life is not identified.

Typically, fatigue life exhibits data scatter as schematically shown in Fig. 9.2 for ferrous and titanium alloys. Nonferrous alloys also exhibit data scatter. Therefore, the difference in failure response of test specimens is due to microstructural defects and machining defects. In addition, a particular material having a fine-grained microstructure exhibit superior fatigue properties over coarse-grained microstructure.

Metal fatigue is a significant engineering problem because it can occur due to repeated or cyclic stresses below the static yield strength; unexpected and catastrophic failure of a vital structural part may occur and rack initiation may start at discontinuities in highly stressed regions of the component. Fatigue failure may be due to discontinuities such as inadequate design, improper maintenance, and so forth.

Fatigue failure can be prevented by

- Avoiding sharp surfaces caused by punching, stamping, and the like
- Preventing the development of surface discontinuities during processing
- Avoiding misuse, abuse, assembling errors, and improper maintenance
- Using proper material and heat treatment procedures
- Using inert environments whenever possible

Normally, the nominal stresses in most structures are elastic or below the static yield strength of the base material. In pertinent cases, crack-free specimens are tested using repeated or cycle loading to determine the strain-life curve (ϵ-N) at low cycles with variable strain amplitude.

On the other hand, the stress-life curve (S-N) in high-cycle fatigue schemes with variable stress ratio (R) is very important for determining the total number of cycles to failure. Consequently, fatigue failure, in general, is a weakening process of the material since internal or external defects are the sources for crack formation.

9.3   Fatigue Crack Initiation

It is not intended here to include a detailed background on crystallography, but a brief explanation on this subject can make the reader be aware of the implications in preventing and understanding fracture initiation at a atomic or nanoscale. With regard to common metallic structural materials, metals and engineering alloys are crystalline in nature since atoms are arranged uniformly forming unique repetitive three-dimensional arrays, which constitute what is known in crystallography as unit cells. As a result, the unit cells repeat themselves within grains (crystals) and their atomic mismatch is known as grain boundary. Figure 9.3a shows a schematic unit cell within the space lattice, and Fig. 9.3b depicts the atomic arrangement in a body-centered cubic (BCC) showing primary slip system $(110)[\bar{1}11]$.

On the other hand, Fig. 9.4 shows two real and different types of microstructures, and therefore, the static and dynamic behavior of the corresponding alloys have distinct mechanical behavior. This means that the microstructure plays a very significant role in mechanical behavior of solid bodies. Figure 9.5 exhibits two different dislocation networks as the representative line defects that develop during permanent deformation [56–58]. These figures are intended to show the different microstructural features of polycrystalline materials responsible for any observable mechanical behavior.

In general, fatigue crack initiation and growth depend on microstructural features, the maximum fluctuating stress, and the environment. Conversely, plastics or polymers are composed of molecules and are also important engineering materials; however, their fatigue mechanism is different from metals.

Consider a polycrystalline solid with a smooth surface being subjected to an elastic-cyclic stress range, in which $\sigma_{max} < \sigma_{ys}$, but σ_{max} is high enough to activate a slip mechanism such as the Cottrell–Hull modified mechanism [16, 20, 26, 54] shown in Fig. 9.6.

Fig. 9.3 Atomic arrangement within a crystal. (**a**) Repetitive array of unit cells and (**b**) body-centered cubic (BCC) showing the slip direction (s), direction of the applied stress (σ), and the most dense plane for shear motion of atoms along the shown direction

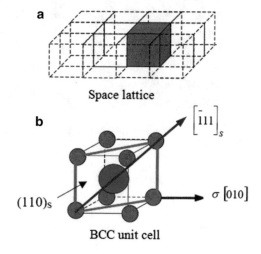

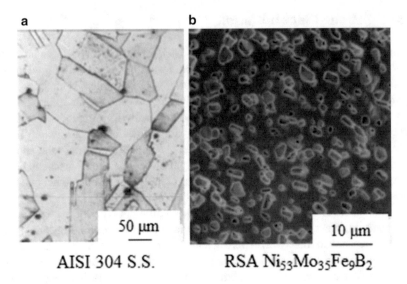

Fig. 9.4 Microstructural features of (**a**) 41 % cold rolled stainless steel type AISI 304 showing austenite grains and (**b**) rapidly solidified alloy exhibiting hard boride particles embedded in an Ni-Mo matrix [56–58]

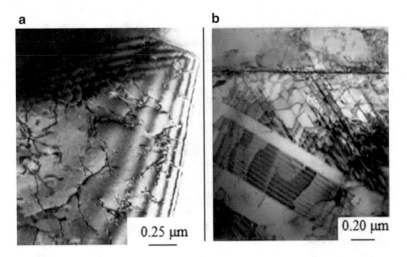

Fig. 9.5 Bright field TEM photomicrographs showing dislocation networks [57]. (**a**) AISI 304 S.S., (**b**) RSA $Ni_{53}MO_{35}Fe_9B_2$

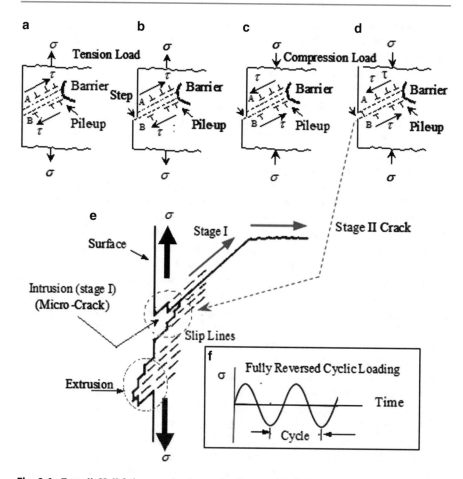

Fig. 9.6 Cottrell–Hull fatigue mechanism in ductile materials [16, 20, 26, 54]

Let the stress ratio be $R = -1$ for a fully reversed cyclic load system causing irreversible damage after many cycles. Take the slip planes A and B for convenience so that dislocation pileups occur on both sides of the planes, but having opposite signs as indicated in Fig. 9.6a. When the slip system $\{hkl\} < uvw >$ is activated due to the local maximum shear stress τ, a surface step is created at $\sigma_{max} > 0$ (Fig. 9.6b). Then dislocation motion is reversed once $\sigma_{min} > 0$ (Fig. 9.6c). Also, the upper part moves toward its original position, leaving an inward step called intrusion (Fig. 9.6d) in the order of the Burger's vector $b = 3$ mm [20]. This mechanism is repeated many times until a deeper intrusion acts as a microcrack (Fig. 9.6e). In polycrystalline materials, microcracks (not visible to the naked eye) interact with the material microstructural features, such as grain boundaries, secondary particles, or phases and dislocation networks induced by previous plastic deformation during cold working or a stress-time history (Fig. 9.6f).

Microcracks can also develop in concrete due to the cement hydration process and compressive loads. Consequently, easy microcrack growth is restricted to an extent during monotonic or fatigue testing.

During this stage I, many life cycles are consumed before crack growth in the direction perpendicular to the local principal normal tension stress, which governs the crack growth behavior, and the Cottrell mechanism no longer applies in a simple manner. This mechanism can take place in a few grains before the crack changes direction. Once this occurs other dislocation mechanisms such as the Frank-Read source may take place. When multiple cross-slip occurs, the Frank-Read source may not complete a loop cycle [16].

Nevertheless, the microcrack becomes long enough, causing an increase in the stress concentration at the crack tip, and subsequently, the local stress is truncated to the yield strength of the solid. This leads to a stage II crack growth during which many grains are deformed making the plastic zone, which eventually reaches a critical size for crack growth continuation in the direction of the local principal stress which governs the stress field at the crack tip.

Since each grain has a different preferred orientation, crack growth may be in a zigzag manner due to different slip directions. As a result, fatigue surface features may consist of curved beach marks and striations in between. It is believed that each striation corresponds to a cycle and many striations are formed between beach marks. Therefore, it is possible that the plastic zone growth is related to sets of striations, and its rupture at a critical size may be attributed to the formation of a beach mark. In addition, solid bodies subjected to a fluctuating load, as opposed to a monotonic load, may develop cracks that may grow very slowly. The fatigue maximum tension load (lower than the monotonic maximum) causes crack opening. On the other hand, the minimum load closes the crack.

Fatigue and fracture in a cyclic mechanical process are strongly influenced by the applied constant and variable stress amplitudes. The variable stress amplitude can be treated as random or semi-random stress amplitude, and it may be more detrimental than the constant stress amplitude, specifically when overloads occur.

Moreover, fatigue is a complex phenomenon since it may initiate in slip zones adjacent to the outer surface or in internal defects, such as voids or inclusions. It should be made clear to the reader that the phrase "fatigue fracture" means rupture or separation of a component. This is basically an overload that makes the crack reach its critical length ($a = a_c$) for rapid growth called crack propagation.

Furthermore, the mechanism of fatigue fracture may be initiated at a microscopic defect since a cumulative displacement between slip planes occurs. Thus, an intrusion, as formed in a slip direction, may be the original source for a microcrack. A triple point at grain boundaries may be a source of crack initiation or cracks may be developed by stress corrosion action, such as hydrogen embrittlement.

However, the direction of fatigue crack growth may change from the slip orientation (stage I) to an average cracking normal to the maximum tension direction corresponding to the stage II process. Thus, the characteristics of crack growth may be transcrystalline either by progressive plastic straining, which causes typical striations, or by cleavage at low temperatures or in the presence of brittle inclusions.

On the other hand, crack growth may be intercrystalline due to bonding deficiency, aggressive environment, or initiation and coalescence of voids within or between grains. Then, the final stage (III) of fatigue fracture is caused by a dynamic crack propagation mechanism, in which the applied maximum stress intensity factor reaches a critical value equals to the plane-stress or plane-strain fracture toughness of the material.

The most common fatigue evaluation is normally focused on tension loading, and it is apparent that crack growth in stage III occurs perpendicularly to the direction of the variable applied stress. As a result, fracture surfaces may include fatigue striations, microvoid coalescence, and cleavage facets (for brittle materials). Hence, fatigue failure analysis must reveal these types of fracture surface features in order to determine the mechanism of fatigue fracture.

9.4 Fatigue Crack Growth Rate

Since fatigue is a cyclic dissipation of energy process related to a cumulative damage process, the elapsed time for damage is expressed in terms of load cycles (N). The control parameter that is used to evaluate this process is the rate of crack growth per cycle (da/dN). Hence, da/dN depends on the applied stress intensity factor range (ΔK) and N is the well-known fatigue life term.

For crack initiation in stage I, the threshold stress intensity factor is

$$\Delta K_{th} = \alpha \Delta \sigma_{th} \sqrt{\pi a} \tag{9.5}$$

Here, α is the geometry correction factor given in Eq. (3.29), and $\Delta \sigma_{th}$ is the threshold stress range, which is analogous to the fatigue limit S_L introduced in Fig. 9.2. Nonetheless, this equation indicates that if $\Delta \sigma < \Delta \sigma_{th}$ crack growth does not occur.

On the other hand, in the early 1960s, Paris [54] empirically expressed the fatigue crack growth rate (da/dN) in stage II as a function of the stress intensity factor range (ΔK). The pertinent relationship became known as the Paris power-law for stage II fatigue crack growth (FCG). Thus, the mathematical model is of the form [54]

$$\frac{da}{dN} = A \left(\Delta K \right)^n \tag{9.6}$$

where $\Delta K = K_{max} - K_{min} = (1 - R) K_{max}$ for $R \geq 0$

$\Delta K = K_{max}$ for $R \leq 0$

$A =$ Constant $(MPa^{-n} \, m^{1-n/2}/cycles)$

$n =$ Exponent

$\Delta \sigma = \sigma_{max} - \sigma_{min} =$ Stress range

Fig. 9.7 Schematic fatigue crack growth curves

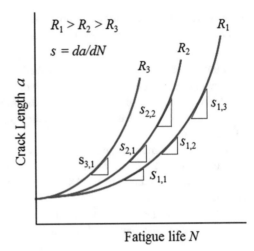

Denote that A and n in Eq. (9.6) are material constants determined empirically from fatigue crack growth rate data and are independent of crack length. On the other hand, ΔK depends on the crack length and can be defined as

$$\Delta K = \Delta\sigma\sqrt{\pi a} \quad \text{(uncorrected)} \qquad (9.7)$$

$$\Delta K = \alpha\Delta\sigma\sqrt{\pi a} \quad \text{(corrected)} \qquad (9.8)$$

Here, α is the specimen geometry correction factor introduced in Chap. 3, and it is a function of normalized crack length, that is, $\alpha = f(a/w)$ for finite plates containing relatively short cracks. It should be clear that Eq. (9.6) through (9.8) exclude the effects of crack closure, related residual stresses at the crack tip [17, 18], and crack tip blunting.

In addition, Fig. 9.7 schematically shows the procedure for estimating the tangential slope da/dN of $a = f(N)$ for a constant stress ratio R. Denote that da/dN increases rapidly as crack growth occurs and significantly approaches a critical stress state. Subsequently, increasing R increases da/dN and accelerates fracture.

Moreover, the Paris law, Eq. (9.6), is the most common mathematical equation used for modeling fatigue crack behavior of structural components, but it can empirically be modified in order to incorporate the effects of cyclic stress ratio R and plane-stress or plane-strain fracture toughness on da/dN. Hence,

- Forman equation [21]:

$$\frac{da}{dN} = \frac{A(\Delta K)^n}{(1-R)K_C - \Delta K} \quad \text{for plane stress and } R > 1 \qquad (9.9)$$

$$\frac{da}{dN} = \frac{A(\Delta K)^n}{(1-R)K_{IC} - \Delta K} \quad \text{for plane strain and } R > 1 \qquad (9.10)$$

- Broek and Schijve equation [9, 19]:

$$\frac{da}{dN} = AK_{max}^2 \left(\Delta K\right)^n \tag{9.11}$$

- Walker equations [77]:

$$\frac{da}{dN} = AK_{max}^m \left(\Delta K\right)^n \tag{9.12}$$

$$\frac{da}{dN} = A\sigma_{max} \left(1 - R\right) \sqrt{\pi a} \quad \text{For } R > 1 \tag{9.13}$$

- Hartman and Schijve equation [25]:

$$\frac{da}{dN} = \frac{A \left(\Delta K - \Delta K_{th}\right)^n}{\left(1 - R\right) K_C - \Delta K} \quad \text{for plane stress and } R > 1 \tag{9.14}$$

9.4.1   Crack Closure

Denote that $da/dN = f\left(\Delta K\right)$ gives a conservative fatigue life prediction by using crack closure-free data. If crack closure is included in a suitable da/dN expression, then ΔK is replaced by an effective stress intensity factor range $\Delta K_{eff} < \Delta K$. Here, $\Delta K_{eff} = K_{max} - K_{op}$ and $K_{op} > K_{min}$ is the stress intensity factor that opens the crack. Thus, it can be assumed that the stress intensity factor for closing the crack is similar to that for opening it, that is, $K_{op} \simeq K_{cl}$. Although the difference between K_{op} and K_{cl} may generally be small and negligible [7,33], the Elber's mathematical model [17,18] for crack closure is adopted in this section which is similar to Paris model. Accordingly,

$$\frac{da}{dN} = C \left(\Delta K_{eff}\right)^m \tag{9.14a}$$

$$\Delta K_{eff} = U\Delta K \tag{9.14b}$$

Thus far, $U < 1$ at low stress ratios and $U \to 1$ at high stress rations [66]. If $U \to 1$, then the effect of crack closure is ineffective. The reader should verify that the factor U in Eq. (9.14b) along with the effective stress ratio $R_{eff} = K_{op}/K_{max}$ can be defined as

$$U = \frac{K_{max} - K_{op}}{K_{max} - K_{min}} = \frac{1 - K_{op}/K_{max}}{1 - K_{min}/K_{max}} \tag{9.14c}$$

$$U = \frac{1 - R_{eff}}{1 - R} \quad \text{For } R \neq 1 \text{ and } R_{eff} \neq 1 \tag{9.14d}$$

Mathematically, $\Delta K = 0$ when $R = 1$ means that there are no cyclic loads and K_{max} becomes a monotonic quantity [74]. Nonetheless, U also depends on K_{max} as shown in Eq. (9.14c). If $K_{min} = 0$, the factor U becomes independent of the nominal stress ratio $R = K_{min}/K_{max}$ and acquires the following definition

$$U = 1 - \frac{K_{op}}{K_{max}} = 1 - R_{eff} \tag{9.14e}$$

It is clear now that the phenomenon of crack closure is less effective when $R_{eff} \to 0$ and $U \to 1$, and the effect of crack closure completely vanishes when $R_{eff} = 0$ because $U = 1$ and $\Delta K_{eff} = \Delta K$.

Further, during crack closure the crack surfaces remain in contact at the crack tip during a portion of the unloading part of the fatigue cycle. Nonetheless, crack closure at the crack tip may significantly influence the fatigue crack growth rate at low stress intensity factors and low stress ratios, and it can be explained through mechanisms such as plasticity-induced crack closure (PICC), roughness-induced crack closure (RICC), oxide-induced crack closure (OICC), and phase transformation-induced crack closure (PTICC) affect fatigue crack growth rate. The reader should consult papers on these specific topics and their related mathematical models available in the literature for relevant theoretical background, experimental observations, and reference citations [41, 48–50, 52, 59].

9.5 Fatigue Life Calculations

The goal here is to develop a mathematical model that predicts fatigue life (N) for a given stress range at a constant load amplitude. Since mode I loading is the most studied, integration of Eq. (9.6) is carried out for this mode for convenience; however, other modes of loadings may be used instead. Nevertheless, the sought fatigue life (N) is

$$\int_{N_o}^{N} dN = \int_{a_o}^{a} \frac{da}{A\,(\Delta K)^n} = \frac{1}{A\left(\alpha\Delta\sigma\sqrt{\pi}\right)^n} \int_{a_o}^{a} \frac{da}{a^{n/2}} \tag{9.15}$$

$\int \frac{da}{a^{n/2}} = -\frac{2a}{(n-2)a^{\frac{1}{2}n}}$

Inserting Eq. (9.8) into (9.15), integrating and arranging the resultant expression yields

$$\Delta N = (N - N_o) = \frac{2\left(a_o^{1-n/2} - a^{1-n/2}\right)}{A\,(n-2)\left(\alpha\Delta\sigma\sqrt{\pi}\right)^n} \qquad \text{for } n \neq 2 \tag{9.16}$$

$$\Delta N = (N - N_o) = \frac{\ln\left(a/a_o\right)}{\pi A\,(\alpha\Delta\sigma)^2} \qquad \text{for } n = 2 \tag{9.16a}$$

Here, ΔN is the constant-amplitude load cycles required for crack growth from the original (initial) crack length (a_o) to a larger size $a > a_o$. Eventually fracture occurs when the crack length reaches a critical size so that $a = a_c$ and $K_{max} = K_{IC}$ or $K_{max} = K_C$.

For plane strain conditions, the critical crack length along with $K_{max} = K_{IC}$ and $\Delta\sigma$ can be predicted using a modified form of Eq. (3.29). Thus,

$$a_c = \frac{1}{\pi}\left(\frac{K_{IC}}{\alpha}\right)^2 \Delta\sigma^{-2} \tag{9.17}$$

Substituting Eq. (9.17) along with $a = a_c$ into (9.16) and arranging terms yields an expression for determining the stress range $\Delta\sigma$ when the final or critical crack length is unknown. Thus,

$$C_1\,(\Delta\sigma)^n + C_2\,(\Delta\sigma)^{n-2} - C_3 = 0 \tag{9.18}$$

The constants C_i with $i = 1, 2$, and 3 are

$$C_1 = A\,(n-2)\,(N-N_o)\,(K_{IC})^n$$

$$C_2 = \frac{2}{\pi}\left(\frac{K_{IC}}{\alpha}\right)^2 \tag{9.19}$$

$$C_3 = 2\,(a)^{1-n/2}\left(\frac{1}{\pi}\right)^{n/2}\left(\frac{K_{IC}}{\alpha}\right)^n$$

Note that Eq. (9.18) is a polynomial of order n, which is a nonnegative integer exponent, and the solutions of this equation are called the roots of the polynomial.

Example 9.1. *A high-strength steel string has a miniature round surface crack of* 0.09 *mm deep and an outer diameter of* 1.08 *mm. The string is subjected to a repeated fluctuating load (*$\sigma_{min} = 0$, $\sigma_{max} > 0$*) at a stress ratio* $R = 0$. *The threshold stress intensity factor is* $\Delta K_{th} = 5$ *MPa* $\sqrt{m}$, *and the crack growth rate equation is given by*

$$\frac{da}{dN} = \left(5\times 10^{-14}\ \frac{MN^{-4}\,m^{-1}}{cycles}\right)(\Delta K)^4$$

*Determine (**a**) the threshold stress* $\Delta\sigma_{th}$ *the string can tolerate without crack growth, (**b**) the maximum applied stress range* $\Delta\sigma$, *(**c**) the maximum (critical) crack length for a fatigue life of* $N = 10^4$ *cycles and (**d**) Will the string fracture? Use the following steel properties:* $K_{IC} = 25$ *MPa* $\sqrt{m}$ *and* $\sigma_{ys} = 880$ *MPa.*

Solution.

It is assumed that the plastic zone with a cyclic range ΔK is smaller than that for K_I applied monotonically and that the surface crack can be treated as a single-edge crack configuration. Note that $N_o = 0$ since a_o already exists.

(a) *From Eq. (3.56),*

$$a = \frac{D - d}{2} \tag{3.56}$$

$$d = D - 2a = 1.08 \text{ mm} - 2\,(0.09 \text{ mm}) = 0.90 \text{ mm}$$

$$\frac{d}{D} = 0.8333 \quad and \quad \frac{D}{d} = 1.20$$

Now, Eqs. (3.54) and (9.5) yield, respectively

$$f\,(d/D) = \frac{1}{2}\sqrt{\frac{D}{d}}\left[\frac{D}{d} + \frac{1}{2} + \frac{3}{8}\left(\frac{d}{D}\right) - \frac{5}{14}\left(\frac{d}{D}\right)^2 + \frac{11}{15}\left(\frac{d}{D}\right)^3\right] \tag{3.55}$$

$$\alpha = f\,(d/D) = 1.198\,9$$

$$\Delta\sigma_{th} = \frac{\Delta K_{th}}{\alpha\,\sqrt{\pi a_o}} = \frac{5 \text{ MPa }\sqrt{m}}{(1.198\,9)\,\sqrt{\pi\,(0.09 \times 10^{-3} \text{ m})}} = 248.02 \text{ MPa}$$

$$\Delta\sigma_{min} < \Delta\sigma_{th}$$

This is a rather large threshold stress range, but it serves the purpose of illustrating the methodology for determining the limit for a minimum stress range.

(b) *Use Eq. (9.19) and subsequently (9.18) to get*

$$C_1 = \left(5 \times 10^{-14}\,\frac{\text{MN}^{-4}\,\text{m}^7}{cycles}\right)(4 - 2)\,(10^4 \text{ cycles})\left(25\,\frac{\text{MN}}{\text{m}^2}\,\sqrt{m}\right)^4$$

$$C_1 = 3.906\,3 \times 10^{-4} \text{ m}$$

$$C_2 = \frac{2}{\pi}\left(\frac{K_{IC}}{\alpha}\right)^2 = \frac{2}{\pi}\left(\frac{25 \text{ MPa }\sqrt{m}}{1.198\,9}\right)^2 = 276.82 \text{ MPa}^2\,\text{m} \tag{9.19}$$

$$C_3 = 2\,(0.09 \times 10^{-3} \text{ m})^{1-4/2}\left(\frac{1}{\pi}\right)^{4/2}\left(\frac{25 \text{ MPa }\sqrt{m}}{1.198\,9}\right)^4$$

$$C_3 = 4.257\,1 \times 10^8 \text{ MPa}^4\,\text{m}$$

and

$$3.906\,3 \times 10^{-4}\Delta\sigma^4 + 276.82\Delta\sigma^2 - 4.257\,1 \times 10^8 = 0$$

Solving the above biquadratic equation yields four roots. The positive root is

$$\Delta\sigma = 864.93 \text{ MPa}$$

$$\sigma_{\max} = \Delta\sigma = 864.93 \text{ MPa} \quad since \quad \sigma_{\min} = 0$$

$$\sigma_{\max} < \sigma_{ys}$$

Thus,

$$K_{\max} = \alpha\sigma_{\max}\sqrt{\pi a} = (1.198\,9)\,(864.93 \text{ MPa})\sqrt{\pi\,(0.09 \times 10^{-3} \text{ m})}$$

$$K_{\max} = 17.44 \text{ MPa}\sqrt{\text{m}}$$

(c) *The critical crack size is calculated from Eq. (9.17)*

$$a_c = \frac{1}{\pi}\left[\frac{K_{IC}}{\alpha}\right]^2 \Delta\sigma^{-2} = \frac{1}{\pi}\left[\frac{25 \text{ MPa}\sqrt{\text{m}}}{(1.198\,9)\,(864.93 \text{ MPa})}\right]^2 \qquad (9.17)$$

$$a_c = 0.185 \text{ mm} = 2.056 a_o$$

$$\Delta a = a_c - a_o = 0.095 \text{ mm}$$

In addition, a_c can also be calculated using the following expression

$$a_c = a\left(\frac{K_{IC}}{K_{\max}}\right)^2 = (0.09 \text{ mm})\left(\frac{25 \text{ MPa}\sqrt{\text{m}}}{17.44 \text{ MPa}\sqrt{\text{m}}}\right)^2$$

$$a_c = 0.185 \text{ mm}$$

(d) *The stress intensity factor range is*

$$\Delta K = \alpha\Delta\sigma\sqrt{\pi a}$$

$$\Delta K = (1.198\,9)\,(864.93 \text{ MPa})\sqrt{\pi\,(0.90 \times 10^{-3} \text{ m})}$$

$$\Delta K = 55.14 \text{ MPa}\sqrt{\text{m}} > 25 \text{ MPa}\sqrt{\text{m}}$$

Therefore, the string fractures because $K_{\max} = \Delta K > K_{IC}$.

Example 9.2. *Consider a hypothetical brittle material subjected to a fatigue load range given next. The objective in this example is to illustrate how the fatigue crack growth rate is affected by the effective stress intensity factor range. (**a**) Derive an expression for the stress intensity factor K_{op} using Elber's model $\Delta K_{eff} = U\Delta K$ [17] and (**b**) calculate K_{op} and da/dN if $U = 0.8$, $R = 0$, $K_{\max} = 100 \text{ MPa}\sqrt{\text{m}}$, $n = 4$, and $A = 5 \times 10^{-16} \text{ MN}^{-4}\text{m}^{-1}/\text{cycle}$.*

Solution.

(a) *Derive an expression for the stress intensity factor K_{op}*

$$\Delta K_{eff} = U \Delta K$$

$$K_{max} - K_{op} = U \left(K_{max} - K_{min} \right)$$

$$1 - \frac{K_{op}}{K_{max}} = U \left(1 - \frac{K_{min}}{K_{max}} \right)$$

$$1 - \frac{K_{op}}{K_{max}} = U \left(1 - R \right)$$

$$K_{op} = K_{max} \left(1 - U + UR \right)$$

(b) *Calculations along with $U = 0.8$, $R = 0$, $K_{max} = 100$ MPa $\sqrt{m}$, $n = 4$, and $A = 5 \times 10^{-16}$ $MN^{-4} m^{-1}/cycle$ give the opening stress intensity factor as*

$$K_{op} = K_{max} \left(1 - U + UR \right) = K_{max} \left(1 - U \right)$$

$$K_{op} = (100 \text{ MPa}) \left(1 - 0.8 \right)$$

$$K_{op} = 20 \text{ MPa}$$

In reality, this value represents the local minimum stress intensity factor on the reverse loading stress spectrum.

The effective and the applied stress intensity factor ranges are

$$\Delta K_{eff} = K_{max} - K_{op} = 100 \text{ MPa } \sqrt{m} - 20 \text{ MPa } \sqrt{m} = 80 \text{ MPa } \sqrt{m}$$

$$\Delta K = K_{max} - K_{min} = 100 \text{ MPa } \sqrt{m} - 0 \text{ MPa } \sqrt{m} = 100 \text{ MPa } \sqrt{m}$$

Thus, the fatigue crack growth rate is calculated using ΔK_{eff} and ΔK individually. The results are

$$\frac{da}{dN} = A \left(\Delta K \right)^4 = 5.00 \times 10^{-6} \text{ m}/cycle \ @ \ \Delta K = 100 \text{ MPa } \sqrt{m}$$

$$\frac{da}{dN} = A \left(\Delta K_{eff} \right)^4 = 2.05 \times 10^{-6} \text{ m}/cycle \ @ \ \Delta K_{eff} = 80 \text{ MPa } \sqrt{m}$$

These results indicate that the fatigue crack growth rate due to the effective stress intensity factor range is the lowest, implying that crack stable growth is the slowest among the two calculated results. Hence, the magnitude of the intensity of the applied stress range has a strong effect on the fatigue crack growth of the hypothetical material in question.

9.6   Crack Growth Rate Diagram

In addition, a detailed fatigue crack growth behavior (sigmoidal curve) is schematically shown in Fig. 9.8 [65].

The ASTM E647 includes steps on how to conduct this type of fatigue test. Stage I is a slow crack growth region in which the fatigue threshold stress intensity factor range is usually less than $DeltaK < 9$ MPa $\sqrt{m}$ [65]. Below this value fatigue crack growth does not occur. In addition, stage I is rather a complex region from a microscale point of view, but it is related to a non-continuum crack growth mechanism, which strongly depends on the material's microstructure, the applied stress ratio R, and the environment.

Analysis of Fig. 9.8

Stage I: This is a slow crack growth process which is related to non-continuum mechanisms and slow fatigue crack growth rate in the order of $da/dN \leq 10^{-6}$ mm/cycle. Thus, this fatigue process strongly depends on the

- Grain size, precipitates, dislocation density, etc.
- Mean tress and stress ratio
- Aggressiveness of the environment
- Surface damage initiation

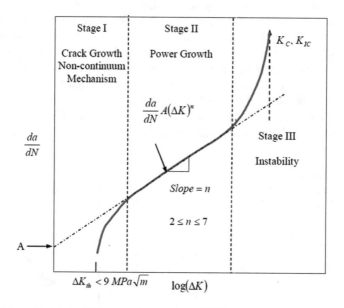

Fig. 9.8 Schematic stages of fatigue crack growth rate [65]

Stage II: This fatigue process is referred to as the power growth behavior usually characterized by the Paris law, and it slightly depends on the parameters in stage I. Particularly, surface characteristics are the formation of beach marks and striations in some metallic materials. Furthermore, the specimen thickness does not strongly influence the fatigue crack growth rate.

Striations are microscopic fatigue features that can be observed with the scanning electron microscope (SEM) and the transmission electron microscope (TEM) at relatively high magnifications. Conversely, each striation indicates a successive advance of one stress cycle. The width of a striation represents the advance of the crack front during one stress cycle, but it depends on the magnitude of the applied stress range. Normally, the appearance of service fatigue striations is irregular due to changeable stress amplitude. Striation is dealt with in the later section. Then, this irregularity in striation configuration is an indication of nonsteady crack growth rate, which may be restricted to geometry dependency and load history, leading to an enhanced or retarded rate.

Stage III: This fatigue process strongly depends on the microstructural parameters cited in stage I and on the specimen thickness. Since the applied stress intensity factor is sufficiently large, the fatigue crack growth rate is high, and the process is under an instable damage process. Therefore, this is the instability region in Fig. 9.8 and fracture occurs when the stress intensity factor reaches a critical value.

Figure 9.9 schematically shows the influence of R on stage I da/dN [22].

Figure 9.10 depicts experimental $da/dN = f(R, \Delta K)$ for a structural steel [5] For instance, the most common engineering metallic materials have a BCC, FCC, or HCP atomic structure; therefore, the fatigue behavior of these materials is expected to be different. Generally, surface crack initiation occurs in this stage I in which the crack growth rate is very slow. In addition, stage II is known as the power crack growth, which is less dependent on the microstructure, stress ratio, and environment than stage I. The log-log linear fatigue behavior is referred to as the Paris Law, which empirically define by Eq. (9.6). The characteristics of the surface fracture appearance may include beach marks and striations that are observable in certain materials such as aluminum and aluminum alloys. Both fatigue fracture features

Fig. 9.9 Schematic effects of the stress ratio (R) on the crack growth rate curves [22]

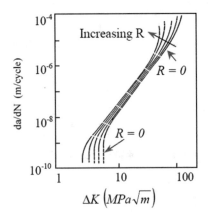

Fig. 9.10 Effect of stress ratio on fatigue crack growth rate of ASTM A533B steel [5]

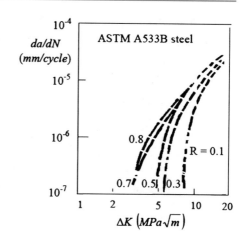

Table 9.1 Fatigue crack growth rate equations for steels [5]

Steel	Paris equation ($m/cycle$)	Strength
Martensitic	$\frac{da}{dN} = 1.35 \times 10^{-10} \, (\Delta K)^{2.25}$ ΔK in MPa $\sqrt{m}$	$\sigma_{ys} > 483 \, \text{MPa}$ $\sigma_{ts} > 621 \, \text{MPa}$
Ferritic Pearlitic	$\frac{da}{dN} - 6.90 \times 10^{-12} \, (\Delta K)^{3}$	$207 \, MPa < \sigma_{ys} < 552 \, \text{MPa}$
Austenitic	$\frac{da}{dN} = 5.60 \times 10^{-12} \, (\Delta K)^{3.25}$	$207 \, MPa < \sigma_{ys} < 345 \, \text{MPa}$

are concentric ridges of circular, semicircular, or semielliptical shape. Beach marks are known as clamshell marks, and from Materials Science point of view, they are macroscopic band-containing striations, represent the position of the crack of length, and are formed due to interruptions of the cyclic stress loading during service.

Despite the experimental data scatter found in steels, Barsom and Rolfe [5] determined conservative crack growth rate expressions as given in Table 9.1 based on Paris equation for fatigue stage II at $R = 0$. If experimental data is not available for a particular steel, these equations can be used in designing with proper caution.

Fatigue crack growth data is quite abundant in the literature, but some selected experimental data are included in Table 9.2 as a reference. Useful experimental data can be found elsewhere [1–3, 6, 8, 10, 13, 15, 17, 18, 24, 28, 30, 31, 35, 37–39, 42–46, 51, 53, 55, 60, 63, 67, 70–73, 79, 81].

With regard to Fig. 9.8, the transition between stage II and III apparently is related to tearing mechanism when the crack tip strain reaches a critical value [3]. For $R = 0$, $\Delta K = \Delta K_{max}$ and the onset of the transition apparently occurs at a constant crack tip opening displacement (δ_t).

According to Barsom's relationship, the crack tip opening displacement for the onset of transition is [4]

$$\delta_{trans} = \frac{\Delta K_{tran}^2}{E \sigma_F} \qquad (9.20)$$

Table 9.2 Fatigue crack growth data for some materials

Materials	σ_{ys} (MPa)	K_{IC}	R	ΔK_{th}	n	A	Ref.
12Cr S.S.	730	102	0	5.00	3.5	1.70×10^{-9}	[71]
304 S.S.	1400	66	0	4.60	3.6	4.77×10^{-11}	[70]
403 S.S.	100	115	0.50	5.52	2.6	1.90×10^{-11}	[1]
Ti-6Al-4V	1172	70	0.40	14.7	4.0	1.00×10^{-12}	[81]
Ti-6Al-4V	880	91	0.02	5.00	3.6	1.91×10^{-13}	[63]
AerMet 100	1724	34	0.5	2.90	3.0	2.00×10^{-11}	[51][a]
Nb-37Ti-13Cr	1107	26	0.10	7.00	5.5	3.40×10^{-14}	[39]
Nb-15Al-49Ti	660	110	0.10	5.00	9.7	5.00×10^{-19}	[43]

K_{IC}, ΔK_{th}, and ΔK in MPa $\sqrt{m}$ and A in MN^{-n} m$^{1-n/2}$/cycle. S.S. = Stainless
steel
[a]$T = -171\,^{\circ}C$

Here, σ_F is called the flow stress and it is simply defined as a arithmetic mean
(arithmetic average)

$$\sigma_F = \frac{\sigma_{ys} + \sigma_{ts}}{2} \tag{9.21}$$

where ΔK_{tran} is the transition stress intensity factor, which can be set equals to the
upper limit of the valid range of ΔK in stage II. Actually, σ_F is the flow stress for
this transition and E is the modulus of elasticity.

In addition, Fig. 9.11 shows recently published da/dN data using compact
tension (CT) specimens according to the ASTM E647 Standard Test Method for
Measurement of Fatigue Crack Growth and a specific software [64]. The testing
material apparently was a steel alloy. From this figure, the threshold stress intensity
factor approximately $\Delta K_{th} \simeq 7.8$ MPa $\sqrt{m}$ agrees with the information given in
Fig. 9.8 that $\Delta K_{th} < 9$ MPa $\sqrt{m}$.

The reason for selecting this particular data is to inform the reader that fracture
mechanics tests can be conducted using state-of-the-art instrumentation accompanied with reliable software to speed up calculations and avoid human errors. Hence,
modern testing equipment and instrumentation are nowadays available as state-of-
the-art closed loop uniaxial and biaxial systems with computer automation and
digital data acquisition. Reliable instrumentation exists for accurate displacement
and strain measurements.

The curve fitting equation for the data given in Fig. 9.11 is a polynomial of second
degree, which represents an average $da/dN = (\Delta K)$ curve.

$$\frac{da}{dN} = 2 \times 10^7 \, (\Delta K)^2 - 10^7 \, (\Delta K) - 4 \times 10^7 \tag{9.22}$$

This numerical approach is common in experimental data analysis, and the resultant
equation does not resemble any of the above empirical and semiempirical models
for fatigue crack growth rate.

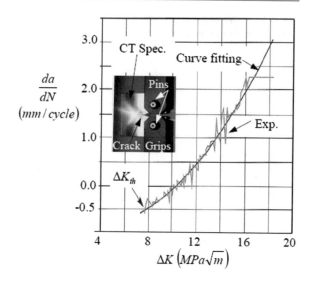

Fig. 9.11 Experimental fatigue crack growth data for a compact tension (CT) steel specimen [64]

   The purpose of this curve fitting is to demonstrate the use of a numerical method for obtaining an average curve fitting curve for fluctuating experimental data. For instance, the ASTM Standard E647 includes a numerical method for determining the da/dN derivatives, as schematically shown in Fig. 9.7. In addition, certain fracture mechanics problems, to a large extent, can be analyzed and solved using numerical methods such as boundary element method (BEM) and finite element method (FEM) which are specialized numerical topics. The reader can find books on BEM and FEM available in the literature.

   Furthermore, from a metallurgical point of view, Paris [53] compared fatigue crack growth rate data for FCC, BCC, and HCP metallic solids. For convenience, only FCC and BCC data are shown in Fig. 9.12. Therefore, it is clear that this data set correlates with the Paris law (Eq. (9.6)).

   An analogous da/dN behavior for polymers is depicted in Fig. 9.13. Therefore, fatigue crack growth behavior is mathematically described by the function $da/dN = f(\Delta K)$, the simplest being the Paris equation.

   The polymeric materials and their da/dN profiles depicted in Fig. 9.13 exhibit significantly different mechanical behaviors when subjected to fatigue loading. It is expected that those polymers showing high fatigue crack growth rates have inferior fatigue strength.

   In general, the fatigue behavior of engineering materials must be characterized using crack-free and cracked specimens for comparing results and determining the degree of fatigue resistance reduction due to the presence of a crack. Thus, assessing structural integrity requires a careful experimental setup to simulate service conditions.

Fig. 9.12 Comparison of
FCC and BCC crack growth
rates [53]

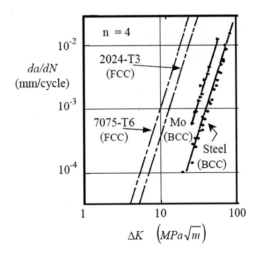

Fig. 9.13 Fatigue crack growth behavior of polymers and two metals [40]

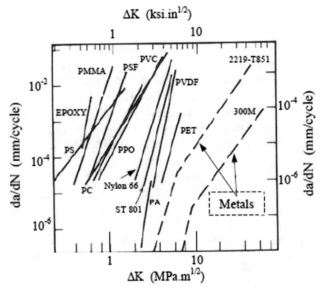

9.7   Weldments

Welding is a fabrication process for joining two or more parts to form one single part. The joining process is a localized metallurgical process related to solidification. The resultant welded area is metallurgically heterogeneous with respect to microstructural features. Nevertheless, the joining process can be achieved by laser beam welding (LBW), arc and gas welding, brazing, and soldering using relatively

Fig. 9.14 Fatigue crack growth behavior of pressure vessel steel weldments in air at 24 °C [32]. This reference was taken from Lancaster [36]

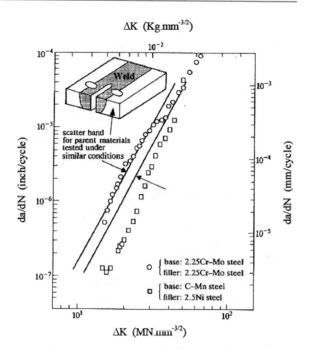

low melting temperature metals (Pb-Sn), explosive welding, ultrasonic welding, and friction stir welding technique [47]. The reader should consult the American Welding Society, the Welding Institute of the United Kingdom, and a new technical article published by the American Society for Metals International. Figure 9.14 shows a typical plot for welded pressure vessel steel specimens [32, 36].

Figure 9.15 depicts Viswanathan's model [76] for microstructural changes in the heat-affected zone (HAZ) adjacent to the weld, which in turn, is related to the Fe-C phase diagram. This schematic representation of the microstructural changes in the base metal with respect to grain size and grain morphology is typical in submerged arc welding (*SMAW*) process of Cr-Mo steels. Consequently, mechanical properties in the HAZ are affected by the cooling rate within the welded region and grain size in the HAZ. In alloy steels, martensite may form as an undesirable phase because it is brittle. The different microstructures are labeled schematically in Fig. 9.15 as zones. These different microstructural zones disturb the microstructural symmetry of the parent metal, and as a consequence, mechanical and corrosion properties vary through the HAZ. One particular solution to this metallurgical problem is to heat treat the welded part whenever possible to obtain a uniform microstructure.

In general, welding cracks or HAZ cracking may occur due to tensile residual stresses that develop upon contraction of the welding bead. Therefore, welding should be done properly in order to avoid welding defects because they have a strongly affect mechanical properties.

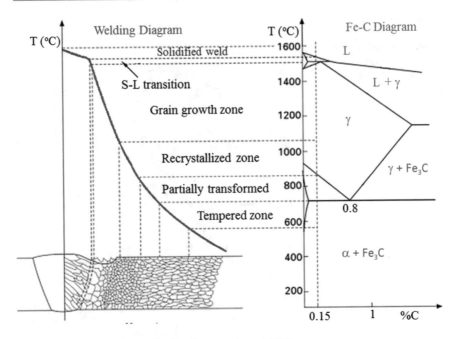

Fig. 9.15 Heat-affected zone (HAZ) adjacent to the weld [76]

The Paris law is the most common approach for correlating da/dN with ΔK in weldments. In fact, the weld bead or solid-phase weld is the part of the joined materials that is tested and characterized. It is apparent that a good correlation exists for $da/dN = f(\Delta K)$ in stage II, where the Paris law describes the observed fatigue behavior of welded specimens [32].

The assessment of welded structures is very complicated due to the complexity of weld deposit, but is important to carry out experiments on welded specimen in order to improve the usage of welded material being dynamically loaded by fatigue loading. Sometimes fracture of a structural component initiates from a defect being introduced by an inadequately welding procedure. Therefore, an optimum fatigue design method for testing welded components is a must in order to integrate appropriate welding procedures in particular welded structures, such as ship hulls and the like.

9.8   Surface Fracture Appearances

This section is focused on the appearance of fracture surfaces that are vital in failure analysis in conjunction with knowledge of the load history in a particular environment. Figure 9.16 shows Schijve's model [67] that provides details of a fracture surface exhibiting typical ductile fracture features encountered in many engineering materials under quasi-static loads. These features are affected by

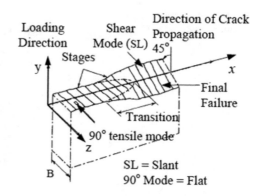

Fig. 9.16 Schematic fracture surface showing the transition from flat to slant appearance during quasi-static loading [67]

applied stress conditions, specimen geometry, flaw size, mechanical properties, and the environment. Subsequent fatigue crack growth or crack extension is associated with increasing applied stress, which in turn increases the stress intensity factor. Some solid materials can show a shear lip as an indicative of ductile fracture, and as a result, the fracture surface exhibits a slant (*SL*) area. According to Fig. 9.16, the crack surface is initially flat, and as crack growth progresses, it continues on its own plane and then changes direction at 45°, leaving a shear lip on the edge. Some materials exhibit double-shear lips which are indications of ductile fracture due to an overload. Conversely, a brittle fracture surface is normally flat without shear lips.

In addition, environmental effects, such as low and high temperatures and corrosive media, can have a significant impact on the mechanical behavior and fracture appearances of solid bodies. For instance, a ductile steel alloy may become brittle at relatively low temperatures.

Stage II Fatigue Failure This is due to a change in crack growth direction of stage I as shown in Fig. 9.6, in which the crack in a polycrystalline material advances along crystallographic planes of high shear stress. The surface characteristics for fatigue failure in stage II are schematically depicted in Fig. 9.17 for a round solid cylinder such as a shaft. In fact, fatigue cracks often produce striations (small ridges) on the fracture surface perpendicular to the direction of crack growth plane.

The possible fatigue fracture surface events during stable crack growth are described as follows:

- Crack growth occurs by repetitive plastic blunting and sharpening of the crack front.
- Shear deformation direction reverts to complete a full cycle in compression. This event may cause cleavage fracture in brittle materials.
- If rapid crack growth rate occurs, then rapid failure takes place, and beach marks and striations may be absent, regardless if the material is ductile or brittle.
- Intercrystalline fracture is possible, particularly at the lower range of stress. When inherent fracture mode of a material takes place during crack growth, a mixture of brittle transcrystalline and brittle intercrystalline fracture might be

Fig. 9.17 Schematic fatigue fracture surface

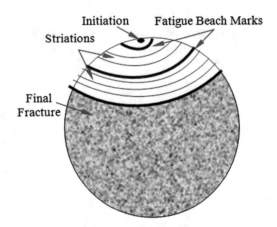

evident. However the latter depends significantly on grain size and secondary phases present in the microstructure since they are obstacles for straightforwardly dislocation motion.

- Formation of striations depends on the nature of the materials, such as aluminum and Al alloys. However, steels may exhibit cleavage mechanism as a dominant fracture mode. Apparently, striations are formed by the instantaneous action of the cyclic stress, stress frequency, and plastic strain.
- Striations indicate the changing position of the crack front with each new cycle of loading. Striations are not visible to the naked eye, but they can be revealed using SEM high magnifications.
- Ripple (annual ring) patterns can form on the fracture surface.
- The domain of high-cycle fatigue prevails during stage II.

Model for the formation of striations: Figure 9.18 shows Broek's possible mechanism [8] for the formation of striations in certain materials, such as aluminum alloys and in some strain-hardened alloys.

The possible stages during the formation of striations are

(1 & 2) Slip formation occurs at the crack tip due to a stress concentration. Slips form in the direction of maximum shear stress as explained by the Cottrell–Hull mechanism (Fig. 9.6), and the crack opens and extends in length (Δa). A particular model for a slip system, $(100)[\bar{1}11]$, in BCC structure is shown in Fig. 9.3b.

(3) Other slip planes are activated and consequently, cross-slip may occur.

(4) Crack tip blunting occurs due to strain hardening, which may activate other slip planes.

(5) The crack re-sharpens due to plastic deformation (plastic zone) embedded in the elastic surroundings. During load release, the elastic surroundings excerpt compressive stresses on the plastic zone. This reversed plastic deformation process closes and re-sharpens the crack tip.

Fig. 9.18 Model for the formation of fatigue striations [8]

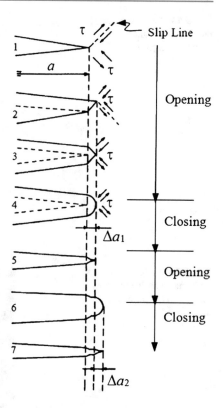

(6 & 7) Crack closure and re-sharpen occur due to repeated loading, leading to more crack growth (extension).

The cyclic opening and closing of the crack develops a typical pattern of ripples called fatigue striations. The above model (Fig. 9.18) of striation formation is a general representation of crack blunting and re-sharpening in ductile or sufficiently ductile materials. However, a cleavage mechanism may involve brittle striations, as opposed to ductile striations.

In addition, Fig. 9.19 shows a unique microphotograph of wavy slip lines in BCC niobium (Nb) [34]. This particular striation morphology can be attributed to cross-slip mechanism since it induces not only the wavy slip bands but tangled dislocations as well. Hence, the wavy slip pattern can be attributed to an easy cross-slip. Nonetheless, cross-slip can produce wavy slip patterns in certain materials and planar slip in other materials.

On the other hand, cross-slip can directly be related to stacking-fault energy associated with plastic deformation of a material. This implies that the higher the stacking-fault energy, the more wavy the slip pattern.

Fig. 9.19 Wavy slip lines in niobium after static deformation [34]

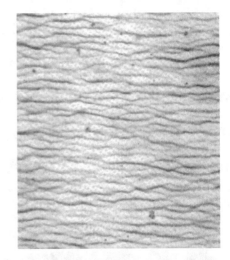

Fig. 9.20 Fatigue striations on a crack surface of an Al alloy. Magnification: 12,000X [30]

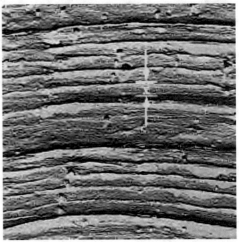

Figure 9.20 shows the characteristic fatigue striations of an Al alloy having a modulus of elasticity approximately equal to $E = 72{,}000$ MPa. These striations are ripples on the fractured surface caused by perturbations in the cyclic stress system. The width of a striation depends on the fatigue stress, but it is in the order of 10^{-4} mm or less.

Furthermore, the apparent stress intensity factor range is related to striation spacing as empirically proposed by Bates and Clark [6]

$$\Delta K = E \sqrt{\frac{x}{6}} \qquad (9.23)$$

where $E = $ Modulus of elasticity (MPa)

$x = $ Average striation spacing (μm)

Certainly, fatigue striation spacing, as well as height, relates to fatigue crack growth, but it seems impossible to determine the minimum and maximum loads from the fatigue fracture surface. Again, it turns out that Eq. (9.23) gives a reasonable ΔK value for a structural component subjected to specific fatigue service conditions.

The striation spacing is a measure of slow crack growth per stress cycle and it may be constant for constant stress amplitude. However, striations may not form when the stress range and the maximum stress are relatively large, leading to fast fatigue crack growth rate

Fatigue fracture surface exhibits features which can aid in understanding the mechanism associated with fracture. For instance, fatigue crack growth, specifically in ductile materials, normally produces thousands of striations which assist in determining the directionality of crack propagation and the location of the initiation site. However, fatigue mechanical damage by friction of the fracture surfaces can occur during opening and closing the crack, and as a result, fine striations are not revealed using SEM and TEM techniques. Consequently, the average striation spacing using the visible striations may not be suitable enough for statistical purposes.

Example 9.3. *Determine the apparent stress intensity factor range and the fatigue crack growth rate using the aluminum alloy fracture surface shown in Fig. 9.20. Use a modulus of elasticity of 72,000 MPa.*

Solution.

The solution to this problem requires that the actual average striation spacing be determined using the magnification given in Fig. 9.20. The average striation spacing is approximately

$$x = \frac{4 \text{ mm}}{12,000} = 3.3333 \times 10^{-4} \text{ mm} = 333.33 \text{ } \mu m$$

Then, using Eq. (9.23) yields the apparent stress intensity factor range

$$\Delta K = E \sqrt{\frac{x}{6}} = (72,000 \text{MPa}) \sqrt{\frac{4 \times 10^{-3} \text{ m}}{(6)(12,000)}} \tag{9.23}$$

$$\Delta K \simeq 17 \text{ MPa } \sqrt{m}$$

Since the striation spacing is a measure of slow crack growth per stress cycle, the estimated fatigue crack growth rate can easily be calculated using the following approximation:

$$\frac{da}{dN} \simeq \frac{\Delta a}{\Delta N} = 333.33 \frac{\mu m}{cycle}$$

Therefore, crack extension for one cycle is simply 333.33 μm.

For convenience, plot the ΔKempirical equation to reveal the ΔK trend with increasing striation spacing x, that is, $\Delta K = f(x)$.

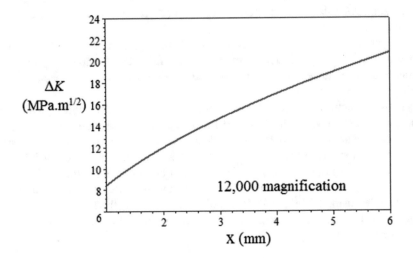

The plot exhibits a slight nonlinearity trend. This implies that ΔK relates to striation spacing in a nearly linear manner.

Example 9.4. *Let us use Fuchs and Stephen Problem 8.13 [22] with reference to a fractograph reported by Crooker et al. [14] to illustrate another approximation technique in fatigue failure analysis. One 17-4 HP stainless steel plate containing a 6-mm single-edge crack was subjected to a constant cyclic loading with a stress ratio of R = 0. The plate was 5-mm thick, 20-mm wide, and sufficiently long. The crack growth rate as per Paris equation is*

$$\frac{da}{dN} = \left(10^{-12} \text{ m/cycle}\right) (\Delta K)^{3.5}$$

Calculate the apparent stress intensity factor range and the maximum load. Explain the results.

Solution.

Given data: $a = 6$ mm, $w = 20$ mm, $B = 5$ mm, $b = w - a = 14$ mm

$$a/w = 0.3, R = K_{min}/K_{max} = 0, K_{min} = 0$$

Firstly, the geometry correction factor for this crack configuration is given in Table 3.1. Thus,

$$\alpha = 1.12 - 0.23 \left(\frac{a}{w}\right) + 10.55 \left(\frac{a}{w}\right)^2 - 21.71 \left(\frac{a}{w}\right)^3 + 30.38 \left(\frac{a}{w}\right)^4$$

$$\alpha = 1.66$$

From the micron marker on the fractograph, linear interpolation allows the determination of the fatigue crack growth rate. Firstly, the average striation spacing is calculated using linear interpolation. The micron bar is approximately 13-mm long and it is equivalent to 1 μm as the magnification factor. Thus,

$$13 \text{ mm} \rightarrow 1 \,\mu\text{m}$$

$$1 \text{ mm} \rightarrow x$$

$$x = \frac{(1 \text{ mm}) (1 \,\mu\text{m})}{13 \text{ mm}} = 7.69 \times 10^{-2} \,\mu\text{m}$$

$$x = 7.69 \times 10^{-8} \text{ m}$$

For one cycle, the average striation spacing represents the crack growth per cycle. Thus,

$$\frac{da}{dN} \simeq \frac{\Delta a}{\Delta N} = \frac{x}{\Delta N} = 7.69 \times 10^{-8} \text{ m/cycle}$$

Solving the Paris equation for the stress intensity factor range, ΔK yields the numerical result equals to K_{max}. Hence,

$$\Delta K = \left(\frac{da/dN}{A}\right)^{1/n} = \left(\frac{7.69 \times 10^{-8}}{10^{-12}}\right)^{1/3.5}$$

$$\Delta K = 24.89 \text{ MPa} \sqrt{\text{m}} = K_{max}$$

For comparison purposes, Eq. (9.23) along with $E = 207$ GPa gives

$$\Delta K \simeq E \sqrt{\frac{x}{6}} = (207 \times 10^3 \text{ MPa}) \sqrt{\frac{7.69 \times 10^{-8} \text{ m}}{6}}$$

$$\Delta K \simeq 23.44 \text{ MPa } \sqrt{m} = K_{max}$$

which is approximately 6 % lower than the previous result. Therefore, both methods can yield reasonable results.

Secondly, taking $\Delta K \simeq 24.17$ MPa $\sqrt{m}$ as the average value, one can calculate the maximum load assuming a zero minimum. Hence, $\Delta K = K_{max}$ and P_{max} is

$$K_{max} = \alpha \sigma_{max} \sqrt{\pi a} = \frac{\alpha P_{max}}{Bw}\bigg|_{max} \sqrt{\pi a}$$

$$\sigma_{max} = \frac{K_{max}}{\alpha \sqrt{\pi a}} = \frac{24.17 \text{ MPa } \sqrt{m}}{(1.66) \sqrt{\pi (6 \times 10^{-3} \text{ m})}} = 106.05 \text{ MPa}$$

$$P_{max} = Bw\sigma_{max} = \left(5 \times 10^{-3} \text{ m}\right)\left(20 \times 10^{-3} \text{ m}\right)\left(106.05 \ \frac{MN}{m^2}\right)$$

$$P_{max} = 10.61 \text{ kN}$$

Despite the undertaken approximation for determining the apparent stress intensity factor range and related striation spacing, the preceding calculations denote acceptable or reasonable results using quantitative fractography, which is a microscopic technique for revealing features on fractured surfaces being related to loading modes and certain characteristics of fracture mechanics.

According to the above results, Eq. (9.23) is a useful empirical relationship because it provides quantitative information for estimating ΔK. Despite that a striation spacing relates to the position of an advancing crack front and it increases with increasing ΔK since $S_K \propto \Delta K^2$. Finally, S_K represents discrete crack advance increments as depicted in Fig. 9.20.

9.9 Mixed-Mode Fatigue Loading

In this section, the physical basis and application conditions of fracture mechanics theory concerned with mixed-mode fatigue loadings are considered for characterizing crack growth rate and for developing a mixed-mode fatigue fracture criterion.

The slow fatigue crack growth ΔK dependency in elastic solids has been investigated under mixed-mode interactions based on remote tension [2, 46] and biaxial cyclic loadings [2, 28, 35, 37, 46]. A remarkable observation is that ΔK_{th} decreases as the stress ratio R increases as in the case of mode I loading depicted in Fig. 9.10. However, if the starting crack size is large and the biaxial stress level is low, da/dN is independent of ΔK [35].

In conducting mixed-mode fatigue studies, a defined effective stress intensity factor ΔK_e may be used in the Paris law for brittle [2,28,35,37,46] or ductile [60,79] materials. Effectively, the definition of $\Delta K_e = \Delta K_{eff}$ depends on the mathematical technique and theoretical background one uses. Nevertheless, the Paris law [53,55] takes the general and empirical form

$$\frac{da}{dN} = A \, (\Delta K_e)^n \tag{9.24}$$

For a mixed-mode I and II interaction, Eqs. (8.17) and (8.18) may be used to defined the effective stress intensity factor range as

$$\Delta K_e^2 = \Delta K_I^2 + \Delta K_{II}^2 \tag{9.25}$$

$$\Delta K_e^2 = \pi a \, (\Delta \sigma)^2 \sin \beta \tag{9.26}$$

For pure mode I at fracture, $\Delta K_e = K_{IC}$ (similar argument was presented in Chap. 8) for $R = 0$ and Eqs. (9.25) and (9.26) become

$$K_{IC}^2 = \Delta K_I^2 + \Delta K_{II}^2 \tag{9.27}$$

$$K_{IC}^2 = \pi a_c \, (\Delta \sigma)^2 \cos^2 \beta \tag{9.28}$$

Combining Eqs. (8.15), (8.16), (9.27), and (9.28) yields an expression similar to Eq. (9.18)

$$D_1 \, (\Delta \sigma)^n + D_2 \, (\Delta \sigma)^{n-2} - D_3 = 0 \quad \text{for } n \neq 0 \tag{9.29}$$

where

$$
\begin{aligned}
D_1 &= A \, (n-2) \, (N - N_o) \, (K_{IC})^n \\[2mm]
D_2 &= \frac{2}{\pi} \left(\frac{K_{IC}}{\alpha \sin \beta} \right)^2 \\[2mm]
D_3 &= 2 \, (a)^{1-n/2} \left(\frac{1}{\pi} \right)^{n/2} \left(\frac{K_{IC}}{\alpha \sin \beta} \right)^n
\end{aligned}
\tag{9.30}
$$

An example can make the above analytical procedure clearly usable for solving mixed-mode problems under cyclic stress systems.

Example 9.5. *Assume that a solid cylinder of 25 mm in diameter has a round surface crack inclined at $\beta = 20°$ and that the material has an average plane-strain fracture toughness and threshold stress intensity factor of 15 MPa $\sqrt{m}$ and 5 MPa $\sqrt{m}$, respectively. If the crack depth is 0.09 mm and the applied cyclic stresses are $\Delta \sigma = \sigma_{max}$ and $\sigma_{min} = 0$, calculate (a) the minimum stress range $\Delta \sigma_{min}$, (b) the applied stress range $\Delta \sigma = \sigma_{max}$, and (c) the critical length a_c for a fatigue life of 10^4 cycles. How much will the crack grow? The Paris equation is*

$$\frac{da}{dN} = \left(5 \times 10^{-12} \, \frac{\mathrm{MN^{-4} \, m^{-1}}}{cycles} \right) (\Delta K)^4$$

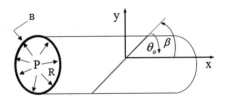

Solution.

(a) *Letting $\Delta K_{th} = \Delta K_e$ gives*

$$\Delta\sigma_{th} = \frac{\Delta K_{th}}{\alpha\sqrt{\pi a_o}} = \frac{5 \text{ MPa }\sqrt{m}}{(1.198\,9)\sqrt{\pi\,(0.09\times10^{-3}\text{ m})}\sin(20)} = 776.25 \text{ MPa}$$

$$\Delta\sigma_{min} < \Delta\sigma_{th}$$

(b) *For maximum stress range, Eq. (9.29) gives*

$$D_1 = A\,(n-2)\,(N-N_o)\,(K_{IC})^n = 5.0625\times10^{-3} \text{ m}$$

$$D_2 = \frac{2}{\pi}\left(\frac{K_{IC}}{\alpha\sin\beta}\right)^2 = 976.16 \text{ MPa}^2\text{ m} \qquad (9.30)$$

$$D_3 = 2\,(a)^{1-n/2}\left(\frac{1}{\pi}\right)^{n/2}\left(\frac{K_{IC}}{\alpha\sin\beta}\right)^n = 5.2939\times10^9 \text{ MPa}^4\text{ m}$$

$$0 = 5.0625\times10^{-3}\,(\Delta\sigma)^4 + 976.16\,(\Delta\sigma)^2 - 5.2939\times10^9$$

$$\Delta\sigma = 964.74 \text{ MPa}$$

(c) *From Eq. (9.28), the critical crack size is*

$$a_c = \frac{K_{IC}^2}{\pi\,(\Delta\sigma)^2\cos^2\beta} = \frac{\left(15 \text{ MPa }\sqrt{m}\right)^2}{\pi\,(964.74 \text{ MPa})^2\cos^2(\pi/9)} = 0.09 \text{ mm}$$

$$\Delta a = a_c - a_o = 0.09 \text{ mm} - 0.09 \text{ mm} = 0$$

Therefore, the crack grew $\Delta a = 0$ which represents a pure brittle material.

9.9.1   Crack Growth Rate Measurements

Characterizing a particular fatigue crack growth behavior involves the determination of uncertainties in crack length (a), stress intensity factor range (ΔK), and the crack growth rate da/dn in a specific environment. The rationalization to organize and integrate an experimental setup or to analyze fatigue data for characterizing fatigue crack growth rate, based on geometry-independent or geometry-dependent specimens, should incorporate elastic and elastic-plastic stress fields in front of a crack. Instead, fatigue crack growth rate assessment is conventionally based on crack extension after N number of cycles are consumed. For instance, ASTM E647, "Standard Method for Measurement of Fatigue Crack Growth Rates," Sect. 9.3, Volume 03.01 covers the recommended general procedures for such a purpose assuming steady-state fatigue crack growth rates.

However, the ASTM E647 recommended experimental procedure includes certain restrictions for conducting fatigue tests. Therefore, details on this matter must be taken into account for providing valid experimental data. Basically, fatigue crack growth is affected by test conditions such as test cyclic frequency and waveform, strain amplitude and strain rate, temperature and aggressive environments, and residual stresses induced by a manufacturing process.

In principle, Griffith strain energy criterion for characterizing fatigue crack growth rate under mode I or even mixed-mode (Eq. (9.27)) loading offers a deterministic mathematical model in which the outcomes are precisely determined through known stress intensity factor(s).

In general, for small increments of the crack length in the order of 1 mm or less, the following procedure gives acceptable results, that is, the fatigue crack growth rate can be approximated by

$$\frac{da}{dN} \simeq \frac{a_i - a_{i-1}}{N_i - N_{i-1}} \quad \text{for } i = 1, 2, 3, 4, 5, \dots \quad (9.31)$$

Then, compute the average crack length and the specimen geometry correction factor, respectively, as

$$\bar{a} = \frac{a_i + a_{i-1}}{2} \quad (9.32)$$

$$\alpha = f(\bar{a}/w) \quad (9.33)$$

For $i = 0$, $a_i = a_o$, and $N_i = N = 0$ since the initial crack length exists. The American Society for Testing Materials (ASTM) E647 Standard Test Method is widely used for measuring the crack length and the elapsed fatigue cycles at constant loading stress amplitude and frequency. This method deals with the procedure for determining low and high steady-state fatigue crack growth rates. This particular test method does not restrict specimen geometry and specimen thickness as long as buckling is precluded and the specimen ligament is large enough.

9.10   Stress-Induced Corrosion

In general, stress-induced corrosion is a phenomenon caused by the combination of quasi-static or cyclic stress and a corrosive environment (hostile chemical solution). If a material is susceptible to deteriorate under these conditions, then a corrosion behavior is established. However, if the material undergoes anodic dissolution at the crack tip, a stress corrosion cracking (SCC) mechanism dominates with the aid of the static or cyclic stress. Secondary cracks may develop on the surface of a component, while the crack tip dissolves and the crack growth rate increases.

The purpose of conducting SCC experiments is to determine the effects of a particular fluid, temperature, applied strain rate, or applied voltage on solid bodies. Hence, the SCC mechanism can prevail since it occurs on ductile and brittle material surfaces containing initially smooth surfaces.

Secondary crack formation on a smooth surface prevails as the source for a major crack to grow statically and cyclically (dynamically). Figure 9.21 shows tensile fracture surfaces of a 304 stainless steel tested in 0.1 N H_2SO_4 solution at room temperature and at strain rate of 5.5×10^{-5} s^{-1} [56]. This material was produced by rapidly solidification and subsequently, consolidated and 50 % cold rolled. Smooth rods were prepared [56] for conducting slow strain rate (SSR) stress corrosion cracking tests in the mentioned environment.

Figure 9.21a also shows secondary cracks tested at −200 mV under the same environmental conditions, but the primary crack grew in a semielliptical manner from opposite sides of the specimen, and the final fracture area due to a stress overload exhibited an elliptical configuration as illustrated by the SEM image in

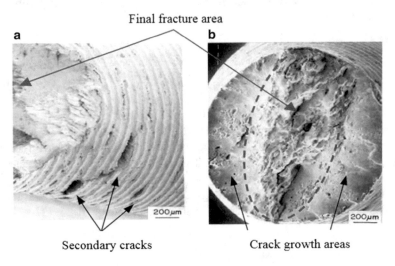

Fig. 9.21 Secondary cracks and overload fracture areas in a 50 % cold rolled (CR) RSA 304 stainless steel [56]

Fig. 9.21b. A typical SCC mechanism is the formation of secondary cracks on the specimen gage length (Fig. 9.21a).

If a component has an initial crack, the fatigue crack growth rate process does not include an incubation period as in testing smooth components. Generally, the corrosion fatigue behavior relates to a high plastic strain at the crack tip [20], which is suitable for anodic dissolution. Apparently, the combination of high plastic strains enhances metal dissolution, which in turn, accelerates fatigue crack growth rates. In addition, if a material is susceptible to develop beach marks and striations as fatigue fracture features in a suitable environment, these features may not be observed or not be cleared enough if metal dissolution takes place on the fracture surface, coating these features with a corrosive product. This corrosion product may be difficult to remove; however, the ultrasonic cleaning technique may be appropriate for this task.

On the other hand, if the corrosive environment, containing hydrogen ions, does not provoke metal dissolution at the crack tip, the phenomenon is called hydrogen embrittlement. This mechanism is highly localized inducing brittle regions that develop at the crack tip [23, 75, 78]. This, then, indicates that the applied cyclic stresses, which induce cyclic strains at the crack tip, and the action of hydrogen atoms, enhance the crack growth rate due to an accelerated breakage of atomic bonds at the crack tip. Thus, atomic hydrogen (H), as oppose to molecular hydrogen (H_2), diffuses into the metal, especially if the amount of hydrogen exceeds that of the solubility limit, at favorable atomic sites at the crack tip. Therefore, the accelerated crack growth rate, specifically in stage II fatigue, may be attributed to this hydrogen diffusion-controlled mechanism since the atomic hydrogen radius is relatively small. These atomic sites are grain boundaries, voids, and inclusions.

However, atomic hydrogen can precipitate in a gaseous or solid form when it reacts with the exposed metal, such as irons and steels, under appropriate environmental (thermodynamics) conditions. Apparently, gaseous precipitation of hydrogen atoms located at these sites react to form hydrogen molecules, which in turn combine themselves to form bubbles at extremely high pressure of 1.3 GPa as an upper limit [20]. Moreover, the solid precipitation of hydrogen with Ti and Zr, among many other elements, is referred to as hydride precipitation, which hardens the material in question [13].

Many investigators have reported the application of linear-elastic fracture mechanics (LEFM) to many material-environment systems, and the literature in this particular engineering field is quite abundant for Al alloys, Fe alloys (steels), Ti alloys, and so on. However, the controlling macroscopic parameter for assessing crack growth has been the stress intensity factor K_I for mode I loading, the most common loading mode in fracture mechanics.

In general, many materials are sensitive to combinations of stress, environments, and microstructure. One particular case is shown in Fig. 9.22 for a Ni-base alloy steel tested in 3 % NaCl saline solution and in air at different test frequencies (cpm = cycles per minute) [29].

Despite that the experimental data obeyed the Paris equation with a common exponent of 2, the environmental effect on the da/dN for the Ni-base alloy steel is evident even at $K_{max} < K_{ISCC}$, which contradicts the stress corrosion cracking

Fig. 9.22 Corrosion fatigue
crack growth behavior as a
function of different test
frequency for 12Ni-5Cr-3Mo
alloy steel in 3 % NaCl
solution and air. All test
conducted at
$K_{max} < K_{ISCC}$ [29]

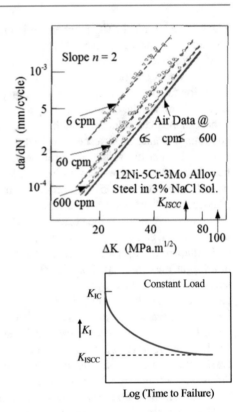

Fig. 9.23 Stress intensity
factor as a function time to
break

behavior depicted in Fig. 9.23. The term K_{ISCC} stands for the stress intensity
factor for mode I loading below which stress corrosion cracking does not occur.
One possible reason for this discrepancy is that crack growth does not occur at
$K_{max} < K_{ISCC}$ for constant load tests due to a protective passive film at the crack
tip. However, this film is sensitive to cyclic stresses and consequently, fatigue
crack growth occurs [27]. The phrase stress corrosion cracking (SCC) is part of
the environment-assisted cracking (EAC) field.

Figure 9.23 illustrates important criteria described in Hertzberg's book [27] when
stress intensity factor is time dependent at constant load. Thus,

- If $K_I < K_{ISCC}$, then failure is not expected in an aggressive or corrosive fluid.
- If $K_{ISCC} < K_I < K_{IC}$, then crack growth and fracture occur after a prolong period
  of time.
- If $K_I > K_{IC}$, then sudden fracture is expected upon loading.

Figure 9.23 represents the possible mechanical behavior of a stressed structural
component exposed to an aggressive environment during service. Therefore, care
should be taken when using this type of data in designing against stress corrosion
cracking.

Fig. 9.24 Experimental crack growth rate data sets for fatigue assessment of $Zr_{41.2}Ti_{13.8}Cu_{12.5}Ni_{10}Be_{22.5}$ under sinusoidal loading and in different common environments [68]

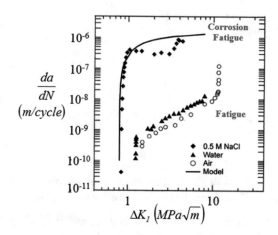

In addition, Fig. 9.24 shows fatigue and corrosion fatigue of a Zr-based glass forming metallic alloy [68, 69].

This alloy exhibits fatigue crack growth (da/dN) behavior comparable to ductile crystalline metals in deionized water and ambient air. However, da/dN is very sensitive in 0.5 M NaCl solution at room temperature. This corrosion fatigue behavior shows a plateau at approximately 2×10^{-7} m/cycle for 1 MPa $\sqrt{m} \leq \Delta K_I \leq 3$ MPa $\sqrt{m}$, and it is an indication of stress-corrosion fatigue. According to Schroeder et al. [68], the plateau in the corrosion fatigue curve occurs about $da/dN = 4 \times 10^{-7}$ m/cycle at a frequency of approximately 25 cycle/s, which is equivalent to a crack velocity of [68]

$$\frac{da}{dt} = CK_{\max}^m = 10^{-5} \text{ m/s}$$

It is likely that high da/dN and da/dt values in the NaCl solution may be attributed to a slow and incomplete repair of stress-induced damage to the oxide film at the crack tip [68]. Nonetheless, fatigue crack growth rate in NaCl can be defined as [68, 69]

$$\frac{da}{dN} = \int_o^{1/f} \left(\frac{da}{dt}\right)_{SCC} dt \qquad (9.34)$$

where f in the integral is the applied frequency.

In addition, the SCC can mathematically be evaluated using an expression for the rate of crack growth or crack velocity (da/dt) defined as

$$\frac{da}{dt} = CK^m \qquad (9.35)$$

where da = Crack growth

dt = Time interval

Fig. 9.25 Crack velocity
profiles

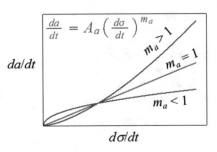

C, m = Material constants which depend on the environment
K = Stress intensity factor for a particular stress mode

Commonly, stress-induced corrosion (SIC) damage under continues action of electrochemical reactions is a dynamic process during which (1) metal atoms on the specimen surface become removable ions from weakest areas at the crack tip and (2) the mechanical surface damage induces the formation of dislocations along preferred slip planes. In this case, inherent damage is due to a combination of stress and anodic reactions at the crack tip. From a macroscale standpoint, the severity of fatigue damage can be characterized by determining the effect of an applied stress rate $d\sigma/dt$, which causes crack growth to behave in a particular manner and induces a strain rate $d\epsilon/dt$ at the crack front.

Generalizing the crack velocity (da/dt) and the strain rate $(d\epsilon/dt)$ as stress rate-dependent entities in the stress-induced corrosion field, one can define them as Paris type equations. For steady crack growth conditions, da/dt and ϵ/dt are defined in closed form as [12, 61, 62]

$$\frac{da}{dt} = A_a \left(\frac{d\sigma}{dt} \right)^{m_a} \tag{9.36}$$

$$\frac{d\epsilon}{dt} = A_\epsilon \left(\frac{d\sigma}{dt} \right)^{m_\epsilon} \tag{9.37}$$

where $A_a = a_c/\left(\sigma_f - \sigma_{th}\right)$ = Constant as per Ref. [61]

a_c = Critical or maximum crack length
σ_f = Fracture stress
σ_{th} = Threshold stress
A_ϵ = Constant
m_a = 1 in Ref. [61]
$m_\epsilon \simeq 1$ in Ref. [12]

Mathematically, $da/dt = f(d\sigma/dt)$ at $0 < m_a < \infty$ can exhibit a unique behavior that depends on the value of M_a. Letting $0.6 \leq m_a \leq 1.5$ in Eq. (9.36) gives the profiles shown in Fig. 9.25. Correspondingly, the strain rate $d\epsilon/dt = f(d\sigma/dt)$ yields similar trends (not shown).

In fact, $da/dt = f(d\sigma/dt)$ and $d\epsilon/dt = f(d\sigma/dt)$ profiles must depend on the properties and microstructural features. Also, A_ϵ must have a meaningful definition dependent on the material's properties.

Plotting Eqs. (9.36) and (9.37) on log-log scale exhibit linearity as the common Paris expression, $da/dN = A(\Delta K)^n$. In light of the above, stress-induced corrosion can be characterized using the effect of steady stress rate, $d\sigma/dt < 1$ MPa/s^{-1}, on crack behavior in a corrosive medium [62].

In order to have a complete assessment on stress-induced corrosion, the fracture surface morphology must be examined to reveal the mechanism of crack growth as intergranular, transgranular, or mixed-mode type. Normally, a stable crack growth is assumed to take place within a range of the applied stress rate for a meaningful characterization of crack growth due to stress-induced corrosion. For instance, Eq. (9.36) predicts that da/dt increases with increasing $d\sigma/dt$. But at very high $d\sigma/dt$, the rate of crack growth, da/dt, is faster than the anodic rate of metal dissolution at the crack tip, and as a result, crack propagation takes place without the effect of the corrosive environment during testing.

9.11 Problems

9.1. Show that Paris equation can take the form

$$da/dN = A\left[E\sigma_{ys}/(1-v^2)\right]^{n/2} \cdot (\Delta\delta_t)^{n/2}$$

where $\Delta\delta_t$ is the change of crack tip opening displacement, E is the modulus of elasticity, and A is a constant.

9.2. (a) Show that $da/dN = C(\Delta K_e)^n$, where $C = A(1-R)^{-n(1-\alpha)}$ and $\Delta K_e = K_{max}(1-R)^\alpha$ is Walker's effective stress intensity factor range. (b) Plot $da/dN = C(\Delta K_e)^n$ and $da/dN = A(\Delta K_e)^n$ for a 4340 steel having $\sigma_{ys} = 1254$ MPa, $\sigma_{ts} = 1296$ MPa, $K_{IC} = 130$ MPa $\sqrt{m}$, $\alpha = 0.42$, $n = 3.24$, $R = 0.7$, and $A = 5.11 \times 10^{-11}$.

9.3. Suppose that a single-edge crack in a plate grows from 2 to 10 mm at a constant loading frequency of 0.020 Hz. The applied stress ratio and the maximum stress are zero and 403 MPa, respectively. The material has a plane-strain fracture toughness of 80 MPa m$^{1/2}$ and a crack growth behavior described by $da/dN = 3.68 \times 10^{-12}(\Delta K)^4$. Here, da/dN and ΔK are in m/cycle and MPa $\sqrt{m}$ units, respectively. Determine the time it takes for rupture to occur. [Solution: $t = 3.87$ h].

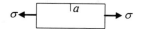

9.4. If a large component is subjected to a cyclic loading under $\Delta\sigma = 300$ MPa and $R = 0$. The material behaves according to Paris law $da/dN = 2 \times 10^{-8} (\Delta K)^{2.45}$, where da/dN and ΔK are in $mm/cycle$ and MPa $\sqrt{m}$ units, respectively. Determine the plane-strain fracture toughness for the component to endure 37,627 cycles so that a single-edge crack grows from 2 mm to a_c.

9.5. Consider a part made of a polycrystalline metal that is stresses in the elastic stress range. If the metal contains inclusions and has an imperfectly smooth exterior surface and natural dislocation, would the metal experience irreversible changes in a microscale? Explain.

9.6. Why most service fatigue fractures are normally not clear?

9.7. What is the physical meaning of the slope of the stage II line in the Paris model?

9.8. Suppose that $d(2a)/dN = 0.001$ mm/$cycle$ and $n = 4$ in the Paris equation for 7075-T6 (FCC), 2024-T3 (FCC), Mo (BCC), and steel (BCC). Determine (a) the constant A and its units and (b) which of these materials will have the higher crack growth rate?

9.9. A *Ti-6Al-4V* large plate containing a 4-mm long central crack is subjected to a steady cyclic loading ($R = 0.10$) The plane strain and the threshold fracture toughness are 70 and 14.7 MPa $\sqrt{m}$, respectively. Determine (a) the minimum stress range, (b) the maximum applied stress range for a fatigue life of 3000 *cycles*, and (c) the critical crack size for 3000 cycles. Let the Paris equation be applicable so that $n = 4$ and $A = 10^{-12}$ MN^4 mm^{-1}/cycle. [Solution: (a) $\Delta\sigma_{min} < 185.45$ MPa, (b) $\Delta\sigma = 642$ MPa, and (c) $a_c = 3.10$ mm]

9.10. Plot $da/dN = f(\Delta K)$ for 403 S.S. using the Paris, Forman, and Broek/Schijve equations. Use the data given in Table 9.2 and a (20 mm)×(300 mm)×(900 mm) plate containing a single-edge crack of 2-mm long. Let 20 MPa $\sqrt{m} \le \Delta K \le 80$ MPa $\sqrt{m}$.

9.11. Plot the data given below and use Eq. (9.6) as the model to draw a curve fitting line on log–log scales. Determine the constants in such an equation. [Solution: $n = 3.50$ and $A = 5.98 \times 10^{-13}$ MPa$^{-3.5}$ m$^{3/4}$]

ΔK	(MPa $\sqrt{m}$)	20	30	40	50	60	70
da/dN	($\times 10^{-7}$ m/$cycle$)	0.241	0.884	2.42	5.29	10	17.20

9.12. A steel plate containing a single-edge crack was subjected to a uniform stress range $\Delta\sigma$ at a stress ratio of zero. Fatigue fracture occurred when the total crack length was 0.03 m. Subsequent fatigue failure analysis revealed a striation spacing per unit cycle of 7.86×10^{-8} m. The hypothetical steel has a modulus of elasticity of 207 GPa. Predict (a) the maximum cyclic stress for a crack length of 0.01 m, (b) the striation spacing per unit cycle when the crack length is 0.02 m, (c) the Paris equation constants, and (d) the plane-strain fracture toughness.

9.13. A 2-cm thick pressure vessel made of a high strength steel welded plates burst at an unknown pressure. Fractographic work using a scanning electron microscope (SEM) revealed a semielliptical fatigue surface crack ($a = 0.1$ cm and $2c = 0.2$ cm) located perpendicular to the hoop stress and nearly in the center of one of the welded plates. The last fatigue band exhibited three striations having an average length of 0.34 mm at 10,000 magnification. The vessel internal diameter was 10 cm. Calculate (a) the pressure that caused fracture and (b) the time it took for fracture to occur due to pressure fluctuations. Assume a pressure frequency of 0.1 cycles per minute (cpm). Given data: $\sigma_{ys} = 600$ MPa, $E = 207$ GPa, $K_{IC} = 75$ MPa $\sqrt{\text{m}}$, and $da/dN = 4.50 \times 10^{-7} (\Delta K)^2$. [Solution: (a) $P = 154.40$ MPa and (b) $t = 85.30$ min]

Crack

9.14. Show that $a_c = a (K_{IC}/K_{max})^2$.

References

1. L. Abrego, J.A. Begley, in *International Corrosion Forum*, Palmer House, Chicago, 3–7 March 1980
2. N.J. Adams, Effect of biaxial stress on fatigue crack growth and fracture. Eng. Fract. Mech. **5**(4), 983–992 (1973)
3. J.M. Barsom, Ph.D. Dissertation, University of Pittsburgh, PA, 1969
4. J.M. Barsom, in *ASTM STP*, vol. 486 (1971), pp. 1–15
5. J.M. Barsom, S.T. Rolfe, *Fracture and Fatigue in Structure: Application of Fracture Mechanics*, 3rd edn. (American Society for Testing and Materials, Philadelphia, PA, 1999)
6. R.C. Bates, W.G. Clark Jr., Fractography and fracture mechanics. ASM Trans. Q. **62**, 380–389 (1969)
7. A.F. Blom, D.K. Holm, An experimental and numerical study of crack closure. Eng. Fract. Mech. **22**, 997–1011 (1985)
8. D. Broek, *Elementary Engineering Fracture Mechanics*, 4th edn. (Kluwer Academic, Boston, 1986)

9. D. Broek, J. Schijve, National Aerospace Laboratory, Amsterdam, Report NLR-TR-M-2111, 1963
10. R.W. Bush et al., Fracture mechanics characteristics of high strength aluminum alloy forging. ALCOA Forged Products, Report Q-395 (1993)
11. J.A. Collins, *Failure of Materials in Mechanical Design: Analysis, Prediction, Prevention*, 2nd edn. (Wiley, New York, 1993), p. 181
12. J. Congleton, T. Shoji, R.N. Parkins, The stress corrosion cracking of reactor pressure vessel steel in high temperature water. Corros. Sci. **25**, 633–650 (1985)
13. T.H. Courtney, *Mechanical Behavior of Materials* (McGraw-Hill, New York, 1990)
14. T.W. Crooker, D.F. Hasson, G.R. Yoder, Micromechanistic interpretation of cyclic crack growth behavior in 17–4 PH stainless steel, in *Fractography: Microscopic Cracking Processes, ASTM STP*, vol. 600 (1976), p. 205
15. M.A. Daeubler, A.W. Thompson, I.M. Bernstein, Influence of microstructure on fatigue behavior and surface fatigue crack growth of fully pearlitic steels. Metall. Trans. A **21A**, 925–933 (1990)
16. G.E. Dieter, *Mechanical Metallurgy* (McGraw-Hill, New York, 1986), p. 178
17. W. Elber, Fatigue crack closure under cyclic tension. Eng. Fract. Mech. 37–45 (1970)
18. W. Elber, The significance of fatigue crack closure, in *ASTM STP*, vol. 486 (1971), pp. 230–242
19. F. Endorgan, *Crack Propagation Theories*. NASA CR-901, National Aeronautics and Space Administration, Springfield (1967), pp. 130–136
20. D.K. Felbeck, A.G. Atkins, *Strength and Fracture of Engineering Solids*, 2nd edn. (Prentice Hall, Upper Saddle River, NJ, 1996), pp. 430–443
21. R.G. Forman, V.E. Kearney, R.M. Engle, Numerical analysis of crack propagation in cyclic-loaded structures. Trans. ASME J. Basic Eng. **89**(3), 459–463 (1967)
22. H.O. Fuchs, R.I. Stephens, *Metal Fatigue in Engineering* (Wiley, New York, 1980)
23. R.P. Gangloff, Crack tip modeling of hydrogen environment embrittlement: application to fracture mechanics life prediction. Mater. Eng. A **A103**(1), 157–166 (1988)
24. W.W. Gerberich et al., Corrosion fatigue, NACE-2, Houston (1972)
25. A. Hartman, J. Schijve, Eng. Fract. Mech. **1** (1970)
26. K. Hellan, *Introduction to Fracture Mechanics* (McGraw-Hill, New York, 1984)
27. R.W. Hertzberg, *Deformation and Fracture Mechanics of Engineering Materials*, 3rd edn. (Wiley, New York, 1989)
28. T. Hoshide, K. Tanaka, A. Yamada, Stress-ratio effect on fatigue crack propagation in a biaxial stress field. Fatigue Eng. Mater. Struct. **4**, 355–366 (1981)
29. E.J. Imhof Jr., J.M. Barsom, Fatigue and corrosion-fatigue crack growth 4340 steel at various yield strengths. ASTM STP **536**, 182–205 (1973)
30. J.M. Barsom, E.J. Imhof Jr., Fatigue and fracture behavior of carbon-steel rails. ASTM STP **644**, 387–413 (1978)
31. L.A. James, Report HEDL-TME-75-82, Westinghouse Hanford Co., Richland, WA, January 1976
32. L.A. James, Elding J. **56**, 386–391 (1977)
33. P. Johan Singh, C.K. Mukhopadhyay, T. Jayakumar, S.L. Mannan, B. Raj, Understanding fatigue crack propagation in AISI 316 (N) weld using Elber's crack closure concept: experimental results from GCMOD and acoustic emission techniques. Int. J. Fatigue **29**, 2170–2179 (2007)
34. A. Kelly, G.W. Groves, *Crystallography and Crystal Defects* (Addison-Wesley, Reading, MA, 1970), p. 212
35. H. Kitagawa et al., in *ASTM STP*, vol. 853 (1985)
36. J.F. Lancaster, *Metallurgy of Welding*, 5th edn. (Chapman & Hall, New York, 1993), p. 356
37. A. Liu et al., in *ASTM STP*, vol. 677 (1979)
38. W.A. Logsdon, P.K. Liaw, Fatigue crack growth rate properties of SA508 and SA533 pressure vessel steels and submerged arc weldments in room and elevated temperature air environments. Eng. Fract. Mech. **22**(3), 509–526 (1985)

39. E.A. Loria, Fatigue crack growth behavior and plane-strain fracture toughness of a multi component Nb-Ti-Al alloy. Mater. Sci. Eng. A254, 63–68 (1998)

40. J.A. Manson et al., in *Advances in Fracture Research*, ed. by D. Francois et al. (Pergamon Press, Oxford, 1980)

41. R.C. McClung, J.C. Newman (eds.), Advances in fatigue crack closure measurement and analysis: second volume, in *ASTM STP*, vol. 1343 (1999)

42. A.J. McEvily, J.L. Gonzalez, Fatigue crack tip deformation processes as influenced by the environment. Metall. Trans. A 23, 2211–2221 (1992)

43. F. Ye, C. Mercer, W.O. Soboyejo, Fracture and fatigue crack growth in niobium aluminide intermetallics. Metall. Mater. Trans. A 29A, 2361–2374 (1998)

44. W.J. Mills, The room temperature and elevated temperature fracture toughness response of alloy A-286. J. Eng. Mater. Technol. Trans. ASME 100(2), 195–199 (1978)

45. W.J. Mills, L.A. James, ASME Publication 7-WA/PUP-3 (1979)

46. K.J. Miller, A.P. Kfouri, The effect of load biaxiality on the fracture toughness parameters J and G. Int. J. Fract. 10(3), 393–404 (1974)

47. R.S. Mishra, Friction stir processing technologies. Adv. Mater. Process. 43–46 (2003)

48. A.R.C. Murthy, G.S. Palani, N.R. Iyer, State-of-the-art review on fatigue crack growth analysis under variable amplitude loading. Inst. Eng. (India) IE (I) J. 85, 118–129 (2004)

49. J.A. Newman, An evaluation of the plasticity-induced crack-closure concept and measurement methods. NASA/TM-1998-208430, The NASA STI Program Office (August 1998)

50. J.A. Newman, The effects of load ratio on threshold fatigue crack growth of aluminum alloys, Dissertation, Virginia Polytechnic Institute and State University, Blacksburg, Virginia, 2000

51. J.A. Newman, S.C. Forth, R.A. Everett Jr., J.C. Newman, W.M. Kimmel, Avaluation of fatigue crack growth and fracture properties of cryogenic model materials. NASA/TM-2002-211673, ARL-TR-2725, NASA Langley Research Center, May 2002

52. J.A. Newman, W.T. Riddell, R.S. Piascik, Analytical and experimental study of near-threshold interactions between crack closure mechanisms. NASA/TM-2003-211755, ARL-TR-2774, The NASA STI Program Office (May 2003)

53. P.C. Paris, *Fracture Mechanics Approach to Fatigue, Fatigue-An Interdisciplinary Approach*, ed. by J.J. Burke et al. Proceedings of 10th Sagamore Army Matl. Research Conf. (Syracuse University Press, Syracuse, NY, 1964)

54. P.C. Paris, F. Erdogan, A critical analysis of crack propagation laws. Trans. ASME J. Basic Eng. 85, 528–534 (1963)

55. P.C. Paris, R.J. Bucci, E.T. Wessel, W.G. Clark, T.R. Mager, Extensive study of low cycle fatigue crack growth rates in A533 and A508 steels, in *ASTM STP*, vol. 513 (1972), pp. 141–176

56. N Perez, Ph.D. Dissertation, University of Idaho, 1989

57. N. Perez, Strengthening mechanism of Ni53Mo35Fe9B2 alloy. J. Mech. Behav. Mater. 10, 123–134 (1999)

58. N. Perez, T.A. Place, X-ray diffraction of a heat treated rapidly solidified alloy. J. Mater. Sci. Lett. 9, 940–942 (1990)

59. R.S. Piascik, J.A. Newman, Accelerated near-threshold fatigue crack growth behavior of an aluminum powder metallurgy alloy. NASA/TM-2002-211676, ARL-TR-2728, The NASA STI Program Office (May 2002)

60. B.L. Boyce, R.O. Ritchie, Effect of Load Ratio and Maximum Stress Intensity on the Fatigue Threshold in Ti-6Al-4V. Eng. Fract. Mech. 68, 129–147 (2001)

61. S. Ramamurthy, A. Atrens, The stress corrosion cracking of As-quenched 4340 and 3.5NiCr-MoV steels under stress rate control in distilled water at 90°C. Corros. Sci. 34(9), 1385–1402 (1993)

62. S. Ramamurthy, A. Atrens, The influence of applied stress rate on the stress corrosion cracking of 4340 and 3.5NiCrMoV steels in distilled water at 30°C. Corros. Sci. 52, 1042–1051 (2010)

63. Report MDC-A0913, Phase B Test Program, McDonnell Aircraft Co., McDonnell Douglas Corp., St. Louis, May 1971

64. A. Riddick, Testing for fatigue crack growth. Adv. Mater. Process. 161(10), 53–55 (2003)

65. R.O. Ritchie, Near-threshold fatigue crack propagation in steels. Int. Met. Rev. **24**(5, 6), 204–230 (1979)
66. A. Saxena, *Nonlinear Fracture Mechanics For Engineers* (CRC Press, New York, 1998), pp. 274–278
67. J. Schijve, *Fatigue Thresholds*, vol. 2 (EMAS, West Midlands, 1982)
68. V. Schroeder, C.J. Gilbert, R.O. Ritchie, Effect of aqueous environment on fatigue crack propagation behavior in a Zr-based bulk amorphous metal. Scr. Mater. **40**(9), 1057–1061 (1999)
69. V. Schroeder, C.J. Gilbert and R.O. Ritchie, A comparison of the mechanisms of fatigue-crack propagation behavior in a Zr-based bulk amorphous metal in air and an aqueous chloride solution. Mater. Sci. Eng. **A317**, 145–152 (2001)
70. P. Shahinian et al., in *ASTM STP*, vol. 520 (1973)
71. M.D. Speidel, NACE stress corrosion cracking and hydrogen embrittlement of iron base alloys, Vilieux, France, June 1973
72. R.L. Tobler, Low temperature effects on the fracture behaviour of a nickel base superalloy. Cryogenics **16**(11), 669–674 (1976)
73. R.D. Pollak, *Analysis of Methods for Determining High Cycle Fatigue Strength of a Material with Investigation of Ti-6Al-4V Gigacycle Fatigue Behavior*. Doctoral Dissertation, Air Force Institute of Technology, Wright-Patterson Air Force Base, Ohio 2005
74. A.K. Vasudeven, K. Sadananda, N. Louat, A review of crack closure, fatigue crack threshold and related phenomena. Mater. Sci. Eng. A **188**, 1–22 (1994)
75. H. Vehoff, H.K. Klameth, Hydrogen embrittlement and trapping at crack tips in Ni-single crystals. Acta Metall. **33**, 955–962 (1985)
76. R. Viswanathan, *Damage Mechanisms and Life Assessment of High-Temperature Components*, 3rd edn. (ASM International, Metal Park, 1995) [ISBN 0-87170-358-0]
77. E.K. Walker, in *ASTM STP*, vol. 462 (1970), pp. 1–14
78. J.S. Wang, Hydrogen induced embrittlement and the effect of the mobility of hydrogen atoms, in *Proceedings of the 1994 5th Inter. Conf. on the Effect of Hydrogen on the Behavior of Materials*, Mora, WY, 11–14 Sept 1994, pp. 61–75
79. W. Wei, H. Qingzhi, in *Proc. Far East Fracture Group Workshop*, Tokyo Institute of Technology, Japan, 28–30 Nov 1988, p. 193
80. W.A. Wood, *Some Basic Studies of Fatigue in Metals, in Fracture* (Wiley, New York, 1959), pp. 412–434
81. J.C. Zola, in *Case Studies in Fracture Mechanics*, ed. by T.P. Rich, D.J. Cartwright (AMMRC MS 77-5 Army Materials and Mechanics Research Center, Watertown, MA, 1996)

Fracture Toughness Correlations **10**

10.1   Introduction

Optimistically, mode I (1) plane-strain fracture toughness (K_{IC}) from the theory of linear-elastic fracture mechanics (LEFM) and (2) critical J-integral (J_{IC}) as per elastic-plastic fracture mechanics (EPFM) theory are properties determined in a laboratory. Hence, K_{IC} and J_{IC} data sets can be used in practical designing procedures. In addition, tension (σ_{ys}) and impact energy (U) and fracture mechanic (K_{IC} and J_{IC}) laboratory tests provide properties of solid materials used in designing. Thus, fracture toughness correlations can be very useful for property conversion.

This chapter is devoted to a brief review of fracture toughness of crack-free and notched specimens. Fracture toughness can be defined as the strain energy absorbed by a material prior to fracture. Thus, this energy is defined as the strain energy density in a tension test, the intensity of the stresses (K_I) ahead of a crack tip, and the strain energy release rate also known as the crack driving force (G_I or J_I) for crack growth or the dynamic strain energy (U) for conventional and instrumented Charpy impact tests. Some useful empirical correlations for determining the plane-strain fracture toughness (K_{IC}) from Vickers microhardness measurements and impact tests are included.

Furthermore, impact testing and microhardness measurement techniques are widely used in materials evaluation since they are simple and cost effective. Thus, fracture toughness correlations have evolved indicating the usefulness of the impact and indentation techniques when proper precautions are taken in conducting experiments. For instance, the dynamic behavior of the Charpy impact test can be understood by modeling the striker and specimen as a spring-mass system.

© Springer International Publishing Switzerland 2017

N. Perez, *Fracture Mechanics*, DOI 10.1007/978-3-319-24999-5_10

10.2   Failure Assessment Diagram

For assessing the safety and reliability of a structure, a failure assessment diagram (FAD) can be developed as a failure curve relating brittle fracture and general yielding of structures containing flaws or cracks [21, 22, 27]. Conventional FAD procedures can be found in the British R6 [60] and BS 7910 [15] standards.

The construction of a FAD failure curve is based on Dugdale's strip yield model [2, 22], also known as Dugdale's approximation, shown in Fig. 5.3 under small-scale yielding conditions and on Burdekin and Stone [16] work. For convenience, the J-integral (J_I) defined by Eq. (6.66) and the crack tip opening displacement (δ_t), Eq. (5.34), are used as the starting point. At fracture,

$$J_{IC} = \delta_{tc}\sigma_{ys} \tag{10.1}$$

$$\delta_{tc} = \frac{8a\sigma_{ys}}{\pi E} \ln\left[\sec\left(\frac{\pi\sigma}{2\sigma_{ys}} \right) \right] \tag{10.2}$$

Inserting Eq. (10.2) into (10.1) at fracture yields

$$J_{IC} = \frac{8a\sigma_{ys}^2}{\pi E} \ln\left[\sec\left(\frac{\pi\sigma}{2\sigma_{ys}} \right) \right] \tag{10.3}$$

Using Eq. (6.61) for Mode I loading only and letting the stress intensity factor be an effective term to account for plasticity effects at the crack tip gives

$$J_{IC} = \frac{K_{IC}^2}{E} \quad \text{(Plane Strain)} \tag{10.4a}$$

$$J_C = \frac{K_C^2}{E} \quad \text{(Plane Stress)} \tag{10.4b}$$

For convenience, combine Eqs. (10.3) and (10.4a) and solve for the plane strain fracture toughness

$$K_{IC} = \sigma_{ys}\sqrt{\pi a}\left\{ \frac{8}{\pi^2} \ln\left[\sec\left(\frac{\pi\sigma}{2\sigma_{ys}} \right) \right] \right\}^{1/2} \tag{10.5}$$

Recall that the nominal stress intensity factor is also define by Eq. (3.29) as

$$K_I = \alpha\sigma\sqrt{\pi a} \tag{10.6a}$$

where the remotely applied stress takes the form

$$\sigma = \frac{P}{wB} \tag{10.6b}$$

The stress σ in Eq. (10.6a) is referred to as the reference stress, P is the applied load, and $A = wB$ is the gross cross-sectional area of a plate. On the other hand, the yield strength σ_{ys} in Eq. (10.5) can be replaced by a collapse stress (σ_c) [62], which is the stress for gross plastic deformation of a structural component containing a crack configuration, but it is common to replace it with the flow stress (σ_F) to account for strain hardening effects [12, 19]. Thus,

$$\sigma_c = \sigma_F = \frac{\sigma_{ys} + \sigma_{ts}}{2} \tag{10.7}$$

Nevertheless, dividing Eq. (10.6a) by (10.5) gives [2]

$$\frac{K_I}{K_{IC}} = \frac{\alpha \sigma}{\sigma_{ys}} \left\{ \frac{8}{\pi^2} \ln \left[\sec \left(\frac{\pi \sigma}{2 \sigma_{ys}} \right) \right] \right\}^{-1/2} \tag{10.8}$$

Let the reference stress intensity factor ratio and the reference load ratio be, respectively,

$$K_r = \frac{K_I}{K_{IC}} \tag{10.9a}$$

$$S_r = \frac{\sigma}{\sigma_{ys}} = \frac{P/[(w-a)B]}{\sigma_{ys}} \tag{10.9b}$$

Denote that $A = (w - a)B$ is the net cross-sectional area of a plate and S_r is found in the literature as L_r. Finally, insert Eq. (10.9b) into Eq. (10.8) to get the stress-based FAD function $K_r = f(S_r)$ defined by

$$K_r = S_r \left\{ \frac{8}{\pi^2} \ln \left[\sec \left(\frac{\pi}{2} S_r \right) \right] \right\}^{-1/2} \tag{10.10}$$

which is plotted in Fig. 10.1. It is clear that K_r and S_r are the two dimensionless parameters that determine the potential for failure of a cracked structural component.

Mathematically, Eq. (10.10) yields $K_r = 0$ at $S_r = 0$ which means that there is no applied load on a cracked component, but the plot predicts that $K_r = 1$ at $S_r = 0$ for brittle fracture and $K_r = 0$ at $S_r = 1$ for ductile or collapse failure by general yielding. In order to plot the FAD curve using Eq. (10.10), it was necessary to use the load ratio inequality $0.001 \leq S_r \leq 0.999$, which gave $K_r = 1.000$ at $S_r = 0.001$ and $K_r = 0.437$ at $S_r = 0.999$. To complete the plot, it was necessary to extrapolate from $K_r = 0.437$ to $K_r = 0$ at $S_r = 1$.

The FAD plot or FAD failure curve can be used as a reliability and system safety analysis technique to assess the integrity of a cracked structure exposed to a particular environment. According to the information in the FAD plot, any assessment point falling on or outside the curve represents failure [2, 19, 62].

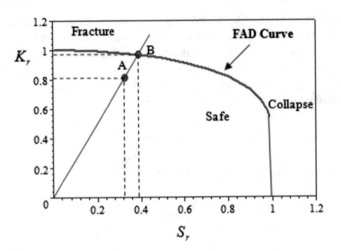

Fig. 10.1 Failure assessment diagram (FAD) using the Dugdale's strip yield model based on Dugdale's approximation, Eq. (10.10)

In addition, $S_r = 1$ is the selected cut-off value used to construct the FAD diagram (Fig. 10.1). Cut-off values larger than unity can be used to construct a FAD plot based on the J-integral approach to account for strain hardening effects [2, 3, 10, 12, 15, 19, 36, 60, 62, 70]. For a FAD based on the J-integral, the driving force and the load ratios are defined by

$$J_r = \frac{J_e(a, P)}{J_e(a_e, P) + J_p(a, n, P)} \tag{10.10a}$$

Here, a is the original crack length, a_e is an effective crack length, $J_e(a, P)$ is the usual elastic J-integral, $J_e(a_e, P)$ is the effective J-integral, and $J_p(a, n, P)$ is the plastic J-integral that accounts for strain hardening effects. In light of the above,

$$K_{rj} = \sqrt{J_r} = \sqrt{\frac{J_e(a, P)}{J_e(a_e, P) + J_p(a, n, P)}} = \sqrt{\frac{J_e(a, P)}{J_n}} \tag{10.10b}$$

$$S_r = \frac{P}{P_o(a, P)} \tag{10.10c}$$

where $K_{rj} = f(S_r) = f(J_e, J_n)$ is the function for a FAD based on the J-integral and strain hardening effects [12].

Example 10.1. *Use the FAD plot to predict failure of a 7075-T6 aluminum plate containing a 6-mm long through-thickness center crack if it is designed to support a load of 200 kN. (a) Will this load cause failure? (b) Calculate the fracture load and the safety factor S_F. Use the data given in Table 3.2.*

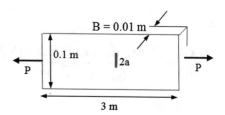

Solution. (a) *From Table 3.2,* $\sigma_{ys} = 572$ *MPa,* $\sigma_{ts} = 641$ *MPa, and* $K_{IC} = 24$ *MPa* $\sqrt{m}$
   Brittle fracture: *From Table 3.1 along with* $x = a/w = 3/100 = 0.03$

$$\alpha = \sqrt{1 + 0.5\,(0.03)^2 + 20.46\,(0.03)^4 + 81.72\,(0.03)^6} = 1.00$$

$$\sigma = \frac{P}{wB} = \frac{(0.200\,\text{MN})}{(0.1\,\text{m})\,(0.01\,\text{m})} = 200\,\text{MPa}$$

$$K_I = \alpha\sigma\,\sqrt{\pi a} = (1.00)\,(200\,\text{MPa})\,\sqrt{\pi\,(3 * 10^{-3}\,\text{m})} = 19.42\,\text{MPa}\sqrt{m}$$

Let $K_{IC} = 24\,\text{MPa}\sqrt{m}$ *so that*

$$K_r = \frac{K_I}{K_{IC}} = \frac{19.42\,\text{MPa}\sqrt{m}}{24\,\text{MPa}\sqrt{m}} = 0.81$$

Collapse: *This is an incipient plastic collapse used as a failure load or stress.*

$$\sigma = \frac{P}{(w-a)\,B} = \frac{(0.200\,\text{MN})}{(0.1\,\text{m} - 0.003\,\text{m})\,(0.01\,\text{m})} = 206.19\,\text{MPa}$$

$$\sigma_c = \frac{\sigma_{ys} + \sigma_{ts}}{2} = \frac{572 + 641}{2} = 606.50\,\text{MPa}$$

$$S_r = \frac{\sigma}{\sigma_c} = \frac{206.19\,\text{MPa}}{606.50\,\text{MPa}} = 0.34$$

Plotting the coordinate $(S_r, K_r)_A = (0.34, 0.81)$ *in Fig. 10.1 defines point A.*
(b) *The failure load is determined by using point B on the FAD curve (Fig. 10.1). This point arises by extrapolating the load line from zero through A to B. Thus,* $(S_r, K_r)_B = (0.38, 0.92)$ *gives the applied stress intensity factor as*

$$K_r = \frac{K_I}{K_{IC}} = 0.92$$

$$K_I = 0.92 K_{IC} = 0.92\,(24\,\text{MPa}\sqrt{m}) = 22.08\,\text{MPa}\sqrt{m}$$

This result implies that elastic-plastic failure occurs at $K_I = 22.08$ *MPa* $\sqrt{m} < K_{IC}$. *Nonetheless, the failure load* P_f *using the calculated* K_I *value is*

$$K_I = \alpha \sigma_f \sqrt{\pi a} = \frac{\alpha P_f}{wB} \sqrt{\pi a}$$

$$P_f = \frac{wBK_I}{\alpha \sqrt{\pi a}} = \frac{(0.1\ m)\ (0.01\ m)\ \left(22.08\ \text{MPa}\sqrt{m}\right)}{(1)\ \sqrt{\pi\ (3 * 10^{-3}\ m)}} = 227.44\,\text{kN}$$

The failure load based on collapse is

$$\sigma = S_r \sigma_c = (0.38)\ (606.50\ \text{MPa}) = 230.47\ \text{MPa}$$

$$P_f = (w - a)\ B\sigma = (0.1\ \text{m} - 0.003\ \text{m})\ (0.01\ \text{m})\ (230.47\ \text{MPa}) = 223.56\,\text{kN}$$

Then, the safety factor can be determined as $S_F = 227.44/223.56 = 1.02$. A similar example problem can be found in Collins' Book, Example 3, page 79 [19].

10.3   Grain Size Refinement

The grain size refinement technique is used for enhancing the yield strength (σ_{ys}) and fracture toughness (K_{IC}), and it has been successfully applied to some body-centered cubic (BCC) steels containing molybdenum (*Mo*), vanadium (*V*), titanium (*Ti*), and aluminum (*Al*). These alloying elements react in the solid-solution state to form either particles or cause microstructural changes that are accountable for pinning grain boundaries and dislocations. The phase transformation mechanism evolved in adding these elements to carbon steels can be found elsewhere [32].

Subsequently, the controlling property, such as yield strength or fracture toughness, is limited to (1) the strength if an applied external load exists, (2) the fracture toughness if absorption of strain energy occurs prior to fracture, and (3) the ductility if metal shaping or forming is required.

From an empirical point of view, monotonic properties and fracture toughness can be correlated using Orowan [48] and the Hahn-Rosenfield [26] plastic constraint factor (λ) at the crack tip. Thus,

$$\lambda = \frac{\sigma_f}{\sigma_{ys}} \qquad \text{(Orowan)} \qquad\qquad (10.11)$$

$$\lambda = 1 + \beta \left(\frac{K_{IC}}{\sigma_{ys}}\right) \qquad \text{(Hahn-Rosenfield)} \qquad (10.12)$$

Here, $\beta = 20\ m^{-1/2}$ for strain hardening mild steels [26]. The yield and fracture strengths as per Hall-Petch-type equation are defined as

$$\sigma_{ys} = \sigma_{oy} + k_y d^{-1/2} \qquad\qquad (10.13)$$

$$\sigma_f = \sigma_{of} + k_f d^{-1/2} \qquad\qquad (10.14)$$

where σ_{oy} = Friction stress due to particle, dislocations, etc. (MPa)

σ_{of} = Stress constant (MPa)
k_y = Dislocation locking term (MPa mm$^{1/2}$)
k_f = Constant (MPa mm$^{1/2}$)
d = Grain size (mm)

Combining Eqs. (10.11) and (10.12) along with Eqs. (10.13) and (10.14) yields the plane-strain stress intensity factor as [20, 56, 59]

$$K_{IC} = \frac{1}{\beta}\left(\sigma_f - \sigma_{ys}\right) \tag{10.15}$$

$$K_{IC} = \frac{1}{\beta}\left(\sigma_{of} - \sigma_{oy}\right) + \frac{1}{\beta}\left(k_f - k_y\right)d^{-1/2} \tag{10.16a}$$

$$K_{IC} = K_o + K_d d^{-1/2} \tag{10.16b}$$

Remarkably, Eq. (10.16b) exhibits a linear correlation $K_{IC} = f(d^{-1/2})$, where K_o is the stress intensity factor due to defect internal stresses and K_d is the slope or the rate of change of $K_{IC} = f(d^{-1/2})$.

Example 10.2.   *Using Stonesifer and Armstrong [72] linear regression analysis for A533B steel at room temperature having an average grain size of 10 μm and modulus of elasticity of 207 GPa, calculate (a) λ and β, and (b) predict the elastic strain energy density (W_e) as a measure of elastic fracture toughness for sound specimens subjected to tension loading. The mechanical properties depend on the average grain size d. Thus,*

$$\sigma_{ys} = 572\,\text{MPa} + \left(0.11\,\text{MPa m}^{1/2}\right)d^{-1/2} = 606.79\,\text{MPa}$$

$$\sigma_f = 1750\,\text{MPa} + \left(3.30\,\text{MPa m}^{1/2}\right)d^{-1/2} = 2793.55\,\text{MPa}$$

$$K_{IC} = 60\,\text{MPa}\sqrt{\text{m}} + (0.16\,\text{MPa m})\,d^{-1/2} = 110.60\,\text{MPa}\sqrt{\text{m}}$$

Solution.

(a) *From Eqs. (10.11) and (10.12), the plastic constraint factor (λ) after localized yielding has occurred at the crack tip and the strain hardening factor (β) is calculated for plane-strain condition:*

$$\lambda = \frac{\sigma_f}{\sigma_{ys}} = \frac{2793.55\,\text{MPa}}{606.79\,\text{MPa}} = 4.60$$

$$\beta = (\lambda - 1)\left(\frac{\sigma_{ys}}{K_{IC}}\right) = (4.60 - 1)\left(\frac{606.79\,\text{MPa}}{110.60\,\text{MPa}\sqrt{\text{m}}}\right)$$

$$\beta = 19.75\,\text{m}^{-1/2}$$

which agree with Orowan and Hahn-Rosenfield [26] approximations given above. In addition, λ represents the local plasticity ahead of defects, and it is also a multiplier for the fracture stress; $\sigma_f = \lambda \sigma_{ys} = 4.6 \sigma_{ys}$.

(b) *From Eq. (6.60), the maximum elastic strain energy density as the area under the typical stress-strain curve is calculated as*

$$W_e = \frac{\sigma_{ys}^2}{2E} = \frac{(606.79\,\text{MPa})^2}{2\,(207 \times 10^3\,\text{MPa})} = 0.89\,\text{MJ/m}^3$$

This is the required strain energy density known as the elastic resilience, which is the maximum elastic energy the material absorbs prior to plastic deformation since the maximum elastic stress is allowed to be equal to the yield strength of the material.

Furthermore, the strain energy density of a material relates to the deformation gradient, and it is the energy stored by a solid material undergoing deformation in a particular environment. In other words, the strain energy density of a material is defined as the strain energy per unit volume, and it is equal to the area under the stress-strain curve of a material. The elastic strain energy is measured from 0 to epsilon$_{ys}$, which is the maximum elastic strain or the transition strain between elastic and plastic deformation.

10.4   Indentation-Induced Cracking

In this section, the theory of indentation is strictly used to measure "indentation hardness" which implies resistance to penetration as depicted in Fig. 10.2 [8]. The most common hardness tests are mechanically static in nature, which have industrial and research applications. Thus, "hardness" can be measured by using the Brinell, Rockwell, and Meyer tests. On the other hand, "microhardness" can be measured using the Knoop and Vickers indentation techniques. The term microhardness indicates the hardness of a very small area such as a grain and/or a particle that constitute the microstructure of a polycrystalline material. Herein, attention is devoted to the Vickers indentation-hardness measurement technique, which has been used very extensively in research for predicting Vickers fracture toughness of brittle materials. The technical procedure for employing the Vickers hardness testing can be found in the ASTM E92 (1997) Standard Test Method. The term microhardness refers to as small micro-indentation hardness due to relatively common light applied load that ranges from 1 kg$_f$ to 120 kg$_f$.

The Vickers indentation is made with a diamond pyramidal-shaped indenter. In fact, the indenter impression schematically shown Fig. 10.2 is a square-based inverted pyramid with a face angle of 136°, and it is so small that it must be observed with a microscope. For brittle materials, such as ceramics (fused, sintered or cemented metallic oxides), cermets (powder metallurgy products containing ceramic particles), polymers, and amorphous metallic materials, indentation-induced cracking overcome difficulties in specimen preparation in a conventional

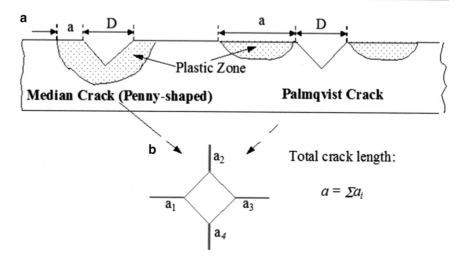

Fig. 10.2 Schematic indentation-induced cracking systems [8]. (**a**) Side view and (**b**) top view showing four cracks

manner as recommended by the American Society for Testing Materials (ASTM) E399 for plane-strain fracture toughness and for the J-integral approach [8]. As a result, an empirical equation may be used to determine the Vickers fracture toughness.

The Vickers hardness (H_v) can be calculated as follows:

$$H_v = \frac{2P \sin(\theta/2)}{D^2} = \frac{1.8544P}{D^2} \qquad (10.17)$$

where P = Applied load (kg)

D = Mean diagonal (mm)
$\theta = 136° =$ Face angle

The advantages of Vickers hardness measurements are (1) its simplicity; (2) it can be applied to microstructural constituents; (3) it does not require fatigue pre-cracking, which is difficult to accomplish in brittle materials; (4) it is cost effective since small specimens are needed; and (5) the tests are considered nondestructive in at macroscale. However, specimen preparation is a slightly time-consuming procedure since a polished surface is required so that uniform indentations are made on a reflective flat plane, which is a must in order to obtain consistent and reproducible results.

In fact, hardness measurements are made for screening materials and characterizing microstructures; subsequent use is made for determining the Vickers fracture toughness, which inevitably shows some degree of inaccuracy when compared with conventional fracture toughness. Nevertheless, this technique has become an

Fig. 10.3 Effects of indentation load on the crack length. The Palmquist fracture toughness for $Al_2O_3 - TiC$ cermet is the inverse of the linear slope [75]

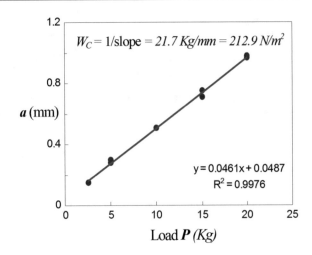

excellent approach for characterizing ceramics, cermets (e.g., $TiC\text{-}Al_2O_3$, $WC\text{-}Co$ composites), and amorphous metals and their alloys.

In the 1950s, Palmquist [50] recognized that indentation-induced cracking observed on cermets was related to fracture toughness, and he developed a procedure to predict fracture toughness. Vickers is the most common and suitable test method due to four possible cracks that may emanate from the corners of the indenter. A typical Palmquist fracture toughness analysis requires a linear plot of total crack length (a) vs. applied load (P); that is, $a = f(P)$. This is shown in Fig. 10.3 for an alumina-titanium carbide cermet ($Al_2O_3\text{-}TiC$). The inverse of the slope of the line is a measure of Palmquist fracture toughness in terms of work done (W_C), which may be taken as the strain energy release rate (G_C) [75].

The prediction of fracture toughness for many brittle materials using empirical formulations can be found in the literature [37, 42, 54, 55, 68]. However, the most common expression is the following general form:

$$K_{IC} = \xi \left(\frac{E}{H_v} \right)^x \left(\frac{P}{a^{3/2}} \right) \tag{10.18}$$

where $\xi =$ Indentation geometry factor ($\xi < 1$)

> $E =$ Modulus of Elasticity (MPa)
> $P =$ Indenter load (MN)
> $H_v =$ Vickers hardness (MPa)
> $x =$ Exponent $a =$ Average crack length (m)

Rearranging and manipulating Eq. (10.18) gives

$$K_{IC} = \sqrt{\pi} \alpha P a^{-3/2} \tag{10.19}$$

$$\alpha = \frac{\xi}{\sqrt{\pi}} \left(\frac{E}{H_v} \right)^x \tag{10.20}$$

Fig. 10.4 Comparison of Vickers and conventional fracture toughness for brittle ceramics and WC-Co cermets. Experimental data taken from Refs. [4, 37]

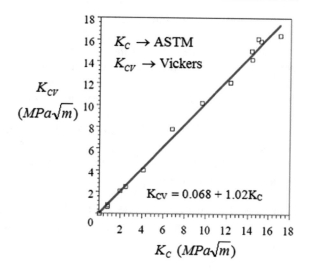

Observe that Eq. (10.19) resembles Eq. (3.29) for the conventional plane stress fracture toughness. Moreover, the exact numerical form of Eq. (10.18) depends on the crack configuration (Fig. 10.2) and the material's properties [4,28,37,54,55,68]. In order to illustrate the usefulness of Eq. (10.18) or (10.19), let us curve fit Laugier [37] and Anstis et al. [4] data for several ceramics and cermets, respectively. Notice the remarkable correlation depicted in Fig. 10.4 for Vickers fracture toughness and conventional fracture toughness testing method (ASTM E399). The pertinent values for each parameter in Eq. (10.18) can be found in the cited references [4, 37].

Despite that Vickers hardness technique has been used for decades, it is still a classical research tool for characterizing materials. For instance, Phelps et al. [53] evaluated toughness of female baboon femurs of 6- to 27-year-old using this technique. On the other hand, Iost and Bigot [33] made use of the Vickers hardness measurements to characterize the brittleness index, which depends on fracture mechanics and hardness, for metallic and ceramics materials, and flux-grown $ErFeO_3$ single crystals [69].

Furthermore, Sridhar and Yovanovich [71] found a power-law relation, $H_v = cD^n$, that fitted hardness data for a tool steel 01, $AISI$ 304 stainless steel, and $Ni200$. In addition, Berces et al. [9] determined the dynamic Vickers hardness as $H_v = P/D^2$ for the characterization of plastic instability of an Al-$3.3Mg$ binary alloy at a loading range of 1.4–70 MN.

The search for finding different avenues to characterize materials continues beyond a researcher's imagination, but apparently, the Vickers hardness measurement technique remains as a research tool.

Recently, Milekhine et al. [42] used Vickers indentation-induced cracking for evaluating the plane-strain fracture toughness of FeSi using Palmquist-type cracks. As a result, $K_{IC}(\text{Vickers}) = K_{IC}(\text{ASTM}) = 2.46\,MPa\sqrt{m}$.

In general, high fracture toughness values are desired for assuring structural integrity and reliability. The ultimate goal is to determine fracture toughness using a cost-effective technique, such as the indentation-induced cracking approach. Apparently, Vickers indentation seems to be the most common experimental approach for determining fracture toughness as accurate as possible. The data shown in Fig. 10.4 is a good example for universally comparing fracture toughness results from other techniques.

Despite that the Vickers indentation technique is limited to small surface areas for measuring fracture toughness, it can provide local fracture toughness data using small amounts of materials. This is an attractive cost-effective experimental technique due to its simplicity.

In addition, the Knoop indentation is also a cost-effective technique for the same purpose. However, it is possible that Vickers and Knoop microhardness and indentation fracture toughness results may differ to an extent due to their load dependency.

10.5   Charpy Impact Testing

In general, impact tests are performed to measure the response of a material to dynamic loading. The most common laboratory test configurations are the pendulum machine and the drop tower. The results obtained from a standard impact tests are usually a single value of the impact energy or energy spent on a single specimen. This is of limited value in describing the dynamic behavior of a particular sample material. Therefore, instrumenting an impact machine yields information on the impact forces, impact velocities, displacements, and strain energies of the striker at any time during the dynamic test. Figure 10.5 shows a conventional Charpy impact testing machine used to measure fracture toughness of a three-point bending specimen (3PB) under an impact loading system at low velocity.

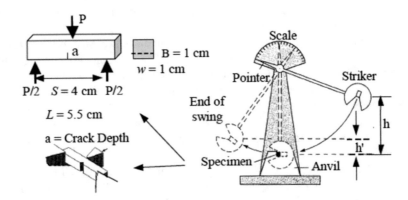

Fig. 10.5 The three-point bending (3PB) specimen and the conventional Charpy impact testing machine [14]

Fig. 10.6 Tinius Olsen Model 84 Instrumented Charpy impact machine equipped with in situ heating and cooling system, optical encoder for measuring the impact velocity, and motorized hammer (striker) return [43]

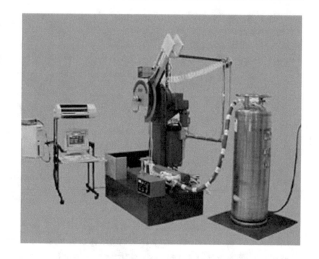

Impact loads generate high strain rates in solid materials. For instance, conventional and instrumented Charpy impact testing machines impart low strain rates at low velocity when compared to ballistic impact velocity. The former technique has been used for characterizing the dynamic behavior of some particular composite materials [1, 13, 18, 25, 29, 34, 39, 49] and the latter technique promotes impact at a high velocity, which varies according to the type of gun projectile being used. Excellent work in the ballistic field can be found elsewhere [17, 24, 31, 77].

The instrumented Charpy impact machine remains a key means for fracture toughness testing due to its low cost, convenience, reliability based on certification standards, and simple use. A particular instrumented Charpy impact machine is shown in Fig. 10.6. Thus, the transient load history during a Charpy test is readily obtained by placing strain gages on the striker so that it becomes the load cell. Using software during an impact, one can record the displacements by integrating the acceleration versus time twice with respect to time. The accuracy of these measurements may be affected by the inertial forces in the striker, variations in the contact force distribution between the striker and the specimen, striker geometry, and strain gage location on the striker [40].

Figure 10.7 illustrates a typical load history for a relevant case [40]. The interpretation of Fig. 10.7 is very vital for characterizing a material dynamic behavior. Denote the characteristic load information and the velocity profile in this figure. This type of plot represents the dynamic load response to dynamic displacement cause by the impact process in a Charpy specimen.

The use of instrumented Charpy impact machine has the advantage of providing information on the toughness characteristics of a specimen [11, 38, 40, 63–67]. In general, the load-time history for a three-point bending (3PB) Charpy specimen can be divided into fracture initiation and fracture propagation regions expressed in terms of areas under the curve (Fig. 10.7).

These areas are measures of the elastic strain energy (U_e) and plastic strain energy (U_p). These energies are strongly dependent on the temperature, specimen size, and impact velocity imparted by the kinetic energy of the striker. Thus, the

Fig. 10.7 Schematic
$P = f(\mu)$ and $v = f(\mu)$
plots, obtainable using MPM
Impact^{Tm} v3.0 software [39]

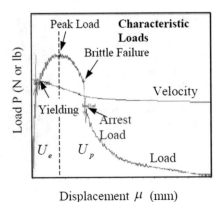

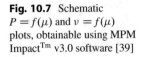

total impact or strain energy a specimen can absorb during impact is known as the
Charpy V-notch (CVN) energy defined by the following general relationship [32]

$$U = U_e + U_p \tag{10.21}$$

$$U = v \int P dt \tag{10.21a}$$

$$v = \sqrt{2gh} \tag{10.22}$$

where v = Striker velocity on contact with the specimen (m/s)

P = Impact load (N)
$\int P dt$ = Impulse (N s)
h = Pendulum initial height (m)
$g = 9.81$ m/s^2

Thus, one must analyze the load-time curve very carefully with respect to the
elastic and plastic strain energies. These energies may be used for classifying solid
materials as $U_e > U_p$ for brittle materials and $U_e < U_p$ for tough and ductile
materials.

The Charpy or Izod notched specimens are used for this purpose, the Charpy
V-notched specimen being the most common. This technique became a conventional
testing method when it was revealed in the 1940s that welded ships, large pipelines,
and other monolithic steel structures fractured at notch roots.

This is a dynamic (impact) testing technique recommended by the ASTM E23
standard test method. The applied impact load (P) is through an impact blow from
a falling pendulum hammer (striker).

The resultant energy measurement is commonly referred to as Charpy impact
energy (U), which is a measure of the fracture toughness of a material at testing
temperatures. Thus, $U = f(T)$ is normally determined experimentally in order to
reveal the effects of impact loads on the dynamic behavior of materials at relatively
low and high temperatures.

In general, structural steels usually have low fracture toughness at relatively low temperatures. In general, hindering the dislocation motion along preferred crystallographic planes and grain-boundary sliding at low temperatures increases strength and decreases the impact energy and ductility, specifically elongation at low temperatures.

In light of the above, relatively ductile materials may exhibit a brittle behavior at low temperatures. Consequently, these types of materials undergo a ductile-to-brittle transition with respect to dynamic or impact behavior.

10.6 Dynamic Effects

The Charpy impact testing machine and its dynamic characteristic can be understood using Williams' [78] one degree of freedom spring-mass model as shown in Fig. 10.8. For a perfectly elastic deformation on impact, the contact stiffness k_1 (= 1/compliance) is high compared with that of the specimen k_2, and consequently, considerable load oscillations are likely over a short period of time.

Firstly, the equation of motion of the system shown in Fig. 10.8 is [78]

$$m\ddot{u} + (k_1 + k_2)\, u = k_1 vt \qquad (10.23)$$

where k_1 = Stiffness of the striker-specimen interface

$k_2 = 1/C$ = Stiffness of the specimen
C = Compliance of the specimen
m = Mass of the specimen

Using the boundary condition $\mathring{u} = u = 0$ yields the solution of Eq. (10.23)

$$u = \frac{v}{\omega}\left(\frac{k_1}{k_1 + k_2}\right)[\omega t - \sin(\omega t)] \qquad (10.24)$$

where the natural angular frequency (ω) and the period of oscillations (τ) are, respectively,

$$\omega = \sqrt{\frac{k_1 + k_2}{m}} \qquad (10.25)$$

$$\tau = \frac{2\pi}{\omega} \qquad (10.26)$$

Fig. 10.8 Spring-mass model of the Charpy impact machine. After Ref. [78]

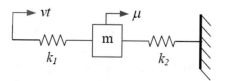

Secondly, assume that the striker slows down insignificantly upon striking the specimen mass (m) so that the spring k_2 expands at velocity v and the spring k_1 compresses. In this case, the impact load (force) can be defined by

$$P = k_1 (vt - u) \tag{10.27}$$

Substituting Eq. (10.24) into (10.27) and rearranging the resultant expression yields

$$P = \frac{vk_2}{\omega} \left(\frac{k_1/k_2}{1 + k_1/k_2} \right) \left[\omega t - \left(\frac{k_1}{k_2} \right) \sin(\omega t) \right] \tag{10.28}$$

Let λ be defined as

$$\lambda = \frac{\omega P}{vk_2} \left(\frac{1 + k_1/k_2}{k_1/k_2} \right) = \left[\omega t - \left(\frac{k_1}{k_2} \right) \sin(\omega t) \right] \tag{10.29}$$

The oscillatory behavior of Eq. (10.29) is shown in Fig. 10.9 as a diagram exhibiting two discrete values of λ. This diagram illustrates that the spring-mass system oscillates as k_1/k_2 increases. The ideal elastic case dictates that $k_1/k_2 = 0$ and load oscillations do not occur as indicated by the straight line. However, real systems are bound to experience load oscillations after the initial contact between the striker and the specimen. These oscillations are likely over a small interval of time.

If loss of contact occurs when the striker bounces in the opposite direction, then the impact load is zero at that instance, but impact reloading resumes in a short interval of time as shown by lines 1 through 5 in Fig. 10.9. In this case, the specimen undergoes free oscillations, and the dynamic behavior that describes this event is defined by the following equation of motion:

$$m\ddot{u} + (k_1 + k_2) u = 0 \tag{10.30}$$

The loss of contact occurs when $P = 0$ and as a result, Eq. (10.28) gives

$$\omega t - \left(\frac{k_1}{k_2} \right) \sin(\omega t) = 0 \tag{10.31}$$

and

$$\omega t \simeq \pi (1 + k_2/k_1) \tag{10.32}$$

Combining Eqs. (10.21a) and (10.28) and integrating the resultant expression yields the impact energy lost by the striker

$$U = \frac{k_1/k_2}{(1 + k_1/k_2)^2} \left\{ \frac{(\omega t)^2}{2} + \frac{k_1}{k_2} [1 - \cos(\omega t)] \right\} mv^2 \tag{10.33}$$

This expression illustrates that the striker kinetic energy (mv^2) is transformed into strain energy U, which is absorbed by the specimen during the impact process. However, this dynamic event exhibits an oscillation behavior to the impact response

Fig. 10.9 Load-time diagram showing oscillations when $k_1 = k_2 > 0$

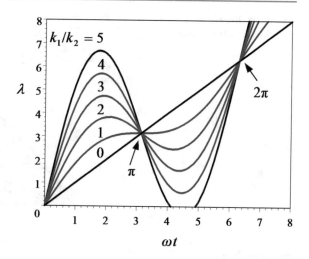

as indicated by the cosine term in Eq. (10.33). Consequently, bouncing generates load oscillations and eventually gives discrete values of energy as illustrated in Fig. 10.9 where all curves coincide when $\omega t = \pi$. Thus, Eq. (10.33) becomes [78]

$$\frac{U}{mv^2} = \frac{k_1/k_2}{(1 + k_1/k_2)^2} \left[\frac{\pi^2}{2} + 2 \left(\frac{k_1}{k_2} \right) \right] \qquad (10.34)$$

For rigid body contact, the stiffness ratios becomes $k_2/k_1 \to 0$ or $k_1/k_2 \to \infty$ and Eq. (10.34) yields a simplified form of the impact strain energy

$$U \simeq 2mv^2 \quad \text{For} \quad \frac{k_1}{k_2} \to \infty \qquad (10.35)$$

Analysis of theoretical and experimental data is a very important issue because erroneous conclusions may be drawn. For example, plotting Eq. (10.33) when $k_1/k_2 \leq 1$ and $k_1/k_2 > 1$ yields different dynamic behavior, while kinetic energy is converted to strain energy as the specimen bends and slows down. The obvious significance of this observation is shown in Fig. 10.10. The curves for $k_1/k_2 > 1$ coincide at $\omega t = \pi$. This is attributed to the first load oscillation since this behavior is also observed in Fig. 10.9. Eventually, a second impact may occur at time predicted by Eq. (10.26) for further kinetic energy transfer.

In summary, an impact is generated by imparting energy from a pendulum carrying large mass that acts as an indenter to the stationary specimen. During this dynamic process, the pendulum swings toward the specimen surface and an impact load-displacement is recorded, and also the pendulum displacement $u(t)$ is recorded as a function of time. Eventually energy loss is inevitable during the specimen deformation. This is dealt with next.

Fig. 10.10 Energy ratio-time diagram for impact loading

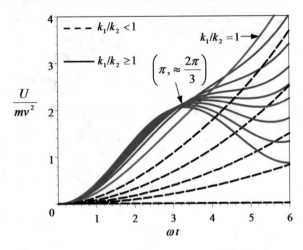

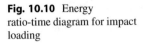

Fig. 10.11 Energy conversion and energy losses

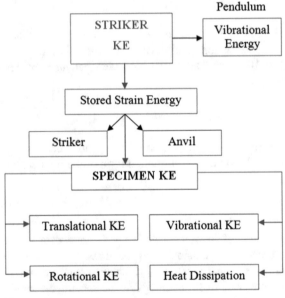

Energy Loss. Despite there are different sources of motion in an impact test, the rigid body motion is considered in the Charpy test analysis. The general overview of the striker kinetic energy (KE) transformation is depicted in Fig. 10.11 [41].

The general equation for kinetic energy is [41, 78]

$$U_{KE} = \frac{1}{2}\left(Mv^2 + I\omega^2\right) = Mgh + \frac{1}{2}I\omega^2 - U_w - U_f \qquad (10.36)$$

where M = Mass of the striker ω = Rotational velocity

   I = Moment of inertia v = Translational velocity

U_w = Windage energy
U_f = Friction energy

This expression takes into account (1) the force transmitted to the points of rotation when the center of percussion (striking) is at an eccentric point from the specimen axis of rotation and (2) the energy losses due to windage and friction of the pendulum [78].

A convenient and realistic model considers a heavy striker (large mass M) and a light specimen (small mass m) in which the displacement is due to relative motion of both bodies. The equation of motion for $M > m$ is [78]

$$m\ddot{u} + \left(1 + \frac{m}{M}\right) f_e(u) + \left(1 + \frac{m}{M}\right) f_d(\overset{\circ}{u}) = 0 \tag{10.37}$$

Here, $f_e(u)$ is the elastic contact stiffness function and $f_d(\overset{\circ}{u})$ is the dissipation function. In fact, elastic stiffness of a solid represents mechanical resistance to elastic deformation under the action of an external loading system, which may be linear or nonlinear. A more exact conceptual treatment of stiffness in a dynamic sense defines it as the stiffness function $f_e(u)$. On the other hand, the dissipation function $f_d(\overset{\circ}{u})$ relates to energy dissipation during dynamic elastic deformation.

The solution of this differential equation, Eq. (10.37), along with the coefficient of restitution (COR) e [78]

$$e = \frac{\text{Rebound velocity}}{\text{Initial impact velocity}} = \frac{v_f}{v_i} \tag{10.38}$$

is the kinetic energy loss of the striker, from which the strain energy absorbed by the specimen and that lost in impact are deduced as [78]

$$U_S = \left(\frac{1+e}{1+m/M}\right)\left[1 - \frac{m}{2M}\left(\frac{1+e}{1+m/M}\right)\right] mv^2 \quad \text{(Striker)} \tag{10.39}$$

$$U_A = \frac{1}{2}\left(\frac{1+e}{1+m/M}\right)^2 mv^2 \quad \text{(Specimen)} \tag{10.40}$$

$$U_A \simeq \frac{1}{2}(1+e)^2 mv^2 \quad \text{For } m/M \ll 1 \tag{10.41}$$

$$U_L = \frac{1}{2}\left(\frac{1-e^2}{1+m/M}\right) mv^2 \simeq \left(\frac{1-e^2}{2}\right) mv^2 \quad \text{(Impact)} \tag{10.42}$$

The coefficient of restitution (COR) e is a measure of the bounciness of the collision between the striker and the specimen. Actually, the coefficient of restitution e is a factor that quantifies the energy loss during impact.

According to Eq. (10.38), e is the ratio of the rebound velocity (v_f) and the initial velocity (v_i) of the striker. In other words, e is the ratio of the final velocity (after) to the initial (before) velocity of the collision between two objects. Hence, $e = 0$ for lack of rebound velocity of the striker, the specimen (object) is plastic, and the kinetic energy of the striker is dissipated as heat during the collision or deformation. Therefore, $v_f = e v_i = 0$.

$e = 1$ for perfect rebound velocity, the specimen (object) is elastic, and no kinetic energy is dissipated as heat. Instead, the kinetic energy is converted to strain energy during collision. Therefore, $v_f = ev_i = 1$.

$0 < e < 1$ for elastic-plastic materials. Some kinetic energy is dissipated as heat and some is transformed to strain energy. For a striker impacting a flexible or soft object COR is low, while impacting a rigid object COR is high. Therefore, $0 < v_f < v_i$.

Example 10.3. *For a linear motion during impact along with* $m/M \ll 1$, *rotational effects are neglected, and consequently, the energy loss of the striker becomes dependent on its kinetic energy and the coefficient of restitution e. Graphically, determine the strain energy* U_S *and* U_A *trends when the kinetic energy of the striker range is* $0 \le mv^2 \le 200$ *mJ and the coefficient of restitution is* $e = 0.5$ *and* 0.8. *Explain.*

Solution.
    From Eq. (10.39),

$$U_S = \left(\frac{1+e}{1+m/M} \right) \left[1 - \frac{m}{2M} \left(\frac{1+e}{1+m/M} \right) \right] mv^2$$

$$U_S = \left(1 - e^2 \right) \left(\frac{1}{2}mv^2 \right) \qquad \text{for } m/M \ll 1$$

$$U_A = \frac{1}{2} \left(\frac{1+e}{1+m/M} \right)^2 \left(\frac{1}{2}mv^2 \right)$$

$$U_A = (1 + e)^2 \left(\frac{1}{2}mv^2 \right) \qquad \text{for } m/M \ll 1$$

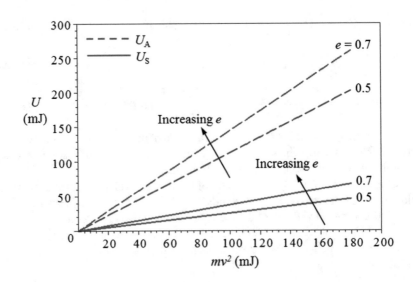

Denote that energies U_S and U_A exhibit increasingly linear trends with slopes $0.5(1 - e^2)$ and $0.5(1 + e)^2$, respectively. These energy straight lines get steeper with increasing e. Therefore, the specimen can bend and slip through the anvil before a fully crack propagates, leading to loss of kinetic energy as the coefficient of restitution increases.

10.7 Dynamic Strain Energy Release Rate

In quasi-static fracture mechanics, a stationary crack under a remotely slow and steady strain rate loading can be characterized by the strain energy release rate as described in previous chapters. In light of this, this energy is the driving force for crack growth or crack propagation, inducing energy dissipation during the formation of new crack surface areas. On the other hand, dynamic fracture mechanics takes into account crack growth including the effects of inertia due to high enough applied strain rates. Subsequently, the term dynamic strain energy release rate appears as the dynamic driving force for assessing the behavior of dynamic crack growth.

This section describes the fracture mechanics of impact testing and how dynamic corrections are derived and demonstrates how impact tests are a coherent part of fracture mechanics. A basic analysis is outlined using elementary linear-elastic fracture mechanics (LEFM) for mode I loading. It is possible to measure G_I concurrently with an impact test by measuring Charpy elastic strain energy U. The mathematical connection between these energies requires a dynamic correction factor ϕ. This is defined by [78]

$$G_I = \frac{U}{\phi Bw} = \frac{K_I^2}{E'} \tag{10.43}$$

The general stress intensity factor and that for a 3PB Charpy specimen are, respectively,

$$K_I^2 = \frac{\pi a f^2 (a/w) P}{Bw} \tag{10.44}$$

$$K_I^2 = f^2 (a/w) \left(\frac{6M}{Bw^2} \right) a = \frac{EP^2}{2B} \frac{dC}{da} \tag{10.45}$$

and the bending moment is

$$M = \frac{PS}{4} \tag{10.46}$$

where E' has been defined in Eqs. (3.5), (6.34), and (8.11).

Specimen variables are given in Fig. 10.5. For brittle materials, this analytical procedure incorporates the Charpy elastic strain energy into G_I as deduced from Williams [78] relationships using a Charpy three-point bending (3PB) specimen.

Fig. 10.12 Charpy impact data on HDPE specimen showing the kinetic energy value. Given data:
$v = 3.36\,\text{m/s},\ \phi \geq 4,$
$\rho = 10^3\,\text{kg/m}^3,\ L = 41\,\text{mm},$
$B = 12\,\text{mm, and}$
$w = 6\,\text{mm}$ [78]

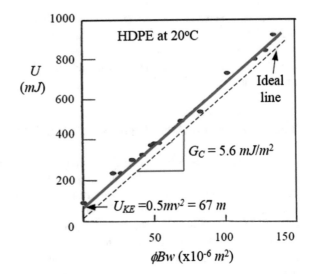

Experimentally, measure $U = f\,(x = a/w)$ and, subsequently, plot $U = f\,(\phi Bw)$ to obtain $G_I = G_{IC}$ from the slope of the straight line.

The significance of this method is shown in Fig. 10.12 for HDPE polymer, which exhibits the apparent linearity of the strain energy. The slope of this linear plot is the critical strain energy release rate as $G_{IC} = 5.6\,\text{mJ/M}^2$, and the intercept is to the kinetic energy of the striker $U_{KE} \simeq 2mv^2 \simeq 67\,\text{mJ}$. The energy absorbed by the HDPE specimen as strain energy is approximately

$$U_{spec} \simeq U_{KE} - U = U_{KE} - \phi BwG_{IC} \simeq U_{KE} \simeq 67\,\text{mJ} \qquad (10.47)$$

Furthermore, rearranging Eq. (10.45) yields the derivative of the compliance with respect to the crack length along with $x = a/w$

$$\frac{dC}{da} = \left(\frac{9S^2}{2Bw^2E'}\right)xf^2\,(x) \qquad (10.48)$$

Integrating Eq. (10.48) gives the total compliance

$$C = \frac{9S^2x^2f^2\,(x)}{4Bw^2E'} + C_o \qquad (10.49)$$

where C_o is an integration constant defined as the compliance for a crack-free specimen.

Thus, [78]

$$C_o = \frac{S}{BwE'} \qquad \text{for tension loading} \qquad (10.50)$$

$$C_o = \frac{S^3}{4Bw^3 E'} \quad \text{for 3-point bending loading} \tag{10.51}$$

From Eq. (10.49), the new compliance equation is

$$\frac{dC}{dx} = \frac{9S^2 x f^2 (x)}{2Bw^2 E'} \tag{10.52}$$

$$C = \int \frac{9S^2 x f^2 (x)}{2Bw^2 E'} dx + C_o \tag{10.53}$$

Now, the energy calibration factor for a 3PB specimen is defined along with $x = a/w$ by

$$\phi = \frac{C}{dC/dx} = \frac{\int x f^2 (x) \, dx}{x f^2 (x)} + \frac{S}{18wx f^2 (x)} \tag{10.54}$$

The geometric calibration factor, $f(x)$, for a Charpy 3PB specimen is given by Brown and Strawley [14] in polynomial form for two span-to-width ratios. The resultant polynomials for $S/w = 4$ and $S/w = 8$ are

$$f_4 (x) = 1.93 - 3.07x + 14.53x^2 - 25.11x^3 + 25.80x^4 \tag{10.55}$$

$$f_8 (x) = 1.96 - 2.75x + 13.66x^2 - 23.98x^3 + 25.22x^4 \tag{10.56}$$

Inserting Eqs. (10.55) and (10.56) into (10.54) and evaluating the resultant expression yields 10th-order polynomials, which can be approximated by the following functions:

$$\phi_4 \simeq \frac{2}{7\pi x} \quad \text{for } S/w = 4 \tag{10.57}$$

$$\phi_8 \simeq \frac{1}{2\pi x} \quad \text{for } S/w = 8 \tag{10.58}$$

Figure 10.13 shows the numerical result for the energy calibration factor *phi*. Denote that the functions defined by Eqs. (10.57) and (10.58) give slight higher results than the polynomials at $a/w \leq 0.18$ for both $S/w = 4$ and $S/w = 8$.

The calibration factor *phi* decreases very rapidly at $x = a/w < 0.1$. It continues decreasing very insignificant at $x = a/w > 0.1$.

10.8   Ductile-to-Brittle Transition

Ideally, Fig. 10.14 shows complete S-shaped curves, $U = f(T)$, for quasi-static (slow-bend) and dynamic testing conditions. Note that the dynamic ductile-to-brittle transition, known as the nil-ductility-transition (NDT) temperature, which is

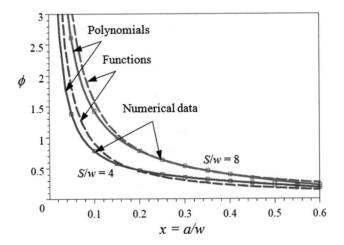

Fig. 10.13 Energy calibration factor ϕ

Fig. 10.14 Schematic
quasi-static and dynamic
impact energy

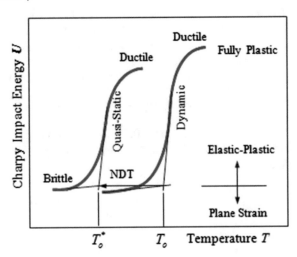

referred to as T_o, is shifted to the left for the quasi-static testing, and the upper region is shifted downward. This temperature shift in Fig. 10.14 is defined as $T_o^* = T_o - \Delta T$, and it defines the upper limit of the plane-strain condition for valid K_{IC} values. This clearly indicates that the loading rate affects the transition region. With respect to the schematic dynamic curve, at $U \le U_{NDT}$ values, elastic behavior prevails since it is a material low-energy consumption process and it is strongly dependent on the notch acuity and thickness of the specimen. In fact, increasing the notch acuity increases the stress concentration at the notch tip and decreases the impact energy.

On the other hand, at $U > U_{NDT}$ values, elastic-plastic behavior occurs in the transition region, while a pure plastic behavior becomes asymptotic accompanied

Fig. 10.15 Typical Charpy
fractured surface showing a
mixture of ductile (*dull and
gray*) and brittle (*shiny*)
fracture modes [51]

by a tearing type of fracture, and the material energy consumption is increased at
relatively high temperatures. All this implies that the material behaves elastically at
low temperatures (below the NDT temperature) and plastically at high temperatures.

Failure analysis on brittle fracture or low-energy fracture would reveal a shiny
and flat surface appearance (Fig. 10.15) since small amount of energy is absorbed
prior to fracture in an elastic deformation fracture mode. This type of fracture is
caused by a cleavage fracture mechanism in which individual grains separate along
definite crystallographic planes.

Brittle fracture literally indicates that only elastic flow occurs and the impact
force provides the absorbed energy. However, in order for the notch to grow as
crack, the structure strain energy must be released so that new crack surfaces are
generated until total separation occurs. On the other hand, the ductile fracture or
shear fracture has different characteristics since yielding occurs due to plastic flow,
which causes a dull fibrous surface appearance as depicted in Fig. 10.15.

In general, the transition from ductile-to-brittle depends on the microstructure,
testing temperature, strain rate, and notch acuity. This may be reflected on ductile
and notch insensitivity steels, which may become brittle if tested at a relatively low
temperature or at a relatively high strain rate [7]. At the transition region, the fracture
appearance is a mixture of brittle and ductile fracture surfaces. If the material has an
extended degree of anisotropy and if it is tested several times at a fixed temperature
in the transition region, then one specimen may exhibit a dominant brittle behavior
and another may show dominant ductile behavior. Therefore, this transition behavior
causes a wide data scatter as encountered in most BCC low-carbon steels.

Moreover, Fig. 10.16 shows a schematic $U = f(T)$ reference curve for a
nonlinear regression procedure. Among several mathematical models that may fit
U data, a model based on the hyperbolic tangent function can give reasonable
results. Oldfield [46, 47] used such a function to interpret the physical meaning of
the nonlinear curve fitting parameters. Thus,

$$U = A + B \tanh\left(\frac{T - T_o}{C}\right) \tag{10.59}$$

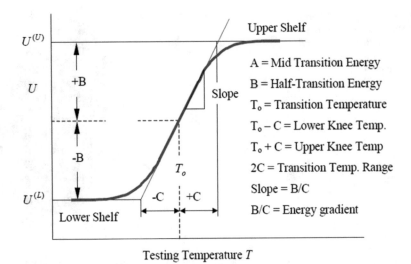

Fig. 10.16 Schematic Charpy V-notch energy, $U = f(T)$, curve showing regression parameters [46]

where A, B, C, T_o = Regression parameters (Fig. 10.16)

T = Testing temperature

The lower shelf energy in Fig. 10.16 represents a brittle behavior in which plane-strain condition should exist. Thus, brittle fracture requires a low consumption of energy. On the other hand, the upper shelf energy is a ductile response and fracture occurs by tearing, which is a fracture process that absorbs a large amount of strain energy.

The slope of the transition region can be defined as an energy gradient and can mathematically be defined by

$$\left.\frac{\partial U}{\partial T}\right|_{T=T_o} = \frac{B}{C} \sec h^2\left(\frac{T - T_o}{C}\right) = \frac{B}{C} \tag{10.60}$$

Additionally, the following functions may give good results as well

$$U = A + B \arctan\left(\frac{T - T_o}{C}\right) \tag{10.61}$$

$$U = \frac{D}{1 + F \exp(-ET)} \tag{10.62}$$

Figure 10.17 shows data for a 25 % cold rolled (CR) ASTM A710 steel, and the nonlinear least-squares fitting curves are drawn as per Eqs. (10.59) and (10.61). The curve fitting equations along with T in °C and U in joules (J) are [52]

$$U = 64.07 + 53.55 \tanh\left(\frac{T - 70.62}{10.67}\right) \tag{10.63}$$

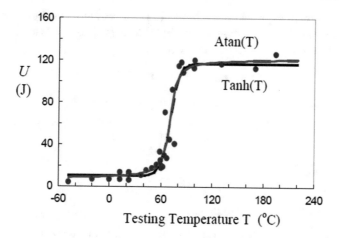

Fig. 10.17 Notch toughness of a 25 %CR A710 steel [2]

$$U = 64.34 + 36.66 \arctan \left(\frac{T - 70.57}{5.38} \right) \tag{10.64}$$

$$U = \frac{119.46}{1 + 23,703.73 \exp(-0.15T)} \tag{10.65}$$

The impact energy data for the A710 steel was fitted to Eq. (10.62) and the resultant expression is given by Eq. (10.65).

Extensive efforts have been devoted to correlate impact fracture toughness (U) and plane-strain fracture toughness (K_{IC}) data [6, 35, 44–47, 58, 61, 73, 74, 76].

Let the plane-strain fracture toughness and the Charpy impact energy be empirically defined by

$$K_{IC} = A + B \tanh \left(\frac{T - T_o}{C} \right) \tag{10.66}$$

$$U = D + E \tanh \left(\frac{T - T_o}{F} \right) \tag{10.67}$$

These two expressions give a similar S-shaped trend, and the transition temperature (T_o) should be the same for a particular data set. Letting $\tanh x \simeq x$ and solving Eq. (10.67) for T and substituting the resultant expression into Eq. (10.66) yields along with $H = F/(CE)$

$$K_{IC} = A + B \tanh \left[H \left(U - D \right) \right] \tag{10.68}$$

where $H = F/(CE)$.

The expressions, Eqs. (10.66) and (10.68), for characterizing the plane-strain fracture toughness (K_{IC}) are confined to the lower shelf region. Thus, valid K_{IC} values are obtainable at $T \leq T_o$.

Other empirical expressions for $U = f(T)$, $K_{IC} = f(T)$, and $K_{IC} = f(U)$ have been successfully used by Nogata and Takahashi [45] for evaluating sound and irradiated materials. One expression is defined by

$$K_{IC} = A + B \exp(CT) \tag{10.69}$$

Combining Eqs. (10.62) and (10.69) and eliminating T yields [45]

$$K_{IC} = A + q\left(\frac{D}{U} - 1\right)^m \tag{10.70}$$

where $m = -C/E$

$q = B/F^m$
$D = $ Upper shelf fracture toughness

This particular correlation, Eq. (10.70), was suitable for evaluating pressure vessels made of ASTM A533B-1 steel [45]. Furthermore, Eqs. (10.62) and (10.69) were fitted to experimental data for evaluating this particular steel, and the resultant empirical equations as functions of testing temperature T (°C) take the form [45]

$$K_{IC} = 20 + 95.5 \exp(0.016T) \quad \text{(in MPa}\sqrt{m}\text{)} \tag{10.71}$$

$$U = \frac{196}{1 + \exp(-0.0297T)} \quad \text{(in Joules)} \tag{10.72}$$

and subsequently, Eq. (10.70) becomes [45]

$$K_{IC} = 20 + 139\left(\frac{196}{U} - 1\right)^{-0.54} \quad \text{(in MPa}\sqrt{m}\text{)} \tag{10.73}$$

This expression, Eq. (10.73), requires that $U \leq 196\,J$; otherwise, K_{IC} cannot be determined. Figure 10.18 shows the response of Eq. (10.73) for the structural steel-type A533B-1.

Observe that there is a semi-linear correspondence in the selected **lower shelf energy**. There is an apparent linearity at $20\,J \leq U \leq 65\,J$. The slope dK_{IC}/dU changes very slightly in this Charpy impact energy range.

Barsom and Rolfe [6] have reported $K_{IC} = f(U, \sigma_{ys})$ for the **upper energy shelf** of ASTM A723 steel. The fitted expression is

$$K_{IC} = \sqrt{0.644U\sigma_{ys} - 0.006\sigma_{ys}^2} \tag{10.74}$$

where U is in Joules, σ_{ys} in MPa and K_{IC} in MPa$\sqrt{m}$.

Other empirical correlations can be found in the literature. Of significance is the Hertzberg's book [30], which contains many compiled expressions in a tabular form. Ideally, K_{IC} and U correlations should correspond to the same loading rate, but quasi-static K_{IC} values can be estimated from dynamic U experimental data, taking into account the related temperature shift.

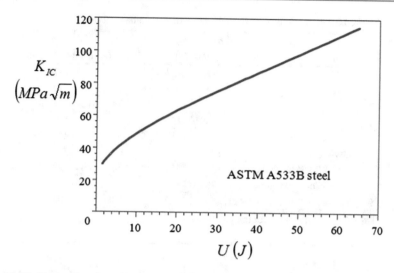

Fig. 10.18 Fracture toughness for A533B steel [45]

In this regard, a simple correlation for ABS-C, A302-B, and A517-F steels in the **transition region** has been reported [5,6] to give conservative results. The reported mathematical expression is of the form

$$K_{IC} = \sqrt{AEU} \quad \text{for } \sigma_{ys} \geq 690\,\text{MPa} \tag{10.75}$$

where E = Modulus of elasticity (MPa)

$A = 37.51$ for dynamic tests (MPa m/J)
$A = 46.89$ for quasi-static tests (MPa m/J)

For the upper shelf energy, which is not strongly dependent on notch acuity and loading rate, Barsom and Rolfe [6] and Rolfe and Novak [57] evaluated several medium-strength high-toughness steels listed in Table 10.1 by normalizing Eq. (10.75) with the room temperature yield strength. Curve fitting such a data set yields an empirical expression with a correlation coefficient of 0.99

$$\left(\frac{K_{IC}}{\sigma_{ys}}\right)^2 = 4.69\left(\frac{U}{\sigma_{ys}}\right) - 0.20 \tag{10.76}$$

The units of K_{IC} and U in Eq. (10.76) are given in Table 10.1. Figure 10.19 shows the curve fitting results.

Table 10.1 Longitudinal mechanical properties of some steels at 27 °C

No	Steel	σ_{ys} (MPa)	σ_{ts} (MPa)	%EL	%RA	U (J)	K_{IC} (MPa$\sqrt{m}$)
1	A517-F	758	834	20	66	84	187
2	4147	945	1062	15	49	35	120
3	HY-130	1027	1096	20	68	121	271
4	4130	1089	1151	14	49	31	110
5	12Ni-5Cr-3Mo	1262	1317	15	61	81	242
6	12Ni-5Cr-3Mo	1282	1324	17	67	88	249
7	18Ni-8Co-3Mo (200 Grade)	1310	1351	12	54	34	123
8	18Ni-8Co-3Mo (190 Grade)	1289	1345	15	66	66	176
9	18Ni-8Co-3Mo (190 Grade)	1696	1772	12	54	22	96

The 0.2 % offset method for the yield strength [7]

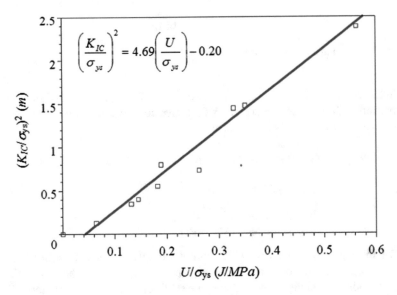

$$\left(\frac{K_{IC}}{\sigma_{ys}}\right)^2 = 4.69\left(\frac{U}{\sigma_{ys}}\right) - 0.20$$

Fig. 10.19 Normalized fracture toughness for steels (Table 10.1)

10.9   Smart Hybrid Composites

This section is devoted to composite materials because of their technological importance in manufacturing lightweight structures. Some theoretical and experimental
results obtained by using the instrumented Charpy impact machine are included. The
effect of low velocity imparted by the Charpy striker may be considered as a non-
penetrating striker. A common low velocity impact is the event of dropping a hard
and sharp tool or a falling bulk and heavy object on a substrate surface. This event
may cause permanent surface damage on metals and alloys due to the irreversible

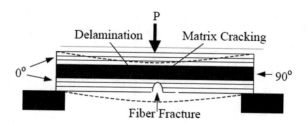

Fig. 10.20 Schematic principal failure mechanisms of a }[0/90]_s bending. Delamination is due to fiber-matrix debonding for weak interface, the matrix cracking occurs due to brittle matrix, and fiber fracture forms a crack. The *dashed lines* represent the bending path of the composite

plastic deformation mechanism. If the substrate is a composite laminated material, then the surface damage may be severe enough for reducing the load carrying capacity of composites [23].

Figure 10.20 schematically shows a three-point bending (3PB) model indicating the possible failure mechanisms encounter in unidirectional composite bars subjected to impact bending loading at low velocity. The model indicates that the impact energy provided by an object traveling at low velocity is absorbed by the composite bar generating defects, and if the plate is thin enough and sufficiently long, some of the impact energy is absorbed by general bending [23]. Thus, the composite specimen damage caused by the impact is represented by delaminations, cracks, and fiber breakage on the opposite side of the impact point P in Fig. 10.20.

Adding tough fibers to the matrix can enhance the impact and fracture resistance of brittle composite materials. For instance, hybrid composites containing embedded shape memory alloy (SMA containing $55Ni$-$45Ti$ or $Ni_{42}Ti_{45}Cu_{13}$) fibers or particulates into the matrix materials, such as polymers, fiber-reinforced polymers, have good impact properties due to the superelastic behavior of SMA materials.

The SMA material undergoes martensitic transformation due to high strain levels. Thus, the stress-induced martensitic transformation mechanism imparts strain energy dissipation, which suppress, to an extent, the formation of defects. Consequently, brittle composites containing SMA fibers become tough to an extent because of the strain energy dissipation upon impact loading. Nevertheless, improving the impact resistance may be accomplished at the expense of material strength.

According to Elber [21], the matrix properties govern the damage initiation and its extent and fiber properties and, on the other hand, control the penetration resistance or the impact resistance. In fact, a superelastic shape memory alloy has a remarkably high strain to failure primarily due to the stress-induced martensitic phase transformation creating a plateau region in the stress-strain curve and a recoverable elastic strain up to 8 % [14, 27]. Consequently, SMA fibers in composites absorb much more strain energy than other fibers before their failure. Thus, SMA hybrid composites become tough.

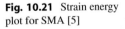

Fig. 10.21 Strain energy plot for SMA [5]

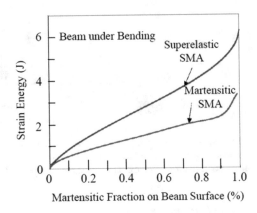

Figure 10.21 shows the relationship between the absorbed strain energy and the martensite fraction on the surface of a SMA beam under bending load. Observe that the martensitic phase transformation absorbs most of the strain energy in the structure. For a martensite fraction of 0.9 %, the superelastic SMA absorbs at least twice the strain energy of the martensitic SMA [14].

10.10 Problems

10.1. (a) Plot the given $U = f(T)$ data for a hypothetical steel. (b) Calculate K_{IC} using the Charpy impact energy U values up to zero °C, and plot K_{IC}-cal. vs. temperature and K_{IC}-exp. vs. temperature. Is there a significant difference between these plots? If so, explain.

$T(°C)$		−125	−100	−60	−40	−12	0	10	25	40	45
$U(J)$		12	18	20	28	40	78	98	110	125	126
$K_{IC} - exp.(MPa\,m^{1/2})$	40	50	80	88	150	210					

10.2. A mild steel plate has a through the thickness single-edge crack, a yield strength of 800 MPa, and a static fracture strength of $\sigma_f = 3.2\sigma_{ys}$. If the plate is loaded in tension and fractures at 600 MPa, calculate the plane-strain fracture toughness of the steel plate and the critical crack length.

10.3. A standard Charpy specimen with $B = 5$ cm, $w = 5$ cm, and $S/w = 4$ was tested at a room temperature. The measured impact energy was 30 J. The tested material has a modulus of elasticity of 70, 000 MPa. Calculate the K_{IC} for this hypothetical specimen having $x = a/w = 0.2$. [$K_{IC} = 42.98$ MPa$\sqrt{m}$].

10.4. Suppose that a design code calls for a Charpy impact energy of 22 J for building a large pressure vessel containing an inert gas. If A533B and A723 ($\sigma_{ys} = 1100$ MPa) steel plates are available for such a purpose, then (a) select the steel that will tolerate the largest critical crack length (depth of a surface semielliptical crack) when the hoop stress is 500 MPa, and **(b)** determine the minimum plate thickness as per ASTM E399 standard for the selected steel.

10.5. Plot the fracture toughness data for a hypothetical polymer and determine the brittle-ductile transition temperature.

T $(°C)$	−160	−120	−90	−80	−75	−70	−60	−50	−30
K_{IC} $(MPa\sqrt{m})$	4.10	4.11	4.00	4.05	4.10	4.5	5.50	6.60	11.00

10.6. For linear motion during Charpy impact tests, the energy lost by the striker (U_S) and the kinetic energy data for some polymers with different masses and spans to depth (S/w) ratios [78] are given below. Let $m/M \ll 1$ and **(a)** plot $U_S = f(mv^2)$ and estimate the coefficient of restitution e for these data. What does $0 < e < 1$ mean? Theoretically, determine **(b)** the $U_S/(mv^2)$ ratio for the first bound which is transformed into specimen strain energy and c) the $U_S/(mv^2)$ ratio when there is no bouncing.

mv^2 (mJ)	0	50	100	150	200
U_S (mJ)	0	75	150	222	297

10.7. A large and thick ASTM A533B-1 steel plate containing a 4-mm-long through-thickness center crack fractures when it is subjected to a tensile stress of 7 MPa. Plot U, K_{IC}, and G_{IC} at $−200°C \leq T \leq 100°C$. Which of the plots is more suitable for determining the transition temperature T_o? Explain. Data: $E = 207,000$ MPa and $v = 1/3$.

10.8. A large plate made of 18Ni-8Co-3Mo Grade 200 alloy is part of structure exposed to relatively high temperature. Charpy impact tests were carried out and the average impact energy is 60 J. Use this information to calculate **(a)** the plane-strain fracture toughness, **(b)** the minimum thickness ASTM requirement. The plate width is at least twice the thickness. Is this thickness practical? **(c)** Assume that a single-edge through the thickness crack develops. What will the critical crack length be? Will its value be reasonable?

References

1. J.C. Aleszka, Low energy impact behavior of composite panels. J. Test. Eval. **6**(3), 202–210 (1978)
2. T.L. Anderson, *Fracture Mechanics: Fundamentals and Applications*, 3rd edn. (CRC Press, Taylor & Francis Group, LLC, New York, 2005), pp. 410–432
3. T.L. Anderson, R.H. Legget, S.J. Garwood, The use of CTOD methods in fitness for purpose analysis, in *The Crack Tip Opening Displacement in Elastic-Plastic Fracture Mechanics* (Springer, Berlin, 1986), pp. 281–313

4. G.R. Anstis, P. Chantikul, B.R. Lawn, D.B. Marshall, J. Am. Cer. Soc. **64**(9), 533–538 (1981)
5. J.M. Barsom, The development of AASHTO fracture toughness requirements for bridge steels. Eng. Fract. Mech. **7**(3), 605–618 (1975)
6. J.M. Barsom, S.T. Rolfe, *Correlations Between K_{IC} and Charpy V-Notch Test Results in the Transition-Temperature Range*, ASTM STP 466 (1970), pp. 281–302
7. J.M. Barsom, S.T. Rolfe, *Fracture and Fatigue in Structure: Application of fracture Mechanics*, 3rd edn. (Butterworth-Heinemann, American Society for Testing and Materials, Philadelphia, PA, 1999)
8. M.W. Barsoum, *Fundamentals of Ceramics* (The McGraw-Hill Companies, Inc., New York, 1997), pp. 404–405
9. G. Berces, N.Q. Chinh, A. Juhasz, J. Lendvai, Kinetic analysis of plastic instabilities occurring in microhardness tests. Acta Mater. **46**(6), 2029–2037 (1998)
10. J.M. Bloom, Prediction of ductile tearing using a proposed strain hardening failure assessment diagram. Int. J. Fract. **6**, R73–R77 (1980)
11. W. Böhme, H.J. Schindler, *Application of Single-Specimen Methods on Instrumented Charpy Tests: Results of DVM Round-Robin Exercises*, ASTM STP 1380 (1999), pp. 327–336
12. H.E. Boyer, T.L. Gall (eds.), *Metals Handbook* (American Society for Metals (ASM), 1985)
13. L.J. Broutman, A. Rotem, *Impact Strength and Toughness of Fiber Composite Material*, ASTM STP 568 (1975), pp. 114–133
14. W.F. Brown, J.E. Strawley, ASTM STP 410 (1966)
15. BS 7910, *Guidance on Some Methods for the Derivation of Acceptance Levels for Defects in Fusion Welded Joints*, (British Standards Institution (August 1991)
16. F.M. Burdekin, D.E. Stone, The crack opening displacement approach to fracture mechanics in yielding materials. J. Strain Anal. **1**, 145–153 (1966)
17. W.J. Cantwell, J. Morton, Comparison of the low and high velocity impact response of CFRP. Composites **20**(6), 545–551 (1989)
18. W.J. Cantwell, J. Morton, Geometrical effects in the low velocity response of CFRP. Compos. Struct. **12**, 39–59 (1989)
19. J.A. Collins, *Failure of Materials in Mechanical Design: Analysis, Prediction, Prevention*, 2nd edn. (Wiley, New York, 1993), 73–80
20. D.A. Curry, J.F. Knott, The relationship between toughness and microstructure in the cleavage fracture of mild steel. Met. Sci. **10**, 1–5 (1976)
21. A.R. Dowling, C.H.A. Townley, The effects of defects on structural failure: a two criteria approach. Int. J. Press. Vessel. Pip. **3**, 77–137 (1975)
22. D.S. Dugdale, *Yielding of Steel Sheets Containing Slits*. J. Mech. Phys. Sol. **8**, 100–104 (1960)
23. W. Elber, Effect of matrix and fiber properties on impact resistance, in *Tough Composite Materials: Recent Developments* (Noyes, Park Ridge, NJ, 1985), pp. 89–110
24. R.L. Ellis, F. Lalande, H. Jia, C.A. Rogers, Ballistic impact resistance of SMA and spectra hybrid graphite composites. J. Reinf. Plast. Compos. **17**(2), 147–164 (1998)
25. Z. Gurdal, R.T. Haftka, P. Hajela, *Design and Optimization of Laminated Composite Materials*, (Wiley-Interscience, New York, 1999)
26. G.T. Hahn, A.R. Rosenfield, Experimental determination of plastic constraint ahead of a sharp notch under plane-strain conditions. Trans. ASM **59**, 909–919 (1966)
27. R.P. Harrison, K. Loosemore, I. Milne, *Assessment of the Integrity of Structures Containing Defects*, CDGB Report R/HRG, CEGB (1976)
28. H.W. Hayden, W.G. Moffatt, J. Wulff, *The Structure and Properties of Materials*, Mechanical Behavior, vol. III (Wiley, New York, 1965)
29. A. Hazizan, W.J. Cantwell, The low velocity impact response of foam-based sandwich structures. Compos. Part B **33**(3), 193–204 (2002)
30. R.W. Herztberg, *Deformation and Fracture Mechanics of Engineering Materials*, 3rd edn. (Wiley, New York, 1989)
31. J.G. Hetherington, P.F. Lemieux, The effect of obliquity on the ballistic performance of two component composite armour. Int. J. Impact Eng. **15**(2), 131–137 (1994)

32. R.W.K. Honeycombe, *Steels: Microstructure and Properties* (Edward Arnold Publisher Ltd. and American Society for Metals, 1982)

33. A. Iost, R. Bigot, *Reply to comment on indentation size effect: reality or artifact?* J. Mater. Sci. Lett. **17** (22), 1889–1891 (1998)

34. R.M. Jones, *Mechanics of Composite Materials*, 2nd edn. (Book News, Inc.®, Portland, OR, June 1998)

35. J.A. Kapp, J.H. Underwood, *Correlations Between Fracture Toughness, Charpy V-Notch, and Yield Strength for ASTM A723 Steel*, ASME, Pressure Vessels & Piping Division (June 21–25, 1992), pp. 219–221

36. V. Kumar, M.D. German, C.F. Shih, *An Engineering Approach To Elastic-plastic Fracture Analysis*, EPRI Report NP-1931, Electric Power Research Institute, Palo Alto, CA (1981)

37. M.T. Laugier, *Comparison of toughness in WC-Co determined by a compact tensile technique with model predictions*. J. Mater. Sci. Lett. **6**, 779–780 (1987)

38. M.P. Manahan, *Advances in Notched Bar Impact Testing*, ASTM Standardization News (October 1996), pp. 23–29

39. M.P. Manahan, *In-Situ Heating and Cooling of Charpy Test Specimens*, ASTM STP 1380 (1999)

40. M.P. Manahan, R.B. Stonesifer, The difference between total absorbed energy measured using an instrumented striker and that obtained using and optical encoder. ASTM STP **1380**, 181–197 (2000)

41. M.P. Manahan, R.B. Stonesifer, The difference between total absorbed energy measured using an instrumented striker and that obtained using and optical encoder. ASTM STP **1380**, 181–197 (2000)

42. V. Milekhine, M.I. Onsoien, J.K. Solberg, T. Skaland, Mechanical properties of $FeSi(c)$, $FeSi_2$ (α) and $Mg_2SiElsevier$. Intermetallics **10**, 743–750 (2002)

43. MPM Technologies, Inc. (2003) www.mpmtechnologies.com

44. B. Nageswara Rao, A.R. Acharya, A computer study on evaluation of fracture toughness from Charpy V-notch impact energy and reduction of area. Eng. Fract. Mech. **41**(1), 85–90 (1992)

45. F.N. Nogata, H. Takahashi, Fracture mechanics evaluation of irradiated embrittlement in reactor vessel steels based on the rate process concept. J. Test. Eval. JTEVA **14**(1), 40–48 (1986)

46. W. Oldfield, *Curve Fitting Impact Test Data: A Statistical Procedure*. ASTM Standardization News, vol. 3(11) (Nov 1975), pp. 24–29

47. W. Oldfield, Fitting curves to toughness data. J. Test. Eval. JTEVA **7**(6), 326–333 (1979)

48. E. Orowan, Notch brittleness and strength of metals. Trans. Inst. Eng. Shipbuild. Scot. **89**, 165–215 (1945)

49. J.S.N. Paine, C.A. Rogers, The response of SMA hybrid composite materials to low velocity impact. J. Intell. Mater. Syst. Struct. **5**, 530–535 (1994)

50. S. Palmquist, Method of determining the toughness of brittle materials, particularly sintered carbides. Jernkontorest Ann. **141**(5), 300–307 (1957)

51. W.S. Pellini, *Symposium on Metallic Materials at Low Temperatures*, ASTM STP 158 (1954), pp. 216–258

52. N. Perez, Ph.D. Dissertation, University of Idaho, Moscow, 1989

53. J. Phelps, J. Shields, X. Wang, C.M. Agrawal, Relationship of fracture toughness and microstructural anisotropy of bone, in *Southern Biomedical Engineering Conference Proceedings*, Sponsored by IEEE, 6–8 Feb 1998, p. 23

54. C.B. Ponton, R.D. Rawlings, Vickers indentation fracture toughness test. Part 1: review of literature and formulation of standardized indentation toughness equations. Mater. Sci. Technol. **5**, 965–872 (1989)

55. C.B. Ponton, R.D. Rawlings, Dependence of the vickers indentation fracture toughness on the surface crack length. Br. Ceram. Trans. J. **88**, 83–90 (1989)

56. R.O. Ritchie, J.F. Knott, J.F. Rice, On the relation ship between critical tensile stress and fracture toughness in mild steel. J. Mech. Phys. Solids **21**, 395–410 (1973)

57. S.T. Rolfe, S.R. Novak, *Slow-Bend K_{IC} Testing of Medium Strength High-Toughness Steels*, ASTM STP 463 (1970), pp. 124–159
58. S.T. Rolfe, W.A. Sorem, G.W. Wellman, Fracture control in the transition-temperature region of structural steels. J. Constr. Steel Res. **12**, 171–195 (1989)
59. A.R. Rosenfield, G.T. Hahn, J.D. Embury, Fracture of steel containing pearlite. Metall. Trans. **3**, 2797–2803 (1972)
60. R6, *Assessment of the Integrity of Structures Containing Defects, Procedure R6, Revision 3* (Nuclear Electric Ltd., Gloucester, 1997)
61. R.H. Sailors, H.T. Corten, Relationship between material fracture toughness using fracture mechanics and transition temperature tests. ASTM STP **514**, 164–191 (1972)
62. A. Saxena, *Nonlinear Fracture Mechanics for Engineers* (CRC Press, New York, 1998), p. 184–187
63. H.J. Schindler, The use of instrumented pre-cracked Charpy specimens, in *Engineering Integrity Assessment Evaluating Material Properties by Dynamic Testing, ESIS 20* (Mechanical Engineering Publications, London, 1996), pp. 45–58
64. H.J. Schindler, Estimation of the dynamic JR curve from a single impact bending test, in *Proceedings of the ECF 11 – Mechanical and Mechanics of Damaged and Failure*, Vol. II, ed. by J. Petit (EMAS Publishing, Warrington, 1996), pp. 2007–2012
65. H.J. Schindler, *Preferable Initial Length for Fracture Toughness Evaluation Using Sub-Sized Bend Specimens*, ASTM STP 1418 (2002), pp. 67–79
66. H.J. Schindler, M. Veidt, *Fracture Toughness Evaluation from Instrumented Sub-Size Charpy-Type Tests*, ASTM STP 1329 (1997), pp. 48–62
67. H.J. Schindler, T. Varga, H. Njo, G. Prantl, Key issues of instrumented pre-cracked Charpy-type testing in irradiation surveillance programs, in *Proceedings of the International Symposium on Materials Aging and Component Life Extension*, vol. II, Milan, 10–13 October 1995, pp. 1367–1377
68. K.K. Sharma, P.N. Kotru, B.M. Wanklyn, Microindentation studies of flux-grown ErFeO3 single crystals. Appl. Surf. Sci. **81**(2), 251–258 (1994)
69. K.K. Sharma, P.N. Kotru, B.M. Wanklyn, Microindentation studies of flux-grown $ErFeO_3$ single crystal. Appl. Surf. Sci. **18**(2), 251–258 (1994)
70. C.F. Shih, M.D. German, V. Kumar, An engineering approach for examining crack growth and stability in flawed structures. Int. J. Press. Vessel. Pip. **9**, 159–196 (1981)
71. M.R. Sridhar, M.M. Yovanovich, Empirical methods to predict vickers microhardness. Wear **193**, 91–98 (1996)
72. F.R. Stonesifer, R.W. Armstrong, Effect of prior austenite grain size on the fracture toughness properties of A533B steel, in Fracture 77, vol. 2A, ed. by D.M.R. Taplin (Pergamon Press, New York, 1977), pp. 1–5
73. S. Tauscher, *The Correlation of Fracture Toughness with Charpy V-Notch Impact Test Data*, USA ARRADCOM Technical Report ARLCB-TR-81012, Benet Weapons Laboratory, Watervliet, NY (March 1977)
74. J.H. Underwood, G.S. Leger, *Fracture Toughness of High Strength Steel Predicted from Charpy Energy or Reduction in Area*, ASTM STP 833 (1984), pp. 481–498
75. G.F. Vander Voort, *Metallography: Principles and Practice* (McGraw-Hill, Inc., New York, 1984), pp. 350–389
76. J.D. Varsik, S.T. Byrne, *An Empirical Evaluation of the Irradiation Sensitivity of Reactor Pressure Vessel Materials*, ASTM STP 683 (1979), pp. 252–266
77. J.D. Walker, C.E. Anderson Jr., The influence of initial nose shape in eroding penetration. Int. J. Impact Eng. **15**(2), 139–148 (1994)
78. J.G. Williams, *Fracture Mechanics of Polymers* (Ellis Horwood Limited Publishers, Halsted Press a division of Wiley, New York, 1984)

Metric Conversions

A

Prefixes		
Factor	Prefix	SI symbol
10^{18}	exa	E
10^{15}	peta	P
10^{12}	tera	T
10^{9}	giga	G
10^{6}	mega	M
10^{3}	kilo	k
10^{-3}	milli	m
10^{-6}	micro	μ
10^{-9}	nano	n
10^{-12}	pico	p
10^{-15}	femto	f
10^{-18}	atto	a

Greek alphabet			
$A\ \alpha$	Alpha	$N\ \nu$	Nu
$B\ \beta$	Beta	$\Xi\ \xi$	Xi
$\Gamma\ \gamma$	Gamma	$O\ o$	Omicron
$\Delta\ \delta$	Delta	$\Pi\ \pi$	Pi
$E\ \epsilon$	Epsilon	$P\ \rho$	Rho
$Z\ \zeta$	Zeta	$\Sigma\ \sigma$	Sigma
$H\ \eta$	Eta	$T\ \tau$	Tau
$\Theta\ \theta$	Theta	$Y\ \upsilon$	Upsilon
$I\ i$	Iota	$\Phi\ \varphi$	Phi
$K\ \kappa$	Kappa	$X\ \chi$	Chi
$\Lambda\ \lambda$	Lambda	$\Psi\ \psi$	Psi
$M\ \mu$	Mu	$\Omega\ \omega$	Omega

Physical constants	
Avogadro's number	$N_A = 6.023 \times 10^{23}$ atom/mol
Boltzmann's constant	$k = 1.38 \times 10^{-23}$ J/atom K
Gas constant	$R = 8.315$ J/K mol
Plank's constant	$h = 6.63 \times 10^{-34}$ J s

© Springer International Publishing Switzerland 2017
N. Perez, *Fracture Mechanics*, DOI 10.1007/978-3-319-24999-5

Length

$1\,m = 10^{10}\,\text{Å}$	$1\,\text{Å} = 10^{-10}\,m$
$1\,m = 10^{9}\,nm$	$1\,nm = 10^{-9}\,m$
$1\,m = 10^{6}\,\mu m$	$1\,\mu m = 10^{-6}\,m$
$1\,m = 10^{3}\,mm$	$1\,mm = 10^{-3}\,m$
$1\,m = 10^{2}\,cm$	$1\,cm = 10^{-2}\,m$
$1\,m = 3.28\,ft$	$1\,ft = 0.3049\,m$
$1\,m = 39.36\,in.$	$1\,in. = 0.0275\,m$
$1\,cm = 10\,mm$	$1\,mm = 0.10\,cm$
$1\,cm = 3.28 \times 10^{-2}\,ft$	$1\,ft = 30.48\,cm$
$1\,cm = 0.394\,in.$	$1\,in. = 2.54\,cm$
$1\,mm = 3.28 \times 10^{-3}\,ft$	$1\,ft = 304.8\,mm$
$1\,mm = 3.94 \times 10^{-2}\,in.$	$1\,in. = 25.4\,mm$

Area

$1\,m^2 = 10^{20}\,\text{Å}^2$	$1\,\text{Å}^2 = 10^{-20}\,m^2$
$1\,m^2 = 10^{18}\,nm^2$	$1\,nm^2 = 10^{-18}\,m^2$
$1\,m^2 = 10^{12}\,\mu m^2$	$1\,\mu m^2 = 10^{-12}\,m^2$
$1\,m^2 = 10^{6}\,mm^2$	$1\,mm^2 = 10^{-6}\,m^2$
$1\,m^2 = 10^{4}\,cm^2$	$1\,cm^2 = 10^{-4}\,m^2$
$1\,m^2 = 10.76\,ft^2$	$1\,ft^2 = 9.29 \times 10^{-2}\,m^2$
$1\,m^2 = 1.55 \times 10^{3}\,in.^2$	$1\,in.^2 = 6.45 \times 10^{-4}\,m^2$

Volume

$1\,m^3 = 10^{27}\,nm^3$	$1\,nm^3 = 10^{-27}\,m^3$
$1\,m^3 = 10^{18}\,\mu m^3$	$1\,\mu m^3 = 10^{-18}\,m^3$
$1\,m^3 = 10^{9}\,mm^3$	$1\,mm^3 = 10^{-9}\,m^3$
$1\,m^3 = 10^{6}\,cm^3$	$1\,cm^3 = 10^{-6}\,m^3$
$1\,m^3 = 35.20\,ft^3$	$1\,ft^3 = 2.83 \times 10^{-2}\,m^3$
$1\,m^3 = 6.10 \times 10^{4}\,in.^3$	$1\,in.^3 = 1.64 \times 10^{-5}\,m^3$
$1\,cm^3 = 3.53 \times 10^{-5}\,ft^3$	$1\,ft^3 = 2.83 \times 10^{4}\,cm^3$
$1\,cm^3 = 6.010 \times 10^{-2}\,in.^3$	$1\,in.^3 = 16.39\,cm^3$
$1\,cm^3 = 2.642 \times 10^{-4}\,gal\ (US)$	$1\,gal\ (US) = 3.79 \times 10^{3}\,cm^3$
$1\,liter\ (l) = 10^{3}\,cm^3$	$1\,cm^3 = 10^{-3}\,liter$
$1\,liter\ (l) = 0.2642\,gal\ (US)$	$1\,gal\ (US) = 3.785\,liters$

Mass

$1\,kg = 10^{3}\,g$	$1\,g = 10^{-3}\,kg$
$1\,kg = 2.205\,lb_m$	$1\,lb_m = 0.454\,kg$
$1\,g = 2.205 \times 10^{-3}\,lb_m$	$1\,lb_m = 454\,g$
$1\,g = 3.53 \times 10^{-2}\,oz$	$1\,oz = 28.35\,g$
$1\,lb_m = 16\,oz$	$1\,oz = 6.25 \times 10^{-2}\,lb_m$

Density

$1\,\text{kg/m}^3 = 10^{-3}\,\text{g/cm}^3$	$1\,\text{g/cm}^3 = 10^3\,\text{kg/m}^3$
$1\,\text{kg/m}^3 = 0.0624\,\text{lb}_m/\text{ft}^3$	$\text{lb}_m/\text{ft}^3 = 16.03\,\text{kg/m}^3$
$1\,\text{kg/m}^3 = 3.61\times10^{-5}\,\text{lb}_m/\text{in.}^3$	$\text{lb}_m/\text{in.}^3 = 2.77\times10^4\,\text{kg/cm}^3$
$1\,\text{g/cm}^3 = 0.0361\,\text{lb}_m/\text{in.}^3$	$\text{lb}_m/\text{in.}^3 = 27.70\,\text{g/cm}^3$

Force

$1\,\text{N} = 1\,\text{kg m/s}^2$	$1\,\text{dyne} = 1\,\text{g cm/s}^2$
$1\,\text{N} = 10^5\,\text{dynes}$	$1\,\text{dyne} = 10^{-5}\,\text{N}$
$1\,\text{N} = 0.2248\,\text{lb}_f$	$1\,\text{lb}_f = 4.448\,\text{N}$
$1\,\text{dyne} = 2.248\times10^{-6}\,\text{lb}_f$	$1\,\text{lb}_f = 4.448\times10^5\,\text{dyne}$

Stress

$1\,\text{MPa} = 0.145\,\text{ksi}$	$1\,\text{ksi} = 6.895\,\text{MPa}$
$1\,\text{MPa} = 145\,\text{psi}$	$1\,\text{psi} = 6.90\times10^{-3}\,\text{MPa}$
$1\,\text{MPa} = 0.1019\,\text{kg}_f/\text{mm}^2$	$1\,\text{kg}_f/\text{mm}^2 = 9.81\,\text{MPa}$
$1\,\text{MPa} - 7.25\times10^{-2}\,\text{Ton}_f/\text{in.}^2$	$1\,\text{Ton}_f/\text{in}^2 = 13.79\,\text{MPa}$

Energy

$1\,\text{J} = 10^7\,\text{ergs}$	$1\,\text{erg} = 10^{-7}\,\text{J}$
$1\,\text{J} = 6.24\times10^{18}\,\text{eV}$	$1\,\text{eV} = 1.60\times10^{-19}\,\text{J}$
$1\,\text{J} = 0.239\,\text{cal}$	$1\,\text{cal} = 4.184\,\text{J}$
$1\,\text{J} = 9.48\times10^{-4}\,\text{Btu}$	$1\,\text{Btu} = 1054\,\text{J}$
$1\,\text{J} = 1.3558\,\text{ft lb}_f$	$1\,\text{ft lb}_f = 0.7376\,\text{J}$
$1\,\text{cal} = 3.97\times10^{-3}\,\text{Btu}$	$1\,\text{Btu} = 252\,\text{cal}$

Fracture toughness

$1\,\text{MPa}\sqrt{\text{m}} = 0.91\,\text{ksi}\sqrt{\text{in.}}$	$1\,\text{ksi}\sqrt{\text{in.}} = 1.10\,\text{MPa}\sqrt{\text{m}}$
$1\,\text{MPa}\sqrt{\text{m}} = 910\,\text{psi}\sqrt{\text{in.}}$	$1\,\text{psi}\sqrt{\text{in.}} = 1.10\times10^{-3}\,\text{MPa}\sqrt{\text{m}}$
$1\,\text{ksi}\sqrt{\text{in.}} = 10^3\,\text{psi}\sqrt{\text{in.}}$	$1\,\text{psi}\sqrt{\text{in.}} = 10^{-3}\,\text{ksi}\sqrt{\text{in.}}$

A.1   Fracture Toughness Data

The average data below can be found elsewhere [1–3] (Table A.1).

Table A.1 Mechanical
properties [1–3]

Material	Temp.	σ_{ys} (MPa)	K_{IC} (MPa$\sqrt{m}$)
Steels			
AISI 1045	$-4\,^\circ$C	269	50
AISI 4340	RT[a]	1567	57
AISI 4340	RT	1408	85
D6AC	RT	1495	93
HP 9-4-20	RT	1295	143
18Ni (200)	RT	1450	110
18Ni (300)	RT	1931	74
ASTM A538	RT	1722	111
Aluminum alloys			
2020-T651	RT	532	25
2024-T351	RT	378	38
2024-T851	RT	450	26
6061-T651	RT	296	28
6061-T651	$-80\,^\circ$C	310	33
7075-T651	RT	538	29
7075-T7351	RT	428	33

[a]Room temperature

References

1. H.O. Fuchs, R.I. Stephens, *Metal Fatigue in Engineering* (Wiley, New York, 1980)
2. J.A. Collins, *Failure of Materials in Mechanical Design: Analysis, Prediction, Prevention* (Wiley, New York,1981)
3. R.W. Hertzberg, *Deformation and Fracture Mechanics of Engineering Materials*, 3rd edn. (Wiley, New York,1989)

Index

A

Aggressive environment, 335, 343
Airy stress function, 16, 58, 82, 262
    biharmonic equation, 14, 16
    cantilever beam, 20
    Cartesian coordinates, 18, 20
    complex form, 28–31, 34, 36, 37, 138
    eigenvalue, 18, 141
    elastic field, 17
    general definition, 18, 141
    hollow cylinder, 17
    plastic field, 263
    polar coordinates, 24, 51, 101, 154, 167
    power series, 18, 19, 143, 154
Alloying elements, 378
Anisotropic material, 123
Anisotropy, 123
Antisymmetric displacement, 149
    mode III, 148–150
Antisymmetric mode II, 234
Antisymmetric mode III, 234
Artificial crack extension
    see Effective crack length, 191
ASTM standard test method
    E1290, 206
    E1820, 119, 251
    E23, 386
    E399, 81, 88, 109, 117, 245, 381, 383
    E561, 241
    E647, 343, 346, 361
    E813, 381
    E92, 380
ASTM thickness requirement, 88
Austenite grain growth, 378

B

Ballistic field, 385
Ballistic impact velocity, 385
BCC materials, 73, 344, 378, 397

BCC slip system, 331, 352
BCC structure, 331
BCC unit cell, 331
Beach marks, 344, 351, 363
Bending stress, 328
Biaxial fatigue stress level, 358
Biaxial loading, 5, 299
Biharmonic equation, 17, 30
    Cartesian coordinates, 16
    complex theory, 30
    polar coordinates, 16
Bipotential equation, 148
Body-force field, 17
Body-force intensity, 12, 14, 17
Boundary function, 44, 47, 160, 161
Brittle fracture, 53, 397
Brittle materials
    ceramics, 380
    cermets, 380
    polymers, 380
Burger's vector, 333

C

Calculus of residues, 43
Cantilever beam, 20
Cauchy integral formula, 43, 47, 159, 160,
    162
Cauchy integral formulae, 43
Cauchy integral theorem, 43
Cauchy–Riemann equations, 27
Cauchy-Riemann condition, 83
Cauchy-Riemann equations, 83
    analytic, 27
    not analytic, 28
Center of percussion, 391
Charpy impact energy
    lower shelf, 398, 400
    upper shelf, 398, 400, 401
Charpy impact machine, 394

© Springer International Publishing Switzerland 2017
N. Perez, *Fracture Mechanics*, DOI 10.1007/978-3-319-24999-5